CONTEMPORARY UNDERGRADUATE MATHEMATICS SERIES
Robert J. Wisner, Editor

MATHEMATICS FOR THE LIBERAL ARTS STUDENT
Fred Richman, Carol Walker, and Robert J. Wisner

INTERMEDIATE ALGEBRA
Edward D. Gaughan

ALGEBRA: A PRECALCULUS COURSE
James E. Hall

MODERN MATHEMATICS: AN ELEMENTARY APPROACH, SECOND EDITION
Ruric E. Wheeler

FUNDAMENTAL COLLEGE MATHEMATICS: NUMBER SYSTEMS AND INTUITIVE GEOMETRY
Ruric E. Wheeler

MODERN MATHEMATICS FOR BUSINESS STUDENTS
Ruric E. Wheeler and W. D. Peeples

ANALYTIC GEOMETRY
James E. Hall

INTRODUCTORY GEOMETRY: AN INFORMAL APPROACH, SECOND EDITION
James R. Smart

AN INTUITIVE APPROACH TO ELEMENTARY GEOMETRY
Beauregard Stubblefield

GEOMETRY FOR TEACHERS
Paul B. Johnson and Carol H. Kipps

MODERN GEOMETRIES
James R. Smart

LINEAR ALGEBRA
James E. Scroggs

AN INTRODUCTION TO ABSTRACT ALGEBRA
A. Richard Mitchell and Roger W. Mitchell

INTRODUCTION TO ANALYSIS
Edward D. Gaughan

A PRIMER OF COMPLEX VARIABLES WITH AN INTRODUCTION TO ADVANCED TECHNIQUES
Hugh J. Hamilton

CALCULUS OF SEVERAL VARIABLES
E. K. McLachlan

PROBABILITY
Donald R. Barr and Peter W. Zehna

THEORY AND EXAMPLES OF POINT-SET TOPOLOGY
John Greever

AN INTRODUCTION TO ALGEBRAIC TOPOLOGY
John W. Keesee

NUMBER THEORY: AN INTRODUCTION TO ALGEBRA
Fred Richman

Introductory Geometry: An Informal Approach

Second Edition

James R. Smart
San Jose State College

Brooks/Cole Publishing Company
Monterey, California

A Division of Wadsworth Publishing Company, Inc.
Belmont, California

This book was edited by Micky Stay and designed by Linda Marcetti. Technical art was drawn by John Foster. The book was typeset by J. W. Arrowsmith, Ltd., Bristol, England, and printed and bound at Kingsport Press, Inc., Kingsport, Tennessee.

L.C. Cat. Card No.: 75-170990
ISBN 0-8185-0021-2
Printed in the United States of America

1 2 3 4 5 6 7 8 9 10—76 75 74 73 72

In the years following the publication of the first edition of this text, many significant trends that directly affect the material of introductory geometry from an informal and intuitive point of view have become apparent. Some of these trends, reflected in the choice of material for this text, are as follows:

—the growing influence on American education of the mathematics curricula of various foreign countries where informal geometry has long been a major subject throughout the elementary and secondary years;

—the recognition that a course in informal geometry may be of greater value to some high school and college students than would be a more formal course whose sole purpose is the teaching of proofs;

—the inclusion of many additional topics in geometry at lower grade levels (transformations, vectors, logic, convexity, and more advanced topics from analytic geometry);

—the ever-stronger emphasis on laboratory experiences in mathematics and the use of constructions as a vehicle for introducing geometric concepts.

This book was written for those who want to study or teach introductory geometry from a modern, informal, and intuitive point of view. It is appropriate for a person with a limited background in formal mathematics yet has been found of interest to many with stronger preparation. Although the text is more comprehensive than are modern presentations of geometry in elementary and junior high schools, it shares their spirit in that it presents geometry with a precise symbolism and vocabulary, emphasizes concepts, and depends on meaning and intuition as well as on informal and formal proofs. It has a carefully developed structure, but its approach is not strictly postulational.

The plan of the book is first to explain fundamental ideas of geometry, such as mathematical models, sets of points, axioms, undefined terms, definitions, theorems, and logic, and then to introduce the geometry of sets of points in a plane, from both a metric and a non-metric standpoint. The study of sets of points in a plane also includes measurement of length and angles, simple closed curves, circles and symmetry, mathematical constructions, area of plane regions, similar figures, and graphs of points in a plane.

Later chapters extend these concepts to sets of points in space. The non-metric ideas of sets of points in space are followed by the use of coordinates and a brief study of errors in measurement. The last chapter introduces some significant ideas from projective geometry, topology, finite geometry, and non-Euclidean geometry.

I have presented this material to many students, including elementary school teachers and administrators, future elementary school teachers, and general college students. The first edition of the text has been used by many thousands of college students. It has also been used successfully by high school students. Many improvements suggested by students have been incorporated into both the first and second editions.

The second edition has retained all the features of the first edition that proved worthwhile. It involves a reorganization of material, not just a tacking on of additional topics. Some specific examples of reorganization are: (1) the combining of two former chapters into a more thorough explanation of the foundations of geometry; (2) the transfer of emphasis to the fundamental concept of geometric transformation to provide greater unity; (3) the increase in the number of pages devoted to constructions and the delay of more difficult concepts from ordinary college geometry until the constructions can be used as a basic vehicle to give the work more meaning and to make it easier to understand; (4) the delay of the chapter on errors in measurement so that examples can be used from area and volume measurements; and (5) the chapter-by-chapter introduction of a modern axiomatic system for elementary geometry so that the student is aware of the assumptions that must be made.

The book may be used with a wide variety of courses. It is suitable for a three-unit college course in introductory geometry, which may come prior to, with, or following a course on number systems. A minimum course, perhaps consisting of two units of material, may include Chapters 1–8, 10, and 12. A shorter course for students with stronger preparation may include a rapid review of the first eight chapters and concentration on Chapters 9–14.

The text may also be used as one of two books in a course designed to provide an introduction to informal algebra and geometry in a single course. In this case, material should supplement what the reader already knows, and Chapters 3, 6, 9, 11, and 14 probably should be emphasized. The text can be used for extension courses and in-service courses for teachers, as well as for individual study. A variety of enrichment topics, such as caroms, modular arithmetic, and Fibonacci numbers, increases the utility for the reader who hopes to make geometry more interesting for others. Moreover, the text can be used for new-type secondary school geometry courses that provide an alternative to traditional plane geometry.

For their helpful criticisms and suggestions in the preparation of the first edition, I thank the following professors: Leon Bowden, of the University of Victoria; Edward J. Farrell, of the University of San Francisco; Francis J. Mueller, Tampa, Florida; Donald E. Myers, of the University of Arizona; Arnold Seiken, of the University of Rhode Island; Robert L. Stanley, of Portland State College; and Robert J. Wisner, of New Mexico State University.

For their suggestions in the preparation of the second edition, I extend my gratitude to C. W. Oakley, of the University of Nevada; Bruce Partner, of Ball State University; Curtis J. Shaw, of the University of Southwestern Louisiana; John J. Callahan, of Boston State College; Bernard Strickmeier, of California State Polytechnic College; Carol H. Kipps, of Pasadena City College; Glyn K. Wooldridge, of the University of Texas; Edward H. Lezak, of Indiana University; and Robert J. Wisner, of New Mexico State University.

I would also like to thank the fine staff of Brooks/Cole Publishing Company for their cooperation and creativity.

James R. Smart

1/Foundations of Geometry
1

An Informal and Intuitive Approach to Geometry *1*
Axioms, Theorems, and Proof *6*
Using Logic in Geometry (Supplementary) *11*

2/Non-Metric Plane Geometry
17

Definitions in Non-Metric Geometry *17*
Basic Axioms for Euclidean Geometry *25*
Angles *30*
Intersections and Unions of Sets of Points *34*

3/Measurement of Length and Angles
41

The Concept of Measurement *41*
The Measurement of Length *47*
Measurement of Angles *54*

4/Simple Closed Curves
63

Basic Concepts *63*
Classification of Triangles *69*
Quadrilaterals *77*
Polygons with More than Four Sides *83*
Perimeter *88*
The Pythagorean Theorem *89*

5/Circles and Symmetry
95

Basic Concepts *95*
Central Angles and Arcs *101*
Circumference *109*
Symmetry *116*

6/Mathematical Constructions
124

Using Construction Instruments *124*
Construction Problems and Designs *131*
Mathematical Laboratory Experiences with Constructions *136*

7/Area of Plane Regions
146

The Concept of Area *146*
Area of Region Whose Boundary Is a Parallelogram or a Triangle *151*
Areas of Other Polygonal Regions *155*
Areas of Circular Regions and Other Plane Regions *159*

8/Similar Figures *166*

Examples of Similar Figures *166*
Scale Drawings *173*
Indirect Measurement *176*
Constructions and Additional Applications of Similar Figures *181*

9/Graphing Points in a Plane *187*

Coordinates of Points *187*
Analytic Geometry *194*
Trigonometric Ratios *203*
Vectors *210*

10/Non-Metric Geometry of Space *218*

Sets of Points in Space *218*
Polyhedrons *223*
Simple Closed Surfaces and Solid Figures *229*

11/Locating Points in Space *234*

Coordinates for Points in Space *234*
Locating Points on the Surface of the Earth *241*

12/Measurements of Surfaces and Solids *245*

Measurements of Prisms and Prismatic Solids *245*
Measurements of Cylinders and Cylindrical Solids *253*
Measurements of Pyramids, Cones, and Related Solids *256*
Measurements of Spheres and Spherical Solids *260*

13/Errors in Measurement *265*

Greatest Possible Error and Precision *266*
Relative Error and Accuracy *269*
Operations with Measures *271*

14/Some Modern Geometries *275*

Introduction to Projective Geometry *275*
Introduction to Topology *285*
Introduction to Finite Geometry *289*
Introduction to Non-Euclidean Geometry *293*

Bibliography *297*

Answers to Selected Exercises *299*

Index *307*

Introductory Geometry: An Informal Approach

Second Edition

1 Foundations of Geometry

Geometry is one of the major branches of mathematics. The ancient Egyptians and Babylonians used geometry as a practical tool. The Greeks developed geometry as a logical science and were responsible for the proofs of many theorems. Since the eighteenth century much of the emphasis in geometry has been on the use of analytic geometry (see Chapter 9) as a tool in calculus. However, in the beginning of the nineteenth century many new, modern geometries (see Chapter 14) were invented that have greatly increased interest in geometry for its own sake.

An Informal and Intuitive Approach to Geometry

One way of studying geometry is a modern, informal approach. Geometry approached from an informal or intuitive point of view may be considered as the study of points in space. An informal approach to geometry makes use of intuition but does not depend on it alone. The student should not imagine, however, that his unaided intuition will always be right. Everyone has often met situations in which his intuition has led him to false conclusions. Examples include estimating distances, judging people, and answering true–false tests. Other examples come from geometric drawings deliberately planned to create an illusion. Study Figure 1.1. In Figure 1.1a, does the vertical line segment seem longer than the horizontal segment?

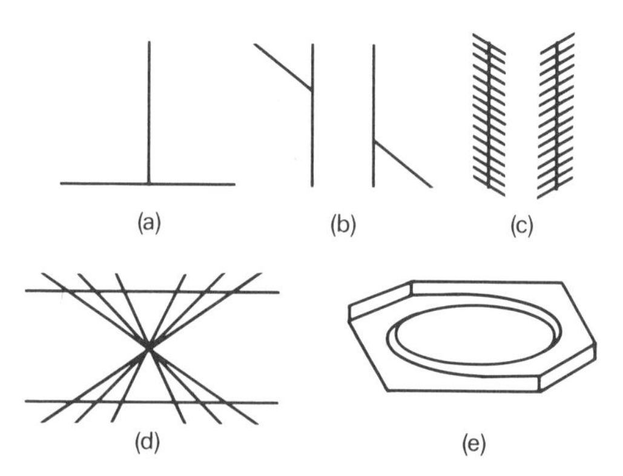

Figure 1.1. Geometric drawings to create illusions

In Figure 1.1b, are the two slanted segments part of the same straight line? In Figure 1.1c, do the segments seem to get closer together at the top? In Figure 1.1d, do the segments seem to get farther apart in the middle? What does your intuition tell you about the physical object pictured in Figure 1.1e? The fact that the geometry in this book is studied from an intuitive approach does not mean that whatever seems obvious is necessarily true.

The approach to the study of the geometry of space in this text depends on abstractions taken from intuitive ideas of space. Think of space as being made up of definite, precise locations. That is, think of each point in space as being a particular location, different from any other. You can then designate the positions of objects by naming specific points of space that they occupy. *Mathematical space is thus made up entirely of points.* Geometric figures consist of collections of these points. Mathematical space is the set of all points—that is, the set of all locations.

The description of mathematical space will become clearer as technical language is introduced. For example, the mathematician, abstracting from the familiar world, talks about sets of points such as lines, planes, circles, and triangles. Mathematical space, then, is a *model* with perfect forms and idealized locations. It should be understood that, although the mathematician sets up an idealized model of space, he can use results obtained in the model in practical ways.

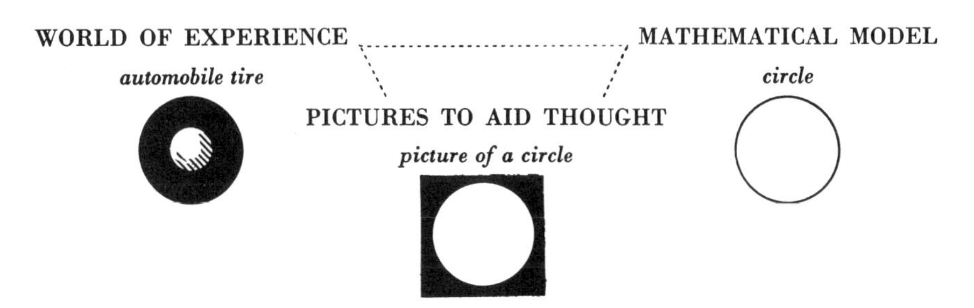

Figure 1.2. Concept of mathematical model

A diagram (see Fig. 1.2) and example will help clarify the distinction between what one might observe and the modern, intuitive approach to geometry. Suppose you look at an automobile tire while the car is parked. Immediately the word *circle* may come to mind. But a car tire is not a circle; a tire is a three-dimensional physical object; a circle is an idea. No person ever saw a circle, since it is an abstraction. You see physical objects that have certain characteristics reminding you of this abstraction and that you can describe using the word *circular*. In the mathematical model, within the realm of pure thought, it is possible to ask questions about the properties of a circle and how circles are related to other geometric figures.

The third part of the diagram in Figure 1.2 is extremely important; yet its significance is often poorly understood. This part is the picture of the circle, which may be considered a part of the mathematical model itself yet which is also a physical object. As an aid to thinking of abstract things, people often draw pictures or diagrams. To help you think about this abstraction called a circle, you draw a picture with a pencil on a piece of paper. Now consider such a picture carefully. This picture of a circle is not the circular tire with which you started; it is not even a good picture of the tire. Yet the two do have something in common. On the other hand, it is not a circle, because a circle is an idea. What is it? It is some lead marking on a piece of paper—a symbol that reminds us of something else. In geometry, it is customary to use pictures. For this reason, geometry is sometimes considered more concrete than is work with numbers. A careful attempt should be made, however, to distinguish between the picture itself, the mathematical idea, and the physical objects that you wish to characterize.

Geometry has a framework consisting of *undefined terms, defined terms, assumptions*, and *theorems* that can be proved. The formal approach to the study of geometry emphasizes the proof of theorems. In approaching geometry informally, it is not always necessary to begin with a complete statement of all the assumptions. Also, it is reasonable in an informal course to state theorems without proof. The emphasis is more on the concepts than on the nature of proof—more on sets of points than on sets of assumptions. For the reasons stated in this paragraph, the modern, intuitive approach to geometry rapidly surveys many aspects of traditional and modern geometry. Proofs are not ignored, but there is less emphasis on a precise form.

In any science, some terms must be undefined or you will find yourself using two words, each in the definition of the other, and hence defining neither. In the intuitive approach to the geometry of space, the mathematician has some freedom to decide which terms he will leave as undefined. Here is a customary beginning list, to which other terms may be added later. Following each term is a description or comment, but it should not be considered as a definition. An understanding of a concept comes from the study of many examples, so that calling a term undefined in no way implies that one cannot learn about it.

1. *Set.* A set is a collection or group of things.

Examples

$A = \{3, 5, 7\}$;

$B =$ {all the past and present Presidents of the United States}.

2. *Element or member of a set.* Each thing in a set is called an element or a member of the set. Elements of a set may be such diverse entities as points, numbers, or physical objects. For set A above, the members are 3, 5, and 7. For set B, each President of the United States is a member. The notation $\in$ is used to state that an element belongs to a set, whereas $\notin$ indicates that it does not.

Examples

$3 \in A$

$6 \notin A$

3. Point. In mathematics, a point is an idea. In the mathematical model for geometry, a point is thought of as a location in space. One exact location in space corresponds to one point. A point in space may be considered as an abstraction or idealization of such physical objects as the point of a pin, a small dot, or the tip of a compass. A point has no width, no length, and no depth. To picture a point, you may draw a dot and write a capital letter near it, as in Figure 1.3a. Why the word *point* is undefined can be understood by noticing the use of more complex words in these possible definitions: "A point is an indivisible part." "A point is something that has no magnitude."

4. Line. In mathematics, a line is a particular set of points. The word *line,* in this text, will always imply *straight line.* A line is considered to be infinite in length, without a beginning point or an endpoint; it has no width. A line is one-dimensional. Furthermore, between any two points on a line are an infinite number of other points. Two common symbols for a line are shown in Figure 1.3b. One uses any two different points, as $\overleftrightarrow{AB}$, and the other simply uses a small letter, such as line m. The notation $D \in \overleftrightarrow{AB}$ means that point D is one of the points on line AB.

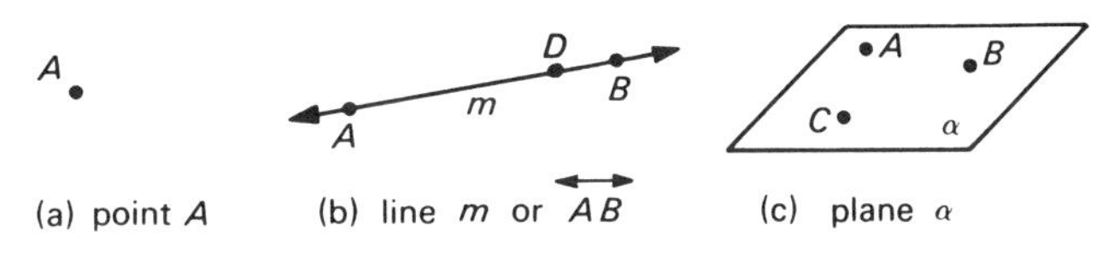

Figure 1.3. Point, line, plane

5. Plane. A plane is also a set of points. It is considered to be infinite in extent, without beginning or end. A plane is two-dimensional. A plane has no thickness, but one may speak of points *on* or *in* a plane. Many other common geometric figures, such as circles, triangles, or squares, lie wholly in a single plane. Often, Greek letters such as α, β, or γ are used to designate planes. (See Fig. 1.3c.) Another notation is plane ABC, where A, B, and C are three points in the plane, not all on the same line.

6. Space. Space is thought of as a set of points. It is considered to be infinite in extent, without beginning or end. A plane and a line are two examples of sets of points in space, as are all other geometrical figures. With the introduction of space as an undefined term, you see finally that the mathematical description "space is a set of points" contains three undefined terms. Within the mathematical model, sets of points form geometric figures that remind you of characteristics of actual physical objects.

Exercise 1.1

1. Can points be seen?
2. Can lines be seen?
3. Criticize each possible definition below by naming one word in it that seems more complex than the word being defined.
   (a) A point is an indivisible part.
   (b) A point is something that has no magnitude.

List in braces the members of each set in Exercises 4 and 5.

4. (a) The set of days of the week.
   (b) The set of labeled points in plane β (Fig. 1.4).

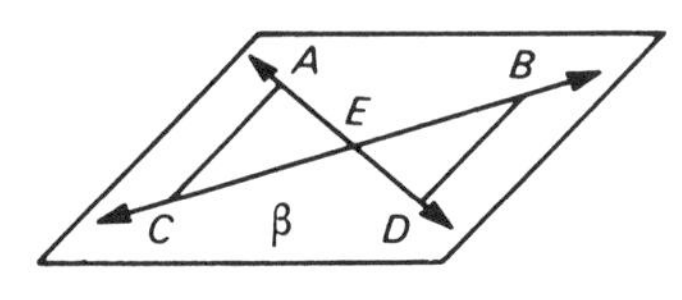

Figure 1.4

5. (a) The set of labeled points in plane β, but not on line CE (Fig. 1.4).
   (b) The set of labeled points (in Fig. 1.4) not on line AC or line BD.
6. Answer yes or no for Figure 1.4.
   (a) $E \in \overleftrightarrow{AD}$ (b) $E \in \overleftrightarrow{BC}$ (c) $E \notin \overleftrightarrow{AB}$ (d) $B \in \overleftrightarrow{AC}$
7. Make a diagram similar to Figure 1.2 (on p. 3) for a postal card and a rectangular region.
8. Draw and label a picture of six points in a plane, such that they lie in sets of three on four different lines.

Axioms, Theorems, and Proof

Axioms (sometimes called *postulates*) are statements that are assumed without proof; they are accepted as a starting point for the development of geometry. Axioms are not proved; theorems are. It is significant that axioms do not need to be even intuitively evident. For example, it is possible to assume that $2 + 3 = 6$, as long as you are willing to accept the conclusions that may result from using this axiom.

The selection of which terms to leave undefined in building a geometry is somewhat arbitrary, as is the selection of statements that you will assume as axioms. Selecting different axioms makes a great deal of difference in the nature of the geometry that results. In Chapter 14 you will study geometries that may seem strange because they are based on axioms that are not familiar and that in some cases may not agree with your common experience.

The geometry in most of this text has the same logical framework as does the ordinary geometry of high school. It is interesting to see first exactly which axioms Euclid stated when he first formalized a presentation of the structure of the kind of geometry that bears his name. His ten basic assumptions are separated into two sets, called the *common notions* and the *postulates*:

Common Notions

1. Things equal to the same thing are equal to each other.
2. If equals are added to equals, the wholes are equal.
3. If equals are subtracted from equals, the remainders are equal.
4. Things that coincide with one another are equal.
5. The whole is greater than the part.

Axioms

1. A straight line may be drawn from any point to any point.
2. A straight line may be produced continuously.
3. A circle may be drawn with any point as center and any distance as radius.
4. All right angles are equal to one another.
5. Through a given point in a plane, only one parallel can be drawn to a given line. (This version of Euclid's fifth postulate is not the actual statement Euclid made, although it is equivalent to it.)

Mathematicians have found that Euclid actually used some assumptions he did not list. For example, he also assumed that a line is infinite in length. Many people have attempted to develop better sets of axioms than those suggested by Euclid, and many modern sets of axioms are in existence. One famous set was developed by David Hilbert and is contained in his book *Foundations of Geometry*, written in 1902. His list of axioms is much longer than Euclid's. Hilbert's are in five groups, concerning connection, order, parallels, congruence, and continuity. Three examples of axioms found necessary by Hilbert are paraphrased here, so you can contrast them with those that Euclid used:

(a) If two points of a straight line lie in a plane, then every point of the line lies in the plane.

(b) If A, B, and C are points on the same straight line, and if B lies between A and C, then B also lies between C and A.

(c) If segment AB is congruent to segment CD and also to segment EF, then segment CD is congruent to segment EF.

In some technical presentations of a geometry, each axiom is *independent,* so that it cannot be established by using the other axioms in the set. For a beginning course in formal geometry, however, the set of axioms chosen often includes some that could be proved as theorems.

A set of axioms, whether each is independent or not, must be *consistent.* If a set of axioms is consistent, contradictions will not result from their use. You would not want to use a set of axioms that proves two contradictory statements. Showing that a set of axioms is consistent is important to mathematicians who investigate the foundations of mathematics. An example of an unfamiliar set of axioms that is both independent and consistent is given in the section on finite geometry in Chapter 14.

Modern lists of axioms appropriate for a first course in geometry are much longer than Euclid's list. The set of axioms given in this text is an adaptation and slight simplification of the axioms from Pearson and Smart's *Geometry* (Ginn and Company, 1971). The axioms are given in smaller groups in the chapters in which they are used. In that way it is possible to discuss the unfamiliar technical terms used and to avoid confusion by a long list.

A theorem is a provable statement. To prove a theorem is to show that it follows logically from the set of axioms, the undefined and defined terms, and the theorems that have been introduced previously in the geometry being studied. Whether the theorem is "true" is not the issue. The question to be settled is whether the theorem is valid—whether it is a logical consequence in the geometry you are studying.

In its simplest form, a theorem is an *if-then* statement consisting of a hypothesis and a conclusion. For example, here is one particular theorem (from Chapter 5) worded to emphasize the two parts: *If a quadrilateral is inscribed in a circle, then the sum of the measures of the opposite angles is 180 in degrees.* To prove a theorem in this form, begin with the *hypothesis* (the "if" statement) and reason step-by-step, supporting each of your reasons by some statement of an axiom, a definition, or some previous theorem, until you have arrived at the *conclusion*. This step-by-step process of establishing a theorem is a *formal deductive proof*.

Some of the differences between a rigorous proof in geometry and the more informal proofs found in this text are:

1. All the basic assumptions must be clear for a formal proof.
2. The order of proofs is extremely important in a rigorous development of geometry.
3. Theorems to be proved may have more than one hypothesis and more than one conclusion, so that the structure of the proof may be more complex.

A second kind of proof often used in geometry is called the *indirect method of proof*. Suppose, for example, that the theorem just stated is to be proved indirectly. There are three alternatives possible (often called trichotomy), exactly one of which must be correct: the sum is greater than 180, equal to 180, or less than 180. If the first and third of these alternatives can be shown to lead to some contradiction, then they must be rejected as incorrect and the remaining alternative must be accepted as the correct one. In Figure 1.5, since inscribed angles such as $\angle ABC$ are measured by half

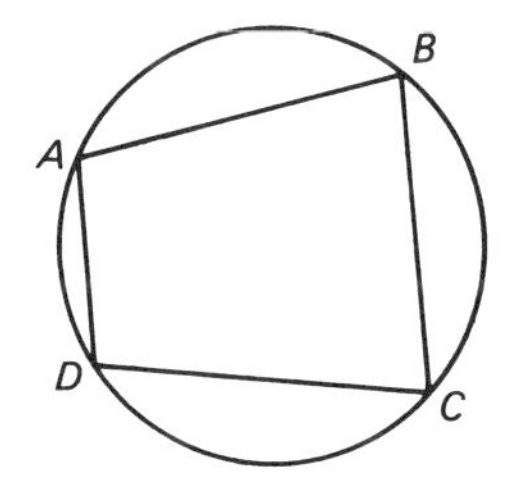

Figure 1.5. Quadrilateral inscribed in a circle

their intercepted arc ($\widehat{ADC}$), the assumption that the sum is greater than 180 or less than 180 will lead to the conclusion that the number of arc degrees in the circle is, respectively, greater than 360 or less than 360, both of which must be rejected as contradicting an established fact. Specifically, if the sum of the measures of the angles at A and C is less than 180 in degrees, then the sum of the measures of the arcs BCD and BAD would be less than 360 in arc degrees, which is a false conclusion.

Although the indirect method of proof is common in the work of Euclid himself, it is often used in modern geometry as well. The method is also used in other branches of mathematics; it is especially useful when only a few possibilities exist, all but one of which can be eliminated.

This brief discussion of theorems has been designed to complete the description of the framework of geometry and to give some notion of the idea of proof. The consideration of the form in which a proof is written can more conveniently be studied when there is a need to prove theorems.

Exercise 1.2

1. Euclid's fifth postulate, as originally stated, was: *"If a straight line falling on two straight lines makes the interior angles on the same side less than two right angles, the two straight lines, if produced indefinitely, meet on that side on which the angles are less than the two right angles."* Draw a figure illustrating this axiom and label the two angles ABC and DEF.
2. Some of Hilbert's axioms are called plane axioms and some space axioms (concerning points in more than one plane).
   (a) Which kind of axioms are the three previously stated?
   (b) Which kind is this axiom: "*There exist at least four points that do not lie on the same plane*"?
3. If you had to choose between two sets of axioms, one of which contained some axioms that were not independent and the other of which contained axioms that were not consistent, which set would it be better to choose?

For Exercises 4–7, answer yes or no.

4. The "if" part of a theorem may be considered as the hypothesis.
5. For a formal proof, knowing what the basic assumptions are is not important.
6. The use of the indirect method of proof is limited to non-metric geometry.
7. If a set of axioms is consistent, then two theorems that are valid in this geometry may be inconsistent.

For Exercises 8–11, list the alternatives, all but one of which would have to be eliminated if the theorem is to be proved by the indirect method. The number of alternatives to list is indicated in parentheses.

8. The exterior angle of a triangle has a measure greater than the measure of an opposite interior angle. (3)
9. The sum of the measures of the three angles of a triangle is 180. (3)
10. The longer of two chords in a circle lies nearer the center. (3)
11. The base angles of an isosceles triangle have the same measure. (2)
12. Three of the four parts in the structure of a geometry might be given at the beginning of a course of study. Which one of the four parts would not be given at that time?

Using Logic in Geometry (Supplementary)

Mathematicians feel that the study of logic can contribute greatly to the study of mathematics, since logic is used whenever theorems are to be proved. Although the study of logic may not be so important in an informal approach to geometry with less emphasis on proof, an introduction to the subject will provide some necessary vocabulary, and the reader will have available a tool that is also of interest for its own sake.

Logic is the study of various forms of arguments. The famous Greek philosopher Aristotle, who lived before Euclid, was one of the first persons to develop the science of logical reasoning.

In logic, a *statement* is a simple sentence that is either true or false. Here are some examples of statements from geometry, with the truth or falsity indicated.

Examples

A triangle is a polygon.	True
A square has five sides.	False
Three points always lie on the same line.	False
Two parallel lines do not have a point in common.	True

The statements of logic are joined by words or phrases called *connectives* to make *compound statements*. In elementary logic, five different types of connectives are used to combine two statements or to negate a simple statement. In logic, it is customary to use the letters p and q to represent individual statements. The examples below show the various connectives and the symbolism normally used for each.

Examples

Connective	*Name of Compound Statement*	*Example*	*Symbolism*
and	conjunction	Two points determine a line, and three non-collinear points determine a plane.	$p \wedge q$
or	disjunction	Two points determine a line, or three non-collinear points determine a plane.	$p \vee q$
not	negation	Two points do not determine a line.	$\sim p$
if, then	implication	If two points determine a line, then three non-collinear points determine a plane.	$p \rightarrow q$
if and only if	equivalence	Two points determine a line if and only if three non-collinear points determine a plane.	$p \leftrightarrow q$

The question of whether a compound statement is true or not is harder to answer than for a simple statement. The truth or falsity of a compound statement is important in geometry, since it is often desirable to present theorems in the form of a compound statement. Logicians have developed *truth tables* to show how the truth of a compound statement depends on the truth of the simple statements of which it is composed. These truth tables should be taken as the *definitions* of the meanings of the connectives, although they may also be explained logically without great difficulty.

From the truth table for a conjunction, it can be seen that a conjunction is true only when both of the simple sentences are true.

Truth Table for Conjunction

p	q	$p \wedge q$
T	T	T
T	F	F
F	T	F
F	F	F

Examples

Two points determine a line, and a triangle is a polygon.	True
Two points determine a line, and a triangle has four sides.	False
A triangle has four sides, and a circle is a polygon.	False

Truth Table for Disjunction

p	q	$p \vee q$
T	T	T
T	F	T
F	T	T
F	F	F

The truth table for a disjunction shows that a disjunction is true unless both the simple statements composing it are false.

Examples

Two points determine a line, or a triangle is a polygon.	True
Two points determine a line, or a triangle has four sides.	True
A triangle has four sides, or a square has five sides.	False

Truth Table for Negation

p	$\sim p$
T	F
F	T

The truth value of a negation is always the opposite of the truth value of the original statement. Although a negation involves only one sentence and a connective, it is still called a compound statement.

Example

Two points determine a line.	True
Two points do not determine a line.	False

Truth Table for Implication

p	q	$p \to q$
T	T	T
T	F	F
F	T	T
F	F	T

An implication is always considered true unless the "if" part is true and the "then" part is false.

Examples

If two points determine a line, then a triangle is a polygon.	True
If two points determine a line, then a triangle is a square.	False
If a triangle is a square, then a square has five sides.	True

Truth Table for Equivalence

p	q	$p \leftrightarrow q$
T	T	T
T	F	F
F	T	F
F	F	T

The truth table for an equivalence shows that an equivalence is true only when both of the parts are true or when both of the parts are false.

Examples

Two points determine a line if and only if a triangle is a polygon.	True
Two points determine a line if and only if a triangle has four sides.	False

Of the five types of compound statements introduced, the most commonly used in geometry is probably the implication, since theorems are normally stated in if-then form. The "if" part of the theorem is the *premise* or *hypothesis,* and the "then" part is the *conclusion.* For a basic implication, there are three other related compound statements that may be formed by interchanges and negations.

Example

Implication $p \rightarrow q$	If one angle of a triangle is a right angle, then the other two angles of the triangle are acute angles.
Converse $q \rightarrow p$	If two angles of a triangle are acute angles, then the third angle is a right angle.
Inverse $\sim p \rightarrow \sim q$	If one angle of a triangle is not a right angle, then the other two angles of the triangle are not acute angles.
Contrapositive $\sim q \rightarrow \sim p$	If two angles of a triangle are not acute angles, then the third angle is not a right angle.

In the example above, the implication is true. The contrapositive is also true, but the converse and the inverse are not. In general, the converse and the inverse do not necessarily have to be true when the implication is true. The contrapositive will have the same truth values as the implication itself; hence it is said to be logically equivalent to the implication. If a theorem has been proved, then the contrapositive is also proved, but not the converse or the inverse. As might be suspected, the converse and the inverse are themselves logically equivalent to each other.

The next truth table shows that an implication and its contrapositive have the same truth tables. This point can be observed by comparing the last two columns, which are identical. The last column is obtained by using the columns for $\sim q$ and $\sim p$ and the truth table for an implication.

p	q	$\sim q$	$\sim p$	$p \rightarrow q$	$\sim q \rightarrow \sim p$
T	T	F	F	T	T
T	F	T	F	F	F
F	T	F	T	T	T
F	F	T	T	T	T

Exercise 1.3

1. Write the negation of each statement.
   (a) A sphere is a three-dimensional figure.
   (b) A disjunction is a compound statement.

For Exercises 2–5, write the compound sentence indicated if
p: A right triangle has one right angle.
q: An acute triangle has three acute angles.

2. Disjunction $p \vee q$
3. Equivalence $p \leftrightarrow q$
4. Implication $p \rightarrow q$
5. Conjunction $p \wedge q$

For Exercises 6–14, use the truth tables to tell whether each compound sentence is true or false if
p: A triangle has four sides.
q: A parallelogram has four sides.

6. $p \wedge q$
7. $p \rightarrow q$
8. $p \leftrightarrow q$
9. $p \vee q$
10. $\sim p$
11. $\sim q$
12. $p \rightarrow \sim q$
13. $q \leftrightarrow \sim p$
14. $\sim p \rightarrow \sim q$

For each implication in Exercises 15–16, write the converse, inverse, and contrapositive.

15. If two points of a line are in a plane, then all the points of the line are in the plane.
16. If two lines are parallel, then they are everywhere equidistant.

This chapter analyzes more carefully some of the basic sets of points of plane geometry. Rather than study any sort of measurement associated with such sets of points, you will study their definitions and the relationships among them, expressed in the language of sets and of operations on sets. As used in this text, non-metric geometry means that measurement is not considered. As you read this chapter, you should continue to think of familiar objects that remind you of the various sets of points being discussed.

Definitions in Non-Metric Geometry

Since the language of sets is used so extensively in non-metric geometry, some of the terms need precise definitions. It is important to understand that the concept of *sets* is not confined to sets whose elements are points. What will be said about sets is just as applicable when the elements are numbers, objects, sets of points, or something else.

1. Equality of sets. Two sets are equal if and only if they have exactly the same members. Saying that set A is equal to set B simply means that A and B are different names for the same set.

Examples

Here are two examples of equal sets: in the first example the members are numbers; in the second the members are points.

(a) $\{1, 2, 3\} = \{3, 2, 1\}$.

(b) $A = B$, where B is the set of labeled points on line l in Figure 2.1, and A is the set of points M, N, O, P.

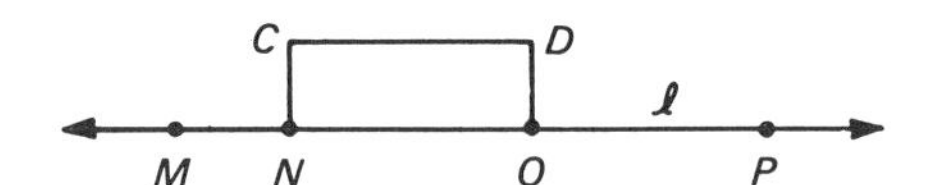

Figure 2.1. Example of equal and unequal sets

Two sets of points are equal only when the members of each set are the same points, which means that the sets include all the same locations in space. In Figure 2.1, it is not correct to say that the set containing points C and D and the points on the line between them is equal to the set of points N and O and the points on the line between them. The two sets of points certainly have something in common, but they are not points on the same line; hence the members could not be the same locations in space.

2. Finite and infinite sets. The number of members of a finite set is a whole number. Whole numbers are 0 and the counting numbers 1, 2, 3, and so on.

Examples of finite sets:

$$A = \{X, Y, Z, W, R\}, \qquad B = \{9, 12\},$$

$$C = \{\text{all states in the United States}\}.$$

Set A has five members, set B has two members, and set C has 50 members.

You cannot use a whole number to describe the number of members of an infinite set. The attempt to actually count each member could go on and on without end. The notation for an infinite set often includes three dots to indicate that the list of members goes on indefinitely.

Examples of infinite sets:

$$A = \{3, 6, 9, 12, \ldots\}$$
$$B = \{\text{all the points on } \overleftrightarrow{AB}\}$$
$$C = \{\text{all the lines in plane } \alpha\}$$

One special finite set needs further discussion, since it is particularly significant in non-metric geometry. This is the set with no members—the *empty set*—symbolized by { } or by Φ.

Examples of empty sets:

$$A = \{\text{all counting numbers between 2 and 3}\}$$
$$B = \{\text{points on both } \overleftrightarrow{MP} \text{ and } \overleftrightarrow{CD} \text{ in Fig. 2.1}\}$$

Two sets are said to have the same *cardinal number*, or to be *equivalent*, if a *one-to-one correspondence* can be established between their members. That is, a matching or pairing can be established in which each member of the first set is paired with exactly one member of the second set and each member of the second set is paired with exactly one member of the first set. In other words, matching pairs can be indicated, with one member of each pair from the first set and one member from the second set. This arrangement includes all the members of both sets. Examples indicating the pairing of members are shown. For finite sets, having the same cardinal number simply means having the same number of members.

$$\begin{matrix} \{4, & 5, & 6\} \\ \updownarrow & \updownarrow & \updownarrow \\ \{3, & 4, & 5\} \end{matrix} \qquad \begin{matrix} \{1, & 2, & 3, \ldots\} \\ \updownarrow & \updownarrow & \updownarrow \\ \{2, & 4, & 6, \ldots\} \end{matrix}$$

3. Universal set. The universal set means the most general and all-inclusive set you are considering at the time. For example, the universal set may be all the counting numbers, or it may be some other more general set, such as the set of rational numbers or real numbers. In non-metric geometry, you have thought of all the points in space as a universal set, with other sets of points all belonging to it. In this chapter, you are studying sets of points in a plane. Here the universal set is all the points of the plane, and all the sets you are studying are included in this universal set.

4. Subsets. Each set is a subset of some universal set. *A given set is a subset of a second set if each member of the first set is also a member of the second set.* For example, each set of points on a line is a subset of the set of all points on the line. The set of points on a line is a subset of the set of all points of any plane containing the line. The set of even numbers is a subset of the set of whole numbers, and the set of counting numbers between 3 and 10 is a subset of the set of all counting numbers.

If A is a subset of B, you can write $A \subseteq B$. This notation allows for the fact that A and B could actually be equal sets. By the definition of a subset, A is a subset of B if $A = B$, since each member of A is also a member of B. If you wish to exclude the possibility that $A = B$—that is, if you wish to specify that B has at least one member not in the subset A—you may use the notation $A \subset B$. In this case, A is called a *proper subset* of B. If $A = B$, A is called an *improper subset* of B.

Examples

$$\{3, 5, 7\} \subset \{1, 2, 3, 4, 5, 6, 7\}$$
$$\{A, B\} \subset \{A, B, C\}$$
$$\{\overleftrightarrow{AB}\} \subseteq \{\overleftrightarrow{AB}\}$$

5. *Intersection of sets.* Two sets may or may not have some members in common. Those members that belong to each of two sets are said to be in the intersection of the two sets. The symbol for the intersection of sets A and B is $A \cap B$. The intersection of two sets is itself a set, and its members are also members of both the original sets.

Examples

$$\{3, 5, 9\} \cap \{4, 9, 11\} = \{9\}$$
$$\{\text{Bob, Mary}\} \cap \{\text{Helen, Bob, Mary}\} = \{\text{Bob, Mary}\}$$

Sometimes two sets have no members in common. In that case, their intersection has no members. A set without members, as you have learned, is called the empty set. If the intersection of two sets is the empty set, then the two sets are said to be *disjoint.*

Examples

$$\{1, 2, 3\} \cap \{4, 5\} = \{\ \}$$
$$\{\text{typewriter, tree}\} \cap \{\text{barn, alligator, persimmon}\} = \{\ \}$$

6. *Union of sets.* The union of two sets is a set whose members belong to either or both of the other sets. Any member of one of the two sets is also a member of the union of the sets.

Examples

$$\{1, 2, 3\} \cup \{4, 5\} = \{1, 2, 3, 4, 5\}$$
$$\{2, 3, 5\} \cup \{2, 3, 4, 6\} = \{2, 3, 4, 5, 6\}$$

Just as you can perform operations such as addition and subtraction on two numbers to arrive at a unique third number, you can also perform operations on two sets to arrive at a unique third set. Intersection and union are operations on two sets that result in a third set, which is uniquely determined by the two original sets.

The idea of intersection and union with sets of points makes it possible to discuss relationships among various sets of points. So far, the discussion in this section has involved only those basic sets of points, such as lines and planes, that were not defined. But the vocabulary of union and intersection of sets also helps make it possible to define other sets of points, which will be recognized as common geometric concepts. The definition in each case consists of a complete description of the specific set of points being designated.

7. *Partition of a set.* Partitioning a set is accomplished by describing any set of non-empty disjoint sets whose union is the original set. Disjoint sets have no members in common, so that each member of the original set is a member of exactly one set in the partition. In general, sets may be partitioned in many different ways. For example, two partitions of the set $\{1, 2, 3, 4\}$ are $\{1, 3\}, \{2, 4\}$ and $\{1, 3, 4\}, \{2\}$.

Any non-empty set of points is called a *geometric figure.* The following definitions explain some of the basic geometric figures that are of great usefulness in the early development of a course in geometry.

8. *Line segments.* Two distinct points of a line and all the points on the line between the two points constitute a set of points named a *segment.* A segment is a part of a straight line; it contains two points of the line, called the *endpoints* of the segment, and all the points on the line between the two endpoints. Two ideas need to be pointed out here. One is that the word *between* is actually considered an undefined term, although its meaning is certainly intuitively clear. The second idea is that, although a segment does

not contain all the points of a line, it does contain an infinite set of points. A segment, then, is a particular infinite set of points, uniquely determined by specifying two endpoints.

The usual notation used for the segment determined by points A and B is $\overline{AB}$. $\overline{AB} = \overline{BA}$, since the order of specifying the endpoints has no significance in the definition. Using the labeled points on $\overline{AB}$ in Figure 2.2, make

Figure 2.2. Line segments on a line

sure that you agree with each of the following statements:

(a) Point D is a member of the set of points on $\overline{AB}$. ($D \in \overline{AB}$)
(b) Point C is not a member of the set of points on $\overline{AB}$. ($C \notin \overline{AB}$)
(c) The intersection of $\overline{AD}$ and $\overline{BC}$ is the empty set. ($\overline{AD} \cap \overline{BC} = \{\,\}$)
(d) The union of $\overline{AD}$ and $\overline{DB}$ is $\overline{AB}$. ($\overline{AD} \cup \overline{DB} = \overline{AB}$)

A segment is *closed,* in the sense that it includes both endpoints. Sometimes, however, it is necessary to use an *open* segment, which consists of all the points on a segment except for the endpoints. The word *segment* used alone refers to a closed segment. The various possibilities and the notation for each follow.

Sets of Points	*Notation*
Closed segment with endpoints A and B.	$\overline{AB}$
Segment without endpoint A.	$\overset{\circ\!-\!-}{AB}$
Segment without endpoint B.	$\overset{-\!-\!\circ}{AB}$
Open segment.	$\overset{\circ\!-\!\circ}{AB}$

9. *Half-plane.* Any line partitions the set of points in a plane into three disjoint subsets of points: the two sets of points on either side of the line and the set of points on the line itself. All the points in a plane on one side of a

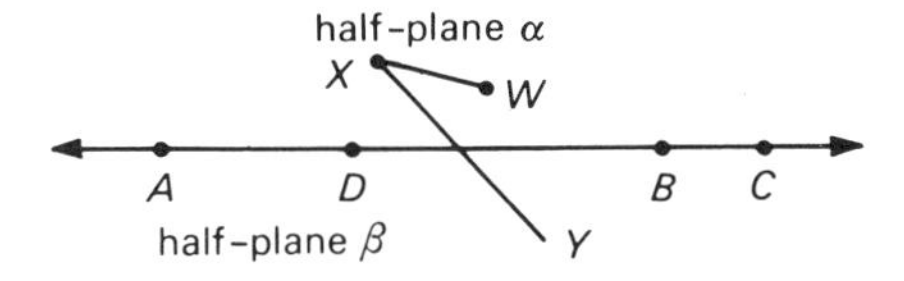

Figure 2.3. Points in half-planes

given line in the plane constitute a *half-plane.* Interestingly enough, the union of the two half-planes of a given line is not all the points in the plane, since neither half-plane includes the points on the line itself. A segment, such as $\overline{XY}$ in Figure 2.3, which has one endpoint in one half-plane and the other endpoint in the opposite half-plane, has exactly one point in common with the line that partitions the plane. If the two endpoints, such as X and W, are in the same half-plane, then the segment determined by them has no points in common with the line that does the partitioning. A half-plane as defined here is sometimes called an *open half-plane.* The union of a half-plane and the line determining it is called a *closed half-plane.*

10. Half-line. One point on a line, such as point B in Figure 2.4, partitions

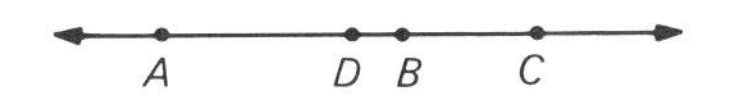

Figure 2.4. Half-lines and rays on a line

the line into three disjoint sets of points. One set consists of the point B itself, and the other two sets contain the points in the two half-lines defined by the point—that is, the points to the left of B and the points to the right of B. If a segment, such as $\overline{AC}$, has one endpoint in one of the half-lines and the other endpoint in the other half-line, it includes the point that defines the particular half-lines. Since both endpoints of $\overline{AD}$ are in the same half-line determined by B, B is not on the segment. An informal definition of half-line is all the points on a line on one side of some particular given point on the line.

11. Ray. A set of points consisting of a point on a line and all the points on one side of it is called a ray. A ray, then, is the union of a point and one of the half-lines formed by it. A ray could be defined as a closed half-line. The symbol for ray AB is $\overrightarrow{AB}$. In this symbolism, the first letter names the endpoint (also called the beginning point), the second letter names one of the points in the particular half-line specified. In Figure 2.4, $\overrightarrow{AD}$ also includes points B and C. $\overrightarrow{DA} \cup \overrightarrow{DB}$ is $\overleftrightarrow{AB}$. The choice of *ray* for the name of this particular set of points agrees with the intuitive association of this set with the common use of ray—for example, rays of the sun or of a ray gun.

In this section, the basic vocabulary of sets has been introduced. This vocabulary includes equality of sets, finite and infinite sets, empty set, universal set, subsets, and intersection and union of sets. The basic terms have been used to help define specific sets of points, such as line segment, half-plane, half-line, and ray. These new terms are a part of the structure of geometry and will be used many times to help develop new concepts in other sections.

Exercise 2.1

Answer yes or no for Exercises 1–14.

1. Two equal finite sets always have the same number of members.
2. A set whose members are points may be equal to a set whose members are numbers.
3. Every member of the intersection of two sets that are not disjoint and whose elements are points is a point.
4. The intersection of two sets can have more members than one of the two sets.
5. The intersection of two sets could be the empty set.
6. The union of two sets can have more members than either of the two sets.
7. The union of two infinite sets is an infinite set.
8. A straight line may have an infinite number of different line segments on it.
9. Two line segments on different lines may have one endpoint in common.
10. The set of points on a line segment is an infinite set of points.
11. An open segment has exactly one endpoint.
12. The two half-planes formed by a particular line include all the points of a plane.
13. A half-line and a ray are both infinite sets of points.
14. $\overrightarrow{AB}$ is the same set of points as $\overrightarrow{BA}$.

Give an example to support your answer to:

15. Exercise 5.
16. Exercise 6.
17. Exercise 9.
18. Exercise 14.
19. Take various examples and then guess whether this statement is always true, sometimes true, or never true: For two finite sets, the sum of the numbers of members in the union and the intersection of the two sets is equal to the sum of the numbers of members in the two sets.

In Exercises 20–21, find the union or intersection of the two sets of points from Figure 2.5 as indicated.

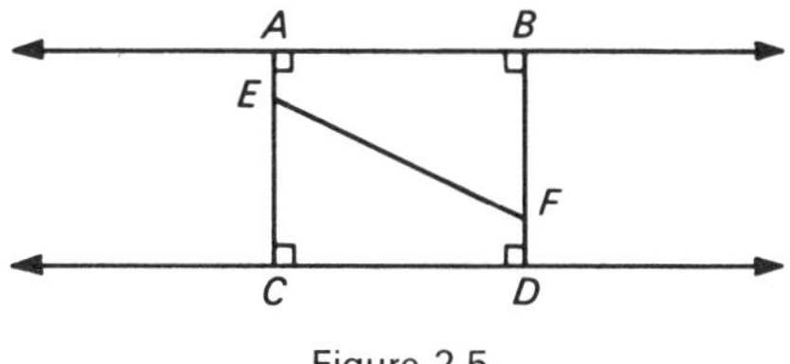

Figure 2.5

20. Find the intersection of $\overline{AB}$ and (a) $\overline{AD}$, (b) $\overleftrightarrow{CD}$, (c) $\overline{EF}$, (d) $\overrightarrow{AB}$, (e) $\overrightarrow{BA}$, (f) B, (g) $\overleftrightarrow{AB}$, (h) half-plane above $\overleftrightarrow{CD}$, (i) half-plane below $\overleftrightarrow{AB}$, (j) $\overset{\circ\!-\!-\!-}{AD}$.
21. Find the union of point E and (a) $\overline{EF}$, (b) $\overline{AD}$, (c) half-plane above $\overleftrightarrow{CD}$, (d) half-plane below $\overleftrightarrow{AB}$.
22. Explain why the empty set can always be considered a subset of any other given set.

Basic Axioms for Euclidean Geometry

The previous section, along with Chapter 1, has given you some of the definitions and undefined terms needed for Euclidean geometry. The following list of assumptions gives the basic list of axioms necessary to begin the study of the geometry of sets of points in a plane. Additional axioms will be given in other chapters as needed.[1]

1. Space contains at least two distinct points.
2. There is exactly one line that contains two distinct points.
3. A line is a set of at least two points.
4. No line contains all of the points of space.
5. If three points are distinct and non-collinear (not on the same straight line), then there is exactly one plane that contains them.
6. If two distinct points of a line lie in a plane, then every point of the line lies in the plane.
7. Any line in a plane separates the points of the plane that are not points of the line into two sets such that each is a convex set (defined below) and every segment that joins a point of one of these sets to a point of the other intersects the line.

[1]From Pearson, H. R., and Smart, J. R., *Geometry*. Boston, Ginn and Company, © 1971. Reprinted by permission of the publisher.

8. If a given point is not in a given line, then there is exactly one line containing the given point and parallel to the given line.

All these beginning assumptions are probably familiar to some extent, and most may be rather self-evident. They are not proved, however, but are used as reasons in proving other statements. The first axiom is an example of what mathematicians call an *existence* axiom. In this case, it guarantees that the entire discussion is not about empty sets. Similarly, axiom 4 guarantees that the geometry will not be confined to the set of points on a single line.

Axiom 7 is related to the definition of half-plane given previously. The idea of a *convex set* will be discussed in the next paragraph. Although concepts not yet defined are used in this discussion, they are illustrated by figures that should help make the meaning clear.

You have used the word *convex* and the idea of convexity outside a mathematics class. A convex lens curves outward, as in Figure 2.6. The mathematical definition of convex sets of points is closely related to the intuitive ideas of convexity.

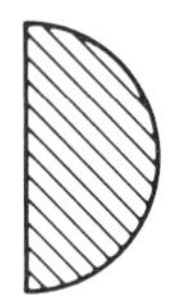

Figure 2.6. A convex set

Try to answer the following question about the sets of points in Figure 2.7a and 2.7b before reading ahead. The sets of points in Figure 2.7a are convex, but those in Figure 2.7b are not. From a mathematical point of view, what is the difference between sets that are convex and those that are not?

The set of points in Figure 2.8a is convex, but the set of points in Figure 2.8b is not. For any two points, such as A and B in Figure 2.8a, does each point of the segment joining them also lie in the set? Does this same property hold for the set in Figure 2.8b? Notice that some of the points in $\overline{A'B'}$ are not included in the original set. This intuitive development leads to a more formal definition of a convex set of points.

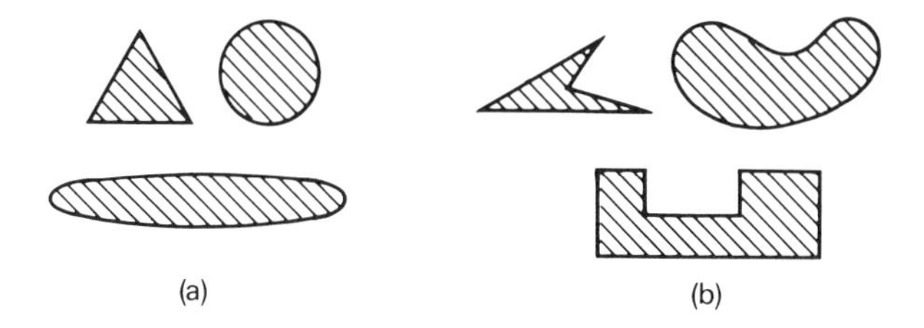

Figure 2.7. Convex and non-convex sets

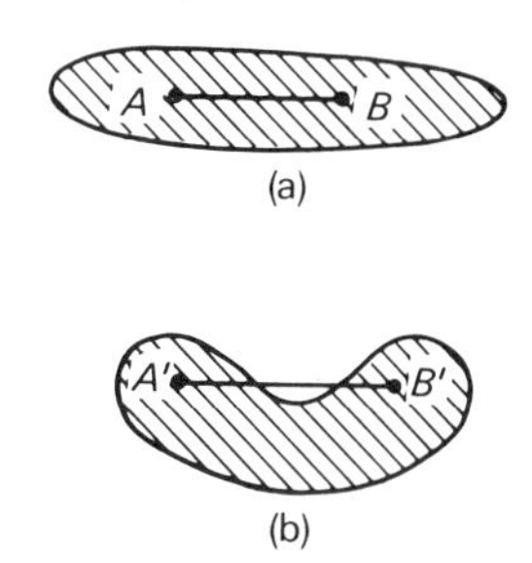

Figure 2.8. A convex and a non-convex set

A convex set of points is a set such that, for any two distinct points of the set, all the points of the segment joining the points are also points of the given set. Also, a single point and the empty set are both considered convex sets.

Examples

To see that the definition is clear, verify the classification of some of these sets with which you are familiar. The concept of region will be discussed later, but it is essential to know that a triangular region is the union of a triangle and its interior.

Convex	*Not Convex*
triangular region	triangle
circular region	circle
half-plane	polygon
half-line	ellipse

The axioms given so far do not tell you everything there is to know about sets of points, but they do make it possible to prove other statements. Two examples of theorems that result from the given axioms follow.

Theorem

If two lines are parallel to the same line in the plane, then they are parallel to each other. (See Fig. 2.9.)

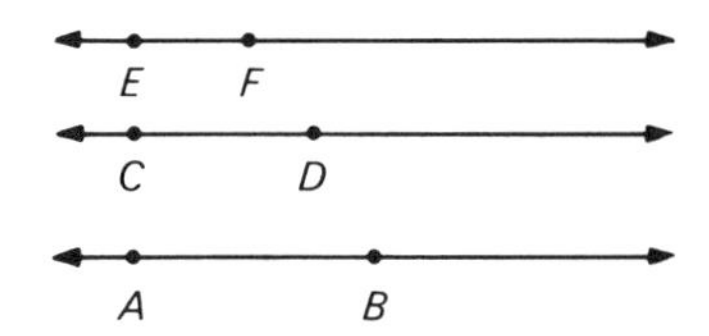

Figure 2.9. $\overleftrightarrow{EF}$ and $\overleftrightarrow{CD}$ parallel to $\overleftrightarrow{AB}$

The contrapositive of this theorem, which is logically equivalent to the theorem, is simpler to prove than the theorem itself. The contrapositive states that if two lines are not parallel to each other, then they are not parallel to the same line in the plane.

In Figure 2.10, if $\overleftrightarrow{GH}$ and $\overleftrightarrow{GI}$ are not parallel, then they have a point in common. By axiom 8, there is exactly one line through G parallel to $\overleftrightarrow{JK}$; hence both $\overleftrightarrow{GH}$ and $\overleftrightarrow{GI}$ cannot be parallel to $\overleftrightarrow{JK}$.

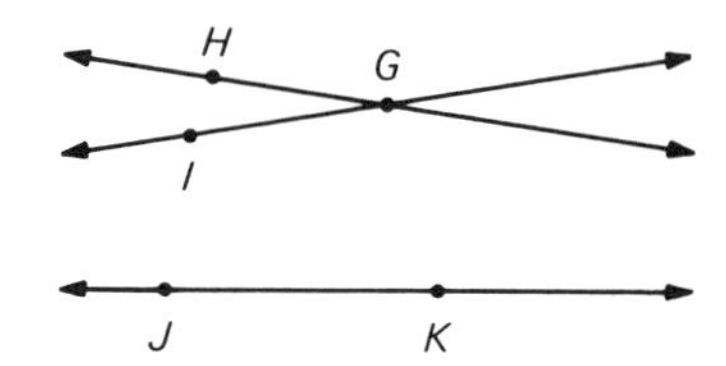

Figure 2.10. Illustration of contrapositive

Theorem

A plane contains at least four points.

The proof of this theorem is given in the traditional two-column form with the statements in the first column and the reasons in the second column. Figure 2.11 will help make the proof clearer.

To prove: A plane contains at least four points.

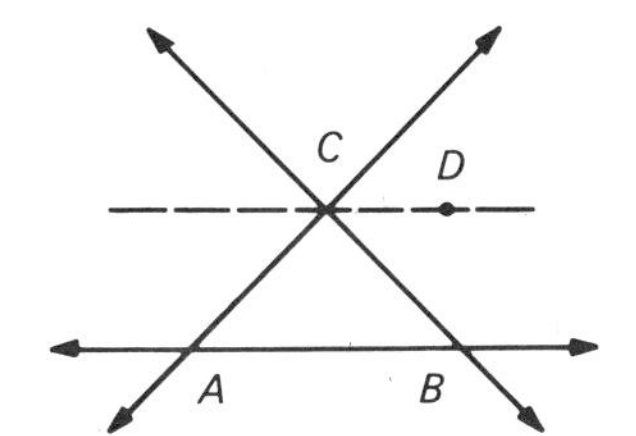

Figure 2.11. Illustration for theorem

1. A plane contains at least three distinct points, not all collinear.	1. Axiom 5
2. These three points determine three lines (lines $\overleftrightarrow{AB}$, $\overleftrightarrow{BC}$, and $\overleftrightarrow{CA}$).	2. Axiom 2
3. None of the three lines are parallel to each other.	3. Definition of parallel
4. There must be a line through one of the points parallel to the line through the other two.	4. Axiom 8
5. The line in step 4 contains a second point (D).	5. Axiom 3
6. The point in step 5 is a fourth distinct point in the plane.	6. Axiom 2

Proofs in this two-column form are rare in mathematics courses beyond the high school level. Instead, the proof is simply written in narrative style. Often the reasons for various steps are not stated explicitly when the author feels that the reader should be able to supply the reason without difficulty. A proof is planned to convince others of the truth of the theorem. Often, in an informal approach to geometry, the reader is asked to accept a theorem without any written proof.

Exercise 2.2

1. On the basis of axioms 1–8, without any additional proof, you can be sure a line contains at least how many points?
2. Axiom 8 is Euclid's postulate on parallels. Name two alternative parallel postulates that might have been stated as contradictory axioms to this one.

For Exercises 3–14, which sets of points are convex and which are not?

3. closed segment
4. open segment
5. open half-plane
6. closed half-plane
7. half-line
8. ray
9. two distinct points
10. three distinct and non-collinear points
11. a finite number of distinct points
12. a line
13. a plane
14. two parallel lines
15. Prove that a line in the plane intersecting one of two parallel lines also intersects the other.
16. Prove that if three lines in a plane have a common point of intersection, a line in the plane parallel to one of these lines intersects the other two.

Angles

The word *angle* is a common word that has some meaning for each person. If you were asked to describe some everyday examples of angles, you might mention the hands of the clock, the arm of a driver signaling to turn, aluminum or steel angles used for brackets and reinforcements, the slope of a roof, or two streets coming together. In all these cases, if you are not careful, you may find that you are thinking of the *measurement* of the angle and not the angle itself. You may have to think carefully to formulate a clear idea of just what you mean by the concept of an angle as a set of points, as distinguished from its measurement. The definition used in this text is only one of several acceptable ways of defining an angle.

Figure 2.12 depicts two angles. Using the language of sets of points, how would you describe the pictures you see? Each angle consists of two different rays. The two rays have the same endpoint, but they do not lie on

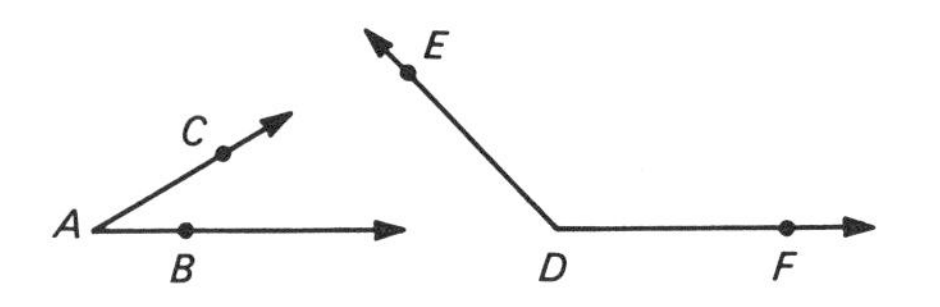

Figure 2.12. Angles

the same line. *An angle is defined as the union of two rays with a common endpoint but not lying on the same line.* The restriction that both rays do not lie on the same line is introduced here to avoid, for the present, considering a line as an angle.

The common endpoint for the two rays is called the *vertex* of the angle. A and D are the vertices of the angles in Figure 2.12. The rays forming an angle are called the *sides* of the angle; they intersect at the vertex. Each point of the angle is a point on one or both of the rays that form it.

The two angles in Figure 2.12 can be called $\angle BAC$ and $\angle FDE$. Notice that three distinct points are used in naming an angle. The second letter in the name refers to the vertex; the first and third letters name a point on each of the two rays. The order of choosing these two rays is not important here. For example, $\angle BAC = \angle CAB$, which is a statement about the equality of two sets of points. Sometimes, if no confusion would result, an angle is named simply by stating the letter for the vertex, as $\angle A$ or $\angle D$.

An angle partitions a plane into three disjoint subsets of points. Disjoint subsets have no members in common. One of these subsets has as members the points of the angle itself. In Figure 2.13, points C and E are in different subsets formed by $\angle BAK$, and point D is a member of the same subset as C.

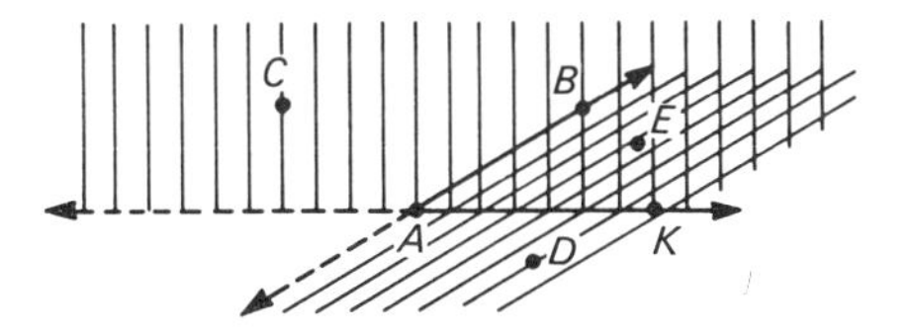

Figure 2.13. Interior of an angle

It seems natural, for angles such as those pictured, to call one of the subsets the *interior* of the angle and the other the *exterior* of the angle. Point E, then, is a point in the interior of $\angle BAK$, whereas points C and D are in the exterior. Any two points in the interior may serve as the endpoints of a segment that has no points in common with the angle. This statement is not true for all points such as C and D in the exterior of the angle.

Another way to distinguish the interior from the exterior uses the idea of half-plane. All points in the interior are in the same half-plane, with respect to $\overleftrightarrow{AB}$, as the point K and the other points on side $\overrightarrow{AK}$. The points in the interior are also in the same half-plane, with respect to $\overleftrightarrow{AK}$, as the point B and the other points on side $\overrightarrow{AB}$. The intersection of these two half-planes is the set of points in the interior of the angle. It has been possible to indicate the interior and exterior of an angle as two sets of points defined in set language, using the intersection of two half-planes. The set of points with cross-hatching in Figure 2.13 is the set of points in the interior of $\angle BAK$. In Figure 2.13, the interior is defined briefly as those points on the K side of $\overleftrightarrow{AB}$ and on the B side of $\overleftrightarrow{AK}$.

The preceding discussion of the interior of an angle can be formalized by stating the needed assumptions as an axiom, using the notation of Figure 2.13.

> Axiom 9
>
> The interior of $\angle BAK$ is equal to the set of the interior points on all rays between $\overrightarrow{AB}$ and $\overrightarrow{AK}$. The interior of the angle is also equal to the intersection of the shaded half-planes. A segment with endpoints B and K in $\overrightarrow{AB}$ and $\overrightarrow{AK}$ is a subset of the interior of the angle.

Now that the concept of angle as a set of points has been developed, it is possible to study pairs of angles whose significance does not depend on measurement. Two special pairs of angles are *adjacent* angles and *vertical* angles. Figure 2.14a shows adjacent angles AOB and BOC. In everyday

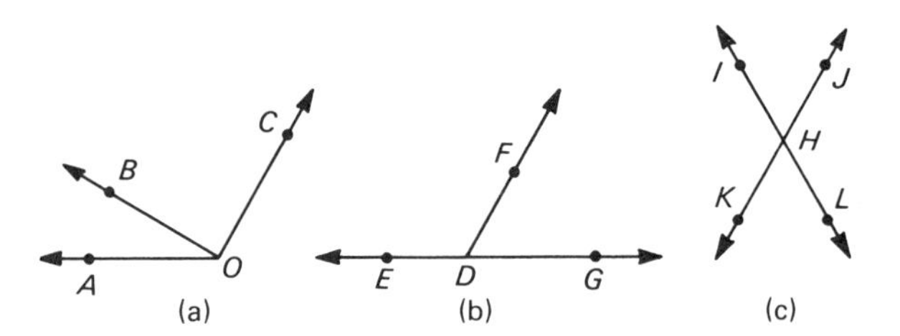

Figure 2.14. Adjacent and vertical angles

language, adjacent angles are two angles that are side by side, with their common side between them. Adjacent angles have the same vertex and also have a ray in common. No points in the interior of one of the angles are in the interior of the other angle. In other words, the sets of points in the interiors are disjoint. Pictured in Figure 2.14b are adjacent angles EDF and FDG. D is the common vertex and $\overrightarrow{DF}$ is the common side.

Figure 2.14c shows vertical angles IHJ and KHL. These two angles also have a common vertex but do not have a common side. Their sides do lie on the same two lines, however. In this same figure, $\angle IHK$ and $\angle JHL$ are another pair of vertical angles. The set of points on two lines that intersect forms four angles. These four angles are two pairs of vertical angles. For any one angle shown, the angle that is not adjacent to it is the other angle of the pair of the vertical angles.

Exercise 2.3

1. Name other common examples of objects that resemble angles.
2. What would always be true of a line segment that had one endpoint in the interior of an angle and one in the exterior of the same angle? Why?
3. How many different angles are shown in Figure 2.15a? 2.15b? 2.15c?

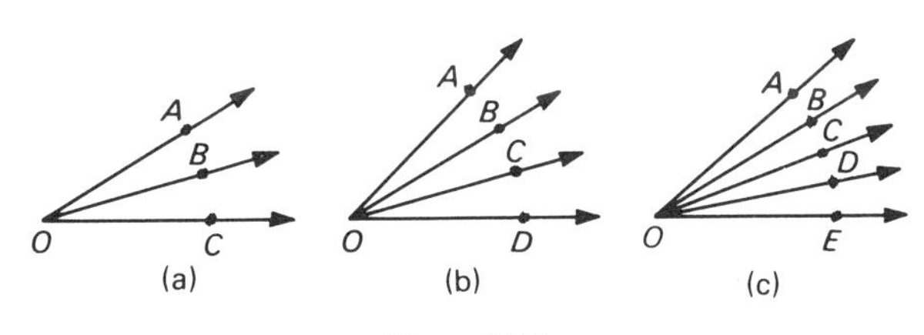

Figure 2.15

4. From Exercise 3, predict the remaining entries needed to complete this table.

number of rays with same vertex but on different lines	*number of angles determined*
2	1
3	3
4	____
5	____
6	____
7	____
8	____

5. Give a formula for finding the missing entries in Exercise 4.
6. How many pairs of adjacent angles are there in Figure 2.15a? 2.15b? 2.15c?

7. How many pairs of vertical angles are there in a picture of:
   (a) three concurrent lines (all lines meeting at the same point)?
   (b) four concurrent lines?
   (c) five concurrent lines?
8. Could two adjacent angles be a pair of vertical angles?
9. Could the angles of a pair of vertical angles be adjacent to one another?
10. Is an angle a convex set of points?
11. Investigate whether or not a line in the plane can be parallel to the lines containing both sides of an angle, and then state a theorem summarizing your findings.

Intersections and Unions of Sets of Points

In this chapter, various sets of points, such as rays and angles, have been defined and illustrated. These sets of points might be considered as the fundamental figures, or building blocks, of non-metric geometry. Many other sets of points can then be explained by using one or more of these basic sets, along with the notation of intersection and union.

The intersection of two sets of points is a figure no more complex than either of the two beginning sets of points. The union of two sets of points, however, is often a more complex figure than either beginning set. For this reason, it is primarily the operation of forming the union of sets of points that makes it possible to use the basic figures defined so far to help define others with more interesting properties.

Selected possibilities for intersection and union of two sets of points in a plane are discussed and illustrated briefly. The concepts of intersection and union will be extended in later chapters as new sets of points are introduced.

1. Intersection of lines in a plane. Two distinct lines may meet at a point in a plane, such as lines l and m meeting at point A in Figure 2.16a. Also, two lines in a plane may not meet, as $\overleftrightarrow{XY}$ and $\overleftrightarrow{EF}$ in Figure 2.16b. Two lines *in a plane* that do not meet are called *parallel* lines. The notation for two parallel lines is two vertical marks—for example, $\overleftrightarrow{XY} \parallel \overleftrightarrow{EF}$. A concise statement of the first possibility is $l \cap m = A$ and the statement of the

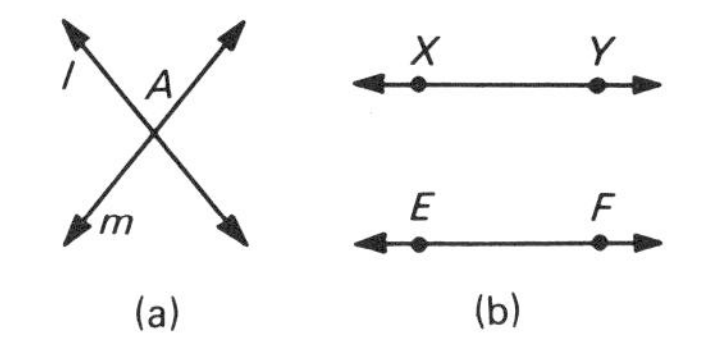

Figure 2.16. Pairs of lines in a plane

second is $\overleftrightarrow{XY} \cap \overleftrightarrow{EF} = \{\ \}$. Two parallel lines have the empty set for their intersection, which is simply another way of saying they have no points in common.

2. *Intersection of two segments.* Study Figure 2.17 to find the intersections of the various segments. Check each of these statements to see that

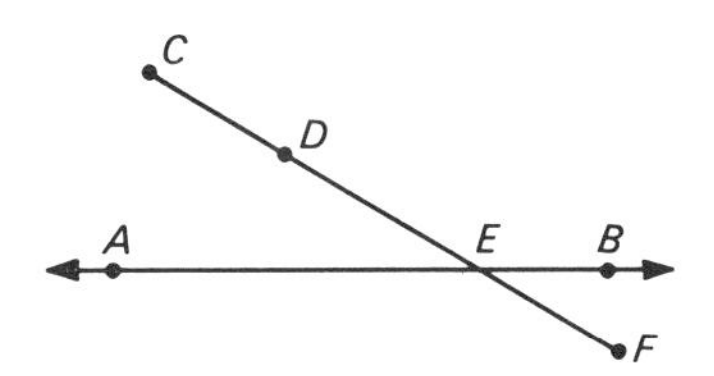

Figure 2.17. Intersection of segments

they describe the relationships shown. $\overline{AE} \cap \overline{ED} = E$ since E is a common endpoint. $\overline{CE} \cap \overline{DF} = \overline{DE}$ since the segments overlap. $\overline{CF} \cap \overline{AB} = E$.

3. *Intersection of a ray and segment.* Study Figure 2.18. Then write, using proper symbols, the intersections of each of these rays and segments: $\overrightarrow{CB}$ and $\overline{DA}$, $\overrightarrow{AB}$ and $\overline{AC}$, $\overline{CD}$ and $\overrightarrow{AB}$, $\overline{EF}$ and $\overrightarrow{CB}$, and $\overline{EG}$ and $\overrightarrow{CB}$.

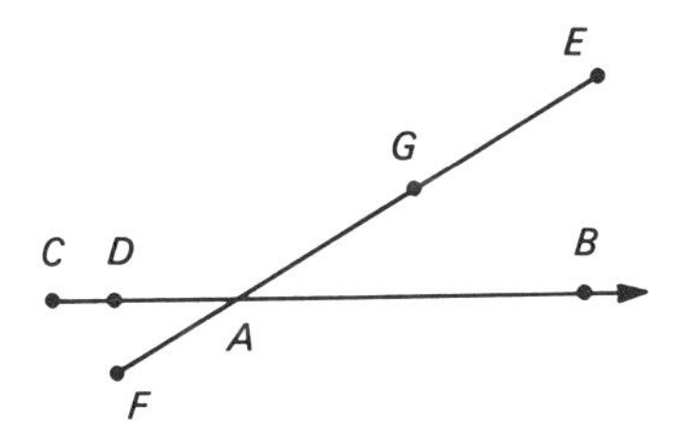

Figure 2.18. Intersection of rays and segments

4. *Intersection of an angle and line.* In Figure 2.19, interpret each of these statements by finding the intersection and explaining the

notation in sentence form: $\angle ABC \cap \overleftrightarrow{BC}$, $\angle ABC \cap \overleftrightarrow{AC}$, $\angle ABC \cap \overleftrightarrow{BE}$, $\angle ABC \cap \overleftrightarrow{DF}$.

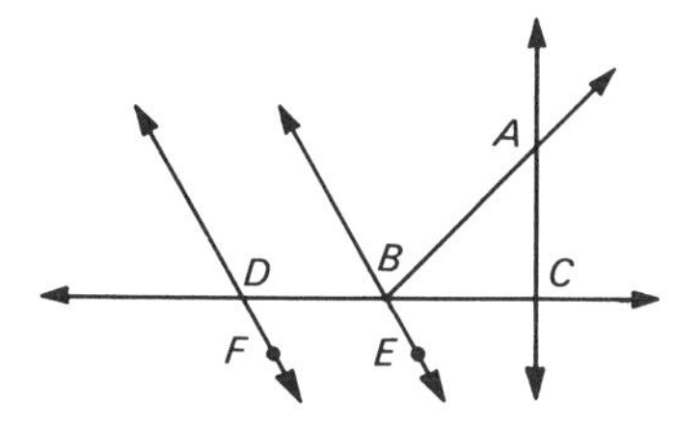

Figure 2.19. Intersection of angles and lines

Some extensions to the study of possibilities for intersection are made in the exercises; however, there are other questions that are not investigated. For example, what sets of points are possible intersections of three different angles? What sets of points are possible intersections for three lines, or for two lines and one segment?

5. *Union of two segments.* Two segments on the same line may have various relationships, which are illustrated in Figure 2.20a. Each of these notations should be explained in words: $\overline{AC} \cup \overline{BC} = \overline{AC}$, $\overline{AB} \cup \overline{BC} = \overline{AC}$, $\overline{AC} \cup \overline{BD} = \overline{AD}$, $\overline{AD} \cup \overline{BC} = \overline{AD}$. Notice that the union of $\overline{AB}$ and $\overline{CD}$ is simply the two separate segments.

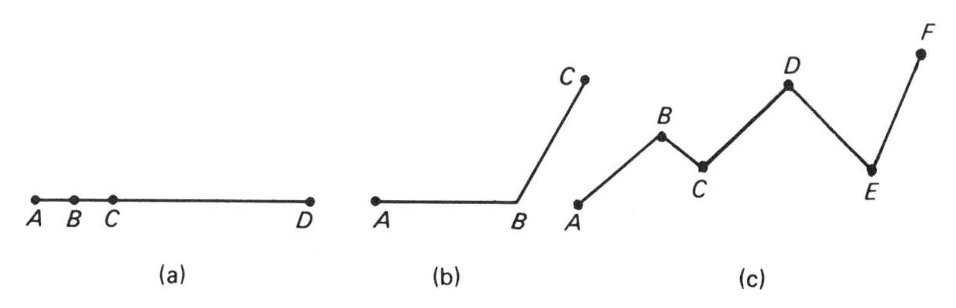

Figure 2.20. Union of segments

Two segments in a plane, not on the same line, may be entirely separate. One of the interesting configurations is pictured in Figure 2.20b. Here the two segments have a common endpoint. The union of a finite number of segments not all on the same line but end-to-end as in Figure 2.20b or 2.20c is called a *broken line.* Notice that a broken line is finite in length, rather than infinite, despite the fact that the word *line* is used in its name.

6. *Union of two rays.* Two rays on the same line may have a union that is a line, a ray, or two separate rays. Each of these possibilities should be

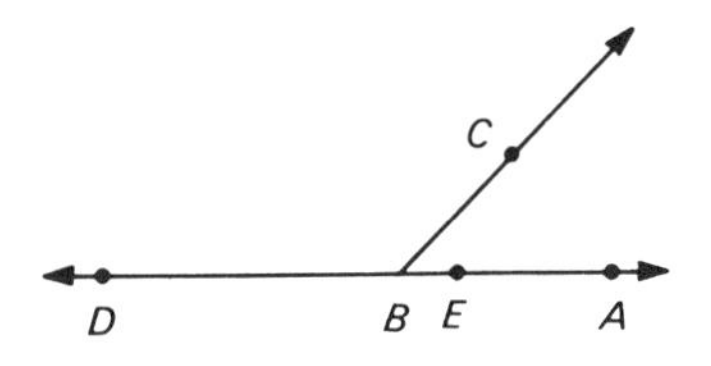

Figure 2.21. Union of rays

named, using the points in Figure 2.21. Two rays not on the same line might be positioned like $\overrightarrow{BC}$ and $\overrightarrow{BA}$ in Figure 2.21. What is the name for this important set of points, which is the union of two rays?

7. *Union of half-lines and half-planes.* Figure 2.22 shows some interesting unions of two sets of points, one of which is a half-line or half-plane. The union of the half-line on $\overleftrightarrow{CD}$ on the right of point C and point C is $\overrightarrow{CD}$. Similarly, the union of that same half-line and $\overline{EC}$ is $\overrightarrow{ED}$. The union of the

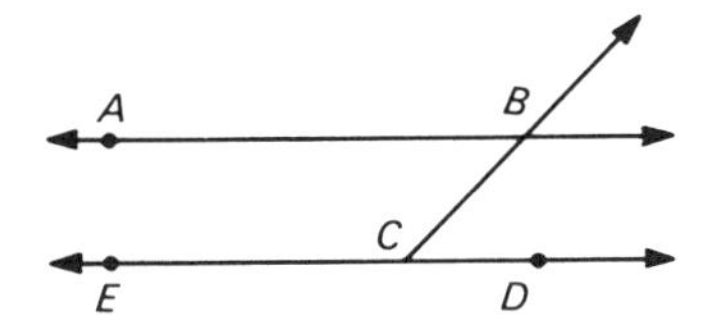

Figure 2.22. Union of half-lines and half-planes

same half-line and $\overrightarrow{CB}$ is angle BCD. The union of two half-lines may be a line, if they are not determined by the same point. For example, the union of the half-line to the right of A and the half-line to the left of B is $\overleftrightarrow{AB}$.

In Figure 2.22, the union of the half-plane below $\overleftrightarrow{AB}$ and the half-plane above $\overleftrightarrow{CD}$ is the entire plane. The union of the half-plane not on the D side of $\overleftrightarrow{BC}$ and not on the B side of $\overleftrightarrow{CD}$ is the exterior of angle BCD. Practice will increase the ability to look at a picture and explain it in terms of the union and intersection of sets of points.

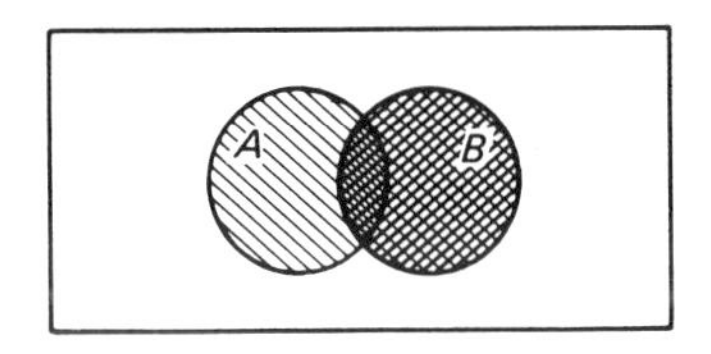

Figure 2.23. Venn diagram

A common way to picture union and intersection of sets is by means of a *Venn diagram*. Venn diagrams are probably used less often with sets of points than with sets of numbers, since it is possible to show union and intersection in geometry without them by showing the actual sets of points. Figure 2.23 shows a Venn diagram. The elements of set A are inside the circle labeled A, and the elements of set B are inside the circle labeled B. The elements in the shaded portion are in the union; the elements in the cross-hatched portion, where the circles overlap, are in the intersection. An example of the use of a Venn diagram with two sets of points is illustrated in Figure 2.24. Set A is the set of labeled points on $\overline{XY}$, and set B is

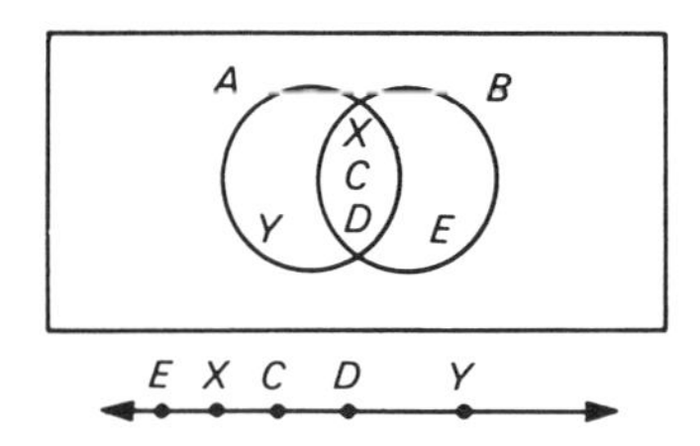

Figure 2.24. Use of Venn diagram

the set of labeled points in $\overline{ED}$. The three points X, C, D are shown in the intersection of the two sets.

Even more so than for intersection, the union of more than two sets of points is of significance in analyzing pictures in geometry. Try drawing various figures showing the union of six segments, for example, other than the two in Figure 2.25, to find some of the different possibilities for this one case.

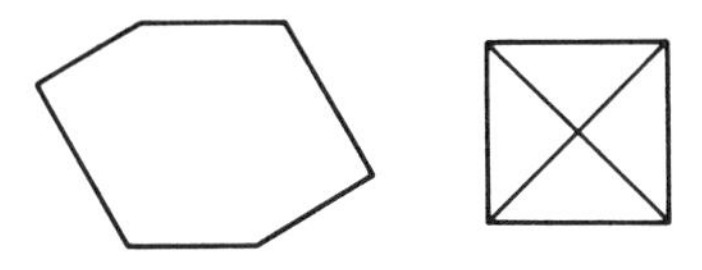

Figure 2.25. Union of segments

Exercise 2.4

1. Which of these sets of points are equal? (Use Fig. 2.26.)

   (a) $\overrightarrow{CA} \cap \overline{BA}$ and $\overrightarrow{AC} \cap \overline{BA}$
   (b) $\overrightarrow{CA} \cap \overrightarrow{DE}$ and $\overrightarrow{CB} \cap \overrightarrow{BD}$
   (c) $\overline{BD} \cap \overline{DE}$ and $\overrightarrow{ED} \cap \overline{DE}$
   (d) $\overleftrightarrow{CA} \cap \overrightarrow{AB}$ and $\overline{BE} \cap \overrightarrow{BD}$

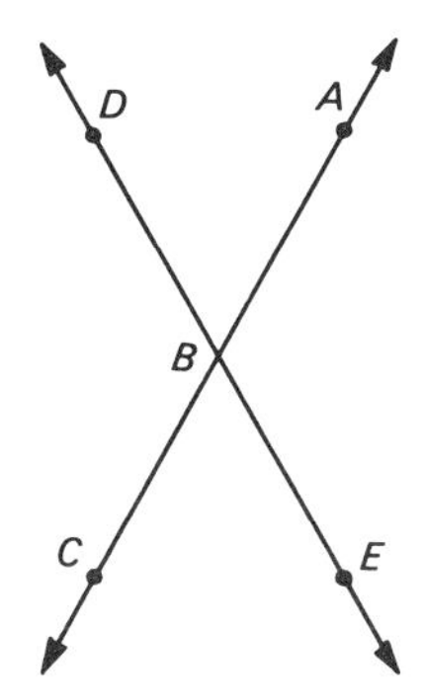

Figure 2.26

2. If possible, draw in one plane two angles whose intersection is:
   (a) three different points (b) the empty set
3. If possible, draw two angles in one plane whose intersection is:
   (a) two points (b) a segment
4. Complete the table by writing yes or no to tell whether it is possible for the intersection of two different sets of points to be the set of points specified. For each yes answer, you should be able to draw a picture to show the relationship.

Two different sets of points in the same plane	*Could the intersection be exactly one:* point?	*line?*	*plane?*	*segment?*	*ray?*	*half-plane?*	*half-line?*	*angle?*
point and line								
point and plane								
point and segment								
point and ray								
point and half-line								
point and angle								
line and plane								
line and segment								
line and ray								
line and half-plane								
line and angle								
plane and segment								
plane and half-plane								
plane and angle								
segment and ray								
segment and half-line								
segment and angle								
ray and half-line								
ray and angle								
angle and angle								

5. Which sets of points from Figure 2.27 are equal?

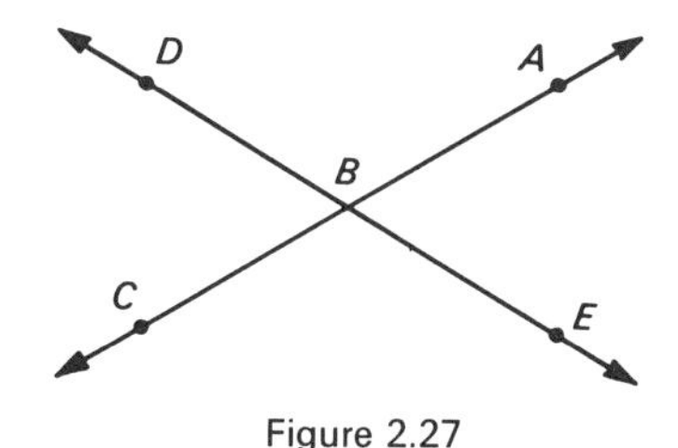

Figure 2.27

(a) $\overrightarrow{CA} \cup \overline{BA}$ and $\overrightarrow{AC} \cup \overline{BA}$ (b) $\overrightarrow{CA} \cup \overline{DE}$ and $\overrightarrow{CB} \cup \overrightarrow{BD}$
(c) $\overline{BD} \cup \overline{DE}$ and $\overrightarrow{ED} \cup \overrightarrow{DE}$ (d) $\overline{CA} \cup \overrightarrow{AB}$ and $\overline{BE} \cup \overrightarrow{BD}$

6. Draw Venn diagrams for (a) and (b).
   (a) $A = \{3, 4, 8, 9, 11\}$ $B = \{1, 3, 5, 7, 9, 11\}$
   (b) $A = \{A, B, C, D, E\}$ $B = \{D, E, F, G\}$
7. Describe each picture as the union of the smallest possible number of segments.
   (a) Figure 2.28a (b) Figure 2.28b

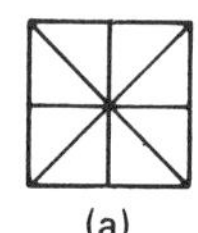

(a)

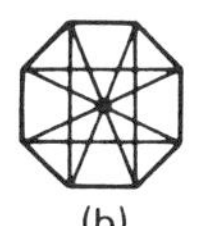

(b)

Figure 2.28

8. Complete a table for union in the same form as the table for intersection. Complete it by writing yes or no to tell whether it is possible for the union of two different sets of points to be exactly the sets of points specified.
9. List the various possibilities for sets of points in the intersection of two open segments.
10. List the various possibilities for sets of points in the union of two open segments.
11. Investigate and then state a theorem about whether or not the intersection of two convex sets is a convex set.
12. Use the definition of convex sets to prove the theorem in Exercise 11.

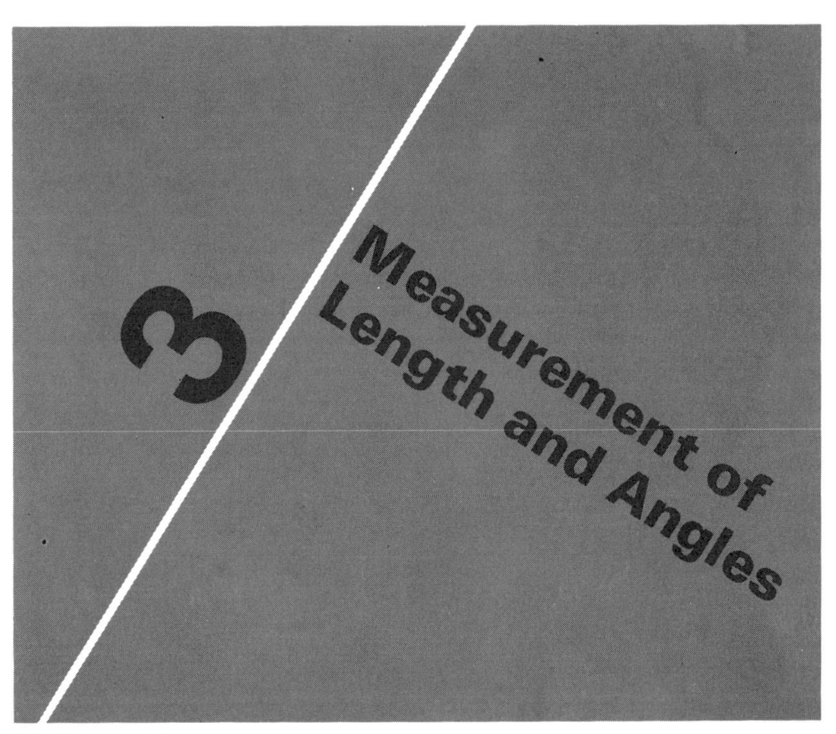

Every person becomes familiar with what is meant by the measurment of physical objects in the sense that he gains some skill in actually making such measurements. This acquaintance with practical measurements provides a beginning for the analysis of the theory of measurement in geometry, which includes considering measurement within the mathematical model, using the language of sets of points, picturing the ideas when desirable, and then applying the results once again to the measurements of actual objects. Additional axioms are needed for the assumptions about measurement. The concept of congruence and transformation is necessary to an understanding of the theory of measurement.

The Concept of Measurement

Measurements of length, angle, area, volume, and capacity are common activities of men. In each of these examples of measurement, certain similarities should be noted, since they are essential for an understanding of the concept of measurement as studied in an informal approach to geometry. The following statements about the measurement of physical objects should agree with your experiences in actually making such measurements.

1. Measurement is used to answer questions such as "How long?" or "How big?" rather than "How many?" The word *measurement* refers both to the process and to the end result. A recorded measurement gives information about a physical property of the object being measured.

2. The results of measuring objects are always approximate. Errors must always be considered in the physical act of measuring.

3. A measurement is an expression, such as 3 inches, that includes a numeral (3 in this case) and specifies some unit of measurement (inch in this case).

4. A measurement compares the object to the unit chosen. If an object is 3 inches long, it is 3 times as long as the unit 1 inch.

5. The unit of measurement must be of the same nature as the object being measured. That is, to measure the length of something, we must choose a unit of length; to measure the area of a surface, we must choose a unit of area. The size of a unit is arbitrary, but standardized units have been adopted for convenience.

6. Although it is possible to compare two objects in size without measuring them, measurement makes a more precise comparison possible, through comparison of both with a unit of measurement.

7. Normally, the physical act of measurement involves placing an instrument such as a ruler or a protractor in the correct position so that the comparison can be made simply and the measurement read from the instrument.

Next, consider abstracting from the measurement of the physical world and describing measurement in the mathematical model. It might seem, on first examining the possibility, that measurement of sets of points is meaningless, since sets and points are nothing but ideas. But it does make sense to speak of measurement in connection with ideas. For example, picture mentally a giant mountain 20,000 feet high. Now this mountain has no existence at all, other than in the world of ideas. The idea certainly is not 20,000 feet high, since it seems to fit nicely into your head as you imagine it. You often use measurements to describe or explain your ideas. You can think of a segment 3 inches long, or of a green Martian 7 feet tall, even though the ideas themselves might not be 3 inches long or 7 feet tall. Furthermore, you can describe your ideas of anything of a certain size so that other people can understand what you mean.

The seven statements previously given will now be interpreted within the mathematical model for sets of points.

1. Measurement is used to answer questions such as "How long is a segment?" or "What is the measure of an angle?" rather than "How many members are in a set of points?" The word *measurement* refers both to the process and to the end result. Measurement gives information about a property of the set of points being investigated.

2. Measurement within the mathematical model does not involve error or approximation. The theory of measurement does not yield estimates; it yields exact answers. It is only when this theory is applied to measurement of physical objects that errors are introduced.

3. A measurement is an expression, such as 3 inches, which includes a numeral (3 in this case) and specifies some unit of measurement (inch in this case). This statement is identical to statement 3 of the previous set.

4. A measurement compares the set of points to the unit.

5. The unit of measurement must be the same kind as the set of points being measured. That is, to measure the length of a segment, we choose a unit of length; to measure area, we choose a unit of area. The size of a unit of measurement is arbitrary, but standardized units have been adopted for convenience.

6. Although it is possible to compare the size of two sets of points without using measurement, the result of measurement makes an exact comparison possible, through comparison of both with a unit of measurement.

7. Normally, the act of measuring in the geometric model is pictured by placing a scale such as a drawing of a number line in the correct position so that the comparison can be made simply and the measurement read on the scale.

The concept of *congruence* is basic to an understanding of the theory of measurement. Congruent sets of points have exactly the same shape and size, but this statement is not meant as a definition. For example, two congruent segments have the same length. Congruent sets of points differ only in location.

Although the intuitive meaning of congruence is easy to develop, a deeper understanding of the concept is related to several fundamental ideas in mathematics: those of mapping, transformation, and isometry.

A *mapping* of one set of points onto another is a pairing of elements of the two sets such that each element of the first is paired with exactly one element of the second, and each element of the second is paired with at least one element of the first. In Figure 3.1, a mapping is shown in which the ordered pairs are (a, b), (c, b), and (d, e). The definition of the word *mapping* in geometry is equivalent to the definition of *function* in mathematics. A function is a set of ordered pairs in which no two different pairs have the same first element.

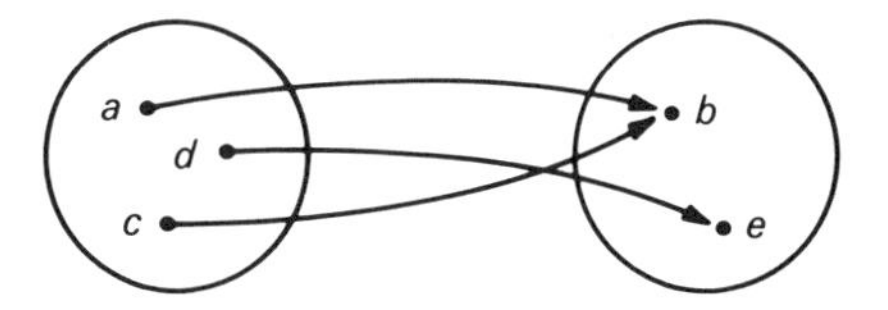

Figure 3.1. Mapping

Example

$\{(1, 3), (4, 2)\}$ is a function, but $\{(1, 3), (1, 4)\}$ is not.

A *transformation* is a mapping of one set onto another such that each element of the second set is the image of exactly one element in the first. This definition means that a one-to-one correspondence exists between the members of the two sets. An example is given in Figure 3.2.

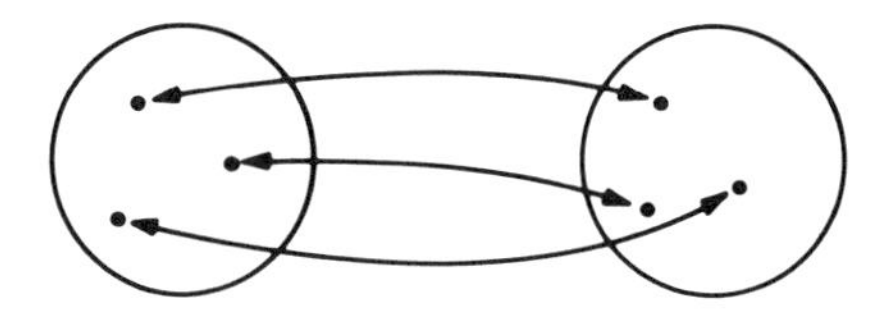

Figure 3.2. Transformation

A transformation that preserves the distance between each pair of points is called an *isometry*. Two sets of points that correspond in an isometry are congruent.

Example

Suppose there is an isometry as illustrated in Figure 3.3 in which $\overline{AB}$ and $\overline{A'B'}$ correspond. The two segments are congruent. The

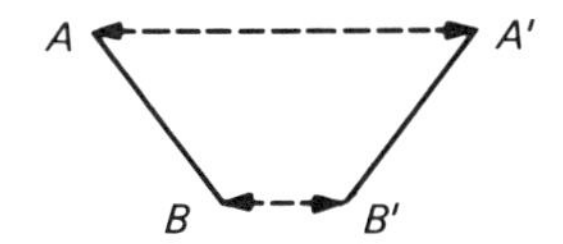

Figure 3.3. First example of isometry

distance between A and B is equal to the distance between A' and B'. Distance is used as an undefined term at this time.

Example

Suppose that the two sets of points in Figure 3.4 are congruent. There is an isometry connecting the two sets in such a way that

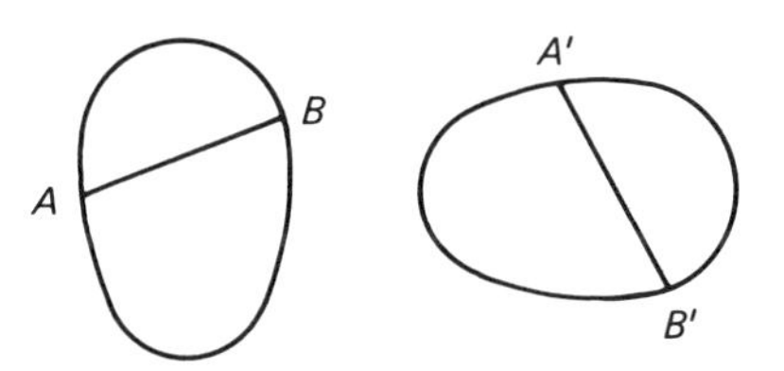

Figure 3.4. Second example of isometry

distance is preserved. If A and A' are corresponding points, and B and B' are also corresponding points in the isometry, then $\overline{AB}$ and $\overline{A'B'}$ are congruent.

Congruence is called an *equivalence relation* because of the following three properties:

1. A set of points is congruent to itself (*reflexive* property).
2. If a first set of points is congruent to a second, then the second is congruent to the first (*symmetric* property).
3. If a first set of points is congruent to a second, and the second is congruent to a third, then the first is congruent to the third (*transitive* property).

Congruence of segments and angles will be studied in more detail in the next two sections; congruence of triangles and other figures will be studied in later chapters.

Exercise 3.1

1. In which of these statements would measurement normally be used to get the results?
   (a) The desk is 29 inches high. (b) The city is 22 miles wide.
   (c) The rectangle is 4 inches high. (d) The cake weighs 2 pounds.
2. In which of these statements would measurement be used to get the results?
   (a) The building has 35 windows.
   (b) The car is traveling 60 miles per hour.
   (c) The intersection of the two sets of points has two members.
   (d) There are five angles pictured in the figure.
   (e) The segment is 5 inches long.
3. For each statement in Exercise 1 involving measurement, tell whether the measurement should be considered as approximate or exact.
4. For each statement in Exercise 2 involving measurement, tell whether the measurement should be considered as approximate or exact.

For Exercises 5–10, answer yes or no.

5. Congruent sets of points have exactly the same size and shape.
6. All mappings are transformations.
7. An isometry is a type of transformation.
8. If two segments are congruent, they are equal.
9. The relation of "is greater than" is an equivalence relation.
10. An isometry preserves the distance between each pair of points.

The Measurement of Length

Two segments that have the same length are *congruent* segments. The concept of congruence, although it does not imply any special unit, is the mathematical basis for the theory of measurement of sets of points. To explain what the length of a particular segment is, a measurement such as 2 inches is used. In the expression *2 inches,* inch is the unit of measurement; the number 2, called the measure, tells how many units.

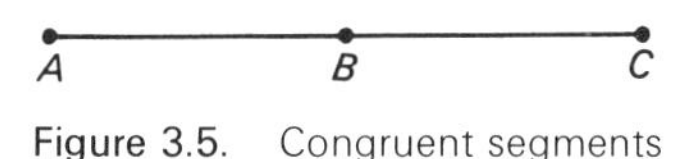

Figure 3.5. Congruent segments

In Figure 3.5, if point C lies on segment AB in such a position that $\overline{AC}$ is congruent to $\overline{CB}$, written $\overline{AC} \cong \overline{CB}$, C is called the *midpoint* of $\overline{AB}$. Then the measure of $\overline{AB}$ is twice the measure of $\overline{AC}$. Segment AB is the union of two congruent segments, $\overline{AC}$ and $\overline{CB}$, on the same line with a common endpoint but with no other points in common. In non-tcchnical language, the two congruent segments are positioned end-to-end without overlapping.

In Figure 3.6, two points, B and C, can be chosen on $\overline{AD}$ such that all three of the segments $\overline{AB}$, $\overline{BC}$, and $\overline{CD}$ are congruent. In this case, $\overline{AD}$ is said to have a measure three times the measure of $\overline{AB}$. Again, notice that $\overline{AB}$, $\overline{BC}$, $\overline{CD}$ are positioned end-to-end along $\overline{AD}$. This process of subdivision of a segment could be continued indefinitely, with a segment partitioned into four, five, or more congruent segments.

The vocabulary of union can be employed to describe the partitioning. If $\overline{AD}$ is the union of three congruent segments positioned as in Figure 3.6, then the measure of $\overline{AD}$ is 3, using the measure of one of the congruent

Figure 3.6. Union of segments

segments as a unit. The measure of a segment is obtained by counting how many times a unit segment can be indicated along the segment, so that the last endpoint coincides with the endpoint of the original segment. Remember that the length of the unit segment is arbitrary.

Figure 3.7 shows $\overline{AB}$ and a unit segment $\overline{XY}$. To find the measure of $\overline{AB}$ in terms of the measure of the unit segment, suppose that a point C is located on $\overline{AB}$ so that $\overline{AC}$ is congruent to $\overline{XY}$, as are $\overline{CD}$, $\overline{DE}$, $\overline{EF}$, $\overline{FG}$, and finally $\overline{GB}$. Then $\overline{AB}$ has a measure of 6, using the measure of $\overline{XY}$ as a unit.

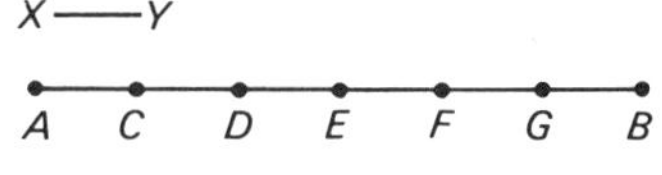

Figure 3.7. Measure of a segment

So far, only counting numbers have been used for measures, but you can easily imagine a situation in which the given segment could not be partitioned into a number of congruent segments for some arbitrary given unit. It might be possible to indicate, for example, 5 congruent segments with some of the original segment left over but not enough for a sixth congruent segment. The use of a number line will help clarify the concept. A *number line* can be thought of as a line with various points indicated by numerals. Sometimes the expression *the number line* is used to imply the existence of *one-to-one correspondence* between all real numbers and all the points on a line.

In addition to the nine axioms introduced in Chapter 2, two additional axioms are necessary for the study of measurement of length.

Axiom 10

There exists a correspondence that associates the number 1 with certain pairs of distinct points and a unique number with every pair of distinct points.

Axiom 10 is illustrated in Figure 3.8. The correspondence that associates the number 1 with the pair A, B associates the number 2 with C, D, the number 3 with E, F, 5 with G, H, and 7 with I, J.

Axiom 11

There is a one-to-one correspondence between the points in a line and the real numbers such that two points of the line correspond to the numbers 0 and 1, and the measure of the distance between any two points in the line is the *absolute value* of the difference of their corresponding numbers.

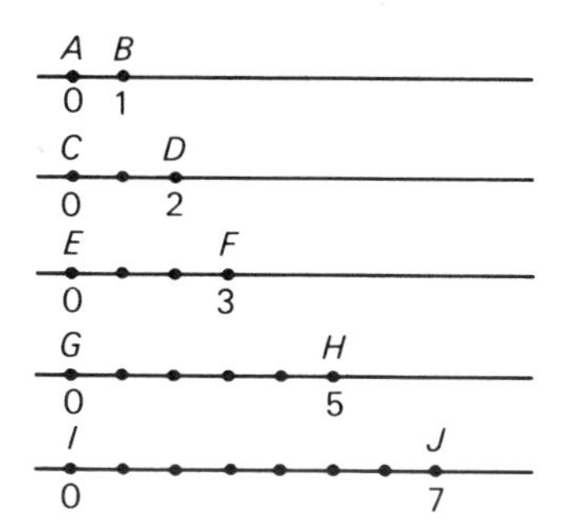

Figure 3.8. Number lines

Recall that the absolute value of a negative number is positive. Thus the absolute value of -5 is 5, written $|-5| = 5$. In this chapter, only measurements with positive numbers are considered.

Example

The distance between points for 6 and 2 is

$$|6 - 2| = |2 - 6| = 4.$$

Example

The distance between points for 7 and 0 is

$$|7 - 0| = |0 - 7| = 7.$$

One-to-one correspondence between real numbers and points on a line implies that, in Figure 3.8, if two distinct points on the line are chosen to correspond to 0 and 1, then exactly one point on the line corresponds to each real number. Also, given any point on the line, exactly one real number corresponds to it.

To find the measure of a segment, such as $\overline{AB}$ in Figure 3.9, think of a number line located so that the number 0 corresponds to one endpoint of the segment. The length from 0 to 1 is designated as the unit length. Then

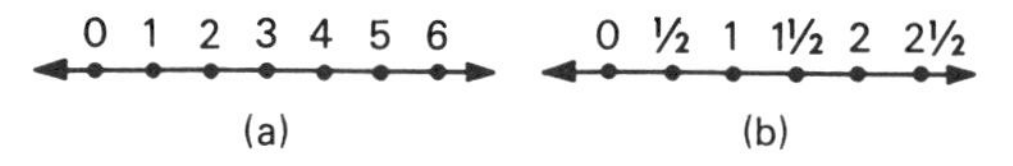

Figure 3.9. Number line and measurement

the number that corresponds to B in Figure 3.9 indicates the measure. In Figure 3.9a, the measure of $\overline{AB}$ is 6. But the use of the number line makes it seem possible that any real number greater than zero can be the measure of a segment. In Figure 3.9b, $2\frac{1}{2}$ is the measure of $\overline{AB}$. It is conceivable that a segment may have as its measure an irrational number such as $\sqrt{5}$ or π.

Measuring the length of a table top can be accomplished by using a foot ruler in an analogous way. The endpoint of the ruler, representing zero, is placed at one edge of the table. In Figure 3.10, the foot ruler is shown in 5

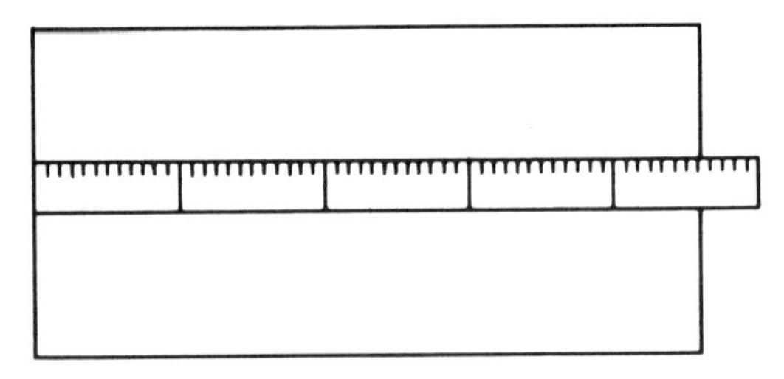

Figure 3.10. Measuring length of table top

different positions, as if laid end to end. In the last position, the reading for 7 inches coincides with the other edge of the table; hence the length of the table is said to be 4 ft 7 in. The analysis of the error involved in physical measurements, including measuring *pictures* of line segments, and of the ways these errors are taken into account to give consistent results in measurement is discussed in Chapter 13.

Throughout the preceding discussion of measurement of length, you probably have noted the importance of the unit of measurement, since it was applied repeatedly to obtain an answer. Units of measurement might be classified as *standardized* and *non-standardized.* A non-standardized unit is not widely recognized or employed regularly by many people. "Stepping off" the distance across a room, or using the distance from your elbow to your fingertip as your unit, or using the width of your thumb, or telling how tall a building is in stories are all examples of the application of non-standardized units. The use of non-standardized units is rarely appropriate in scientific or technological applications. Indeed, the history of measurement is in part the story of the development of useful standardized units.

Today, two common systems for measurement of length are in general use: the *English* system and the *metric* system. Four familiar linear units in the

English system are the inch, the foot, the yard, and the mile. The well-known relationships among these units are: 1 ft = 12 in., 1 yd = 3 ft, and 1 mile = 5280 ft. One other unit sometimes used is 1 rod, defined as $16\frac{1}{2}$ ft or $5\frac{1}{2}$ yd.

Of increasing importance in everyday life in the United States, of paramount importance in measurement in science and industry, and the truly international system on which the English system is now based is the metric system. Great Britain expects to convert to the metric system by 1980. A study of the cost of total conversion has been undertaken in the United States. The metric system of measurement originated in France, becoming legal in that country in 1799. It was intended that a meter have a length of one ten-millionth of the distance from the North Pole to the Equator. The distance between two scratches on a platinum–iridium bar, approximating the earth measurement above, was actually established as the standard unit. The International Bureau of Weights and Measures, at Sèvres, France, serves as a depository for standards of length and other measurements, although other bureaus of standards, such as the Bureau of Standards in Washington, D. C., maintain duplicates. The science of measurement is called *metrology*. Governments and large industries maintain laboratories in which expensive equipment is used to make the extremely precise measurements necessary in an advanced technology.

Recently, by international agreement, the meter has been defined as having a measure 1,650,763.73 times the measure of the wavelength of the orange-red light emitted by a lamp containing krypton 86. (A krypton lamp is shown in Fig. 3.11.) This definition allows the meter to be reproduced in case the official standard should be destroyed. Actually, the metric system has been legal in the United States since 1866. Today, an inch is defined as 2.54 centimeters; a yard is 3600/3937 meters.

One of the chief advantages of the metric system is its employment of powers of ten to define the relationships among common units. A brief table summarizes some important equivalences of linear units in the metric system.

10 millimeters (mm)	= 1 centimeter (cm)
10 centimeters	= 1 decimeter (dm)
10 decimeters	= 1 meter (m)
1000 meters	= 1 kilometer (km)

Changing among units in the metric system is relatively simple, as these two examples illustrate.

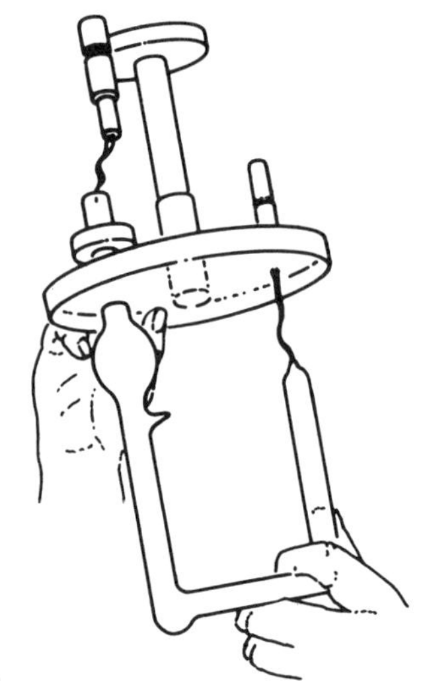

Figure 3.11. A krypton lamp

Examples

1. $54 \text{ cm} = ?\text{ mm}$

Since $1 \text{ cm} = 10 \text{ mm}$

$$54 \text{ cm} = (54 \times 10) \text{ mm}$$
$$= 540 \text{ mm}.$$

2. $7 \text{ cm} = ?\text{ m}$

Since $1 \text{ cm} = {}^{1}/_{100} \text{ m}$

$$7 \text{ cm} = (7 \times {}^{1}/_{100}) \text{ m}$$
$$= {}^{7}/_{100} \text{ m}.$$

Today, scientists customarily use other metric units of length in addition to those named. Two examples of very small units of measurement of length are the *micron,* which is one-thousandth of a millimeter, and a *millimicron,* which is one-thousandth of a micron, or one-millionth of a millimeter.

Exercise 3.2

1. Could two different segments (a) have the same measure? (b) be congruent? (c) be equal?
2. Could the union of two different segments (a) have the same measure as one of the segments? (b) be congruent to one of the segments? (c) be equal to one of the segments?

For Exercises 3–4, find the distance between points on the number line for:

3. 7 and 0.

4. 8 and 3.
5. Draw a number line as it would be used to find the measure of a segment if the measure is $3\frac{1}{5}$.
6. Draw a number line as it would be used to find the measure of a segment if the measure is $5\frac{1}{4}$.
7. Name other examples of non-standardized linear units of measurement based on measurements of the human body.
8. Name other examples of non-standardized linear units of measurement not based on measurements of the human body.
9. Answer each question for the English system.
   (a) How many yd equal one mi? (b) How many rd equal one mi?
   (c) How many in. equal one rd? (d) How many mi equal 32,340 ft?
10. Find a number to replace the question mark to make each statement true.
   (a) 1 m = ? cm (b) 1 km = ? cm (c) 1 dm = ? m
   (d) 4.5 km = ? m (e) 9 mm = ? m (f) 300 m = ? km
   (g) 4.73 m = ? cm (h) 2.34 m = ? dm
11. How many microns equal one cm?
12. How many millimicrons equal one meter?
13. Answer yes or no for each statement relating the English to the metric system.
   (a) 1 yd is longer than 1 m.
   (b) 1 km is shorter than 1 mi.
   (c) 1 cm is longer than 2 in.
   (d) 1 mm is shorter than 1 in.
   (e) 2 cm is shorter than 1 in.
   (f) 1 m is longer than 39 in.
14. Some common distances for races are 100 and 800 meters. Approximately how many yards are the races?
15. Other common distances for races are 200 and 1500 meters. Approximately how many yards is each?

Exercises 16–20 are mathematical laboratory experiences with the metric system.

16. Find your own height in cm.
17. How long is an ordinary pencil in cm?
18. What are the dimensions of a 3-by-5 card, using metric units?
19. What are the dimensions of a sheet of typing paper, using metric units?
20. How many pennies must be put in a stack that is 3 cm tall?

Measurement of Angles

The concept of measurement of representations of angles, such as the two hands of a clock or the intersection of two streets, can be approached by considering an angle in the geometric model as a set of points that is the union of two distinct rays in a plane, not on the same line, but with a common endpoint. In this section notice the similarities between measuring an angle and measuring a segment.

Two congruent segments are segments with the same length. What is meant by two congruent angles? In Figure 3.12, suppose that points $A, C, D,$ and F

Figure 3.12. Congruent angles

are chosen so that $\overline{AB}, \overline{BC}, \overline{DE},$ and $\overline{EF}$ are all congruent segments. The two angles ABC and DEF will then be congruent if $\overline{AC} \cong \overline{DF}$. This figure, using informal language, makes it clear that two angles are congruent if they have the same size. The concept of congruent angles is basic to an understanding of the theory of measurement of angles.

To find the measure of an angle, it is necessary to use a *unit angle,* just as it was necessary to use a unit segment for measuring a segment. In Figure 3.13, suppose that $\angle ABC$ and a unit angle are given. Then $\angle ABC$ has a measure of 2, using that unit angle, which is true because $\angle ABC$ and its interior can be partitioned into two adjacent angles and their interiors by $\overrightarrow{BD}$, and each of these two angles is congruent to the unit angle.

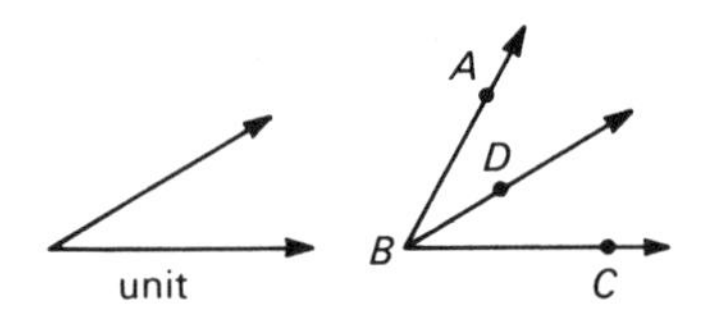

Figure 3.13. Measuring an angle

In Figure 3.14, the measure of $\angle AOB$, using the given unit angle, is 6, since $\angle AOB$ and its interior can be partitioned as shown into 6 angles, each congruent to the unit angle.

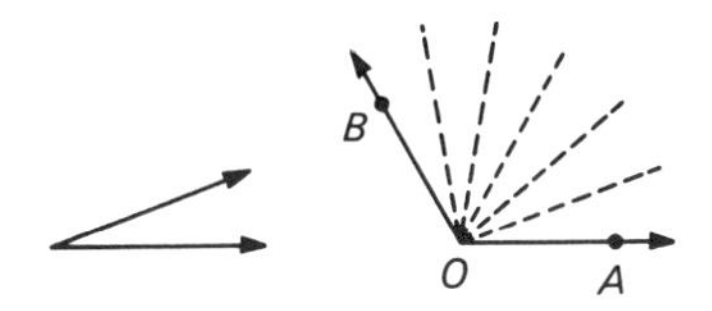

Figure 3.14. Angle with measure of six

Although several common standard units for measurement are in use, the one customarily chosen in everyday measurements of angles has a measurement of one degree. In Figure 3.15, consider the half-plane above $\overleftrightarrow{AB}$. Suppose that rays are drawn, each with O as endpoint, so that $\overleftrightarrow{AB}$ and the

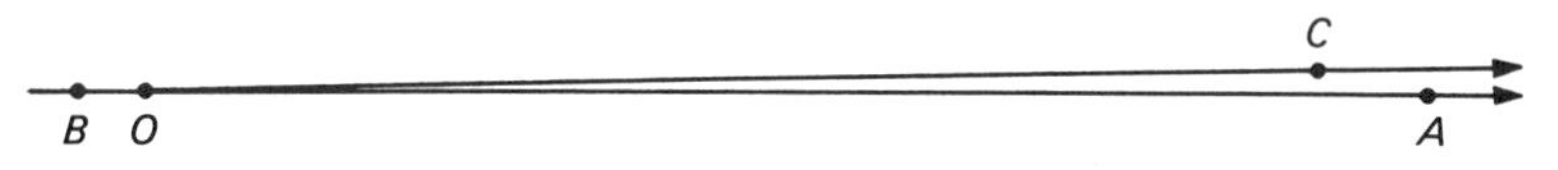

Figure 3.15. One-degree angle

entire half-plane is partitioned into 180 congruent angles and their interiors. If $\angle AOC$ is one of these angles, then its measurement would be 1 degree, written 1°. A second common standard unit angle is introduced in Chapter 5.

Now that a standard unit angle has been chosen, a scale is needed so that the number of times this standard angle may be duplicated by congruent angles in the interior of the given angle can be determined. Figure 3.16

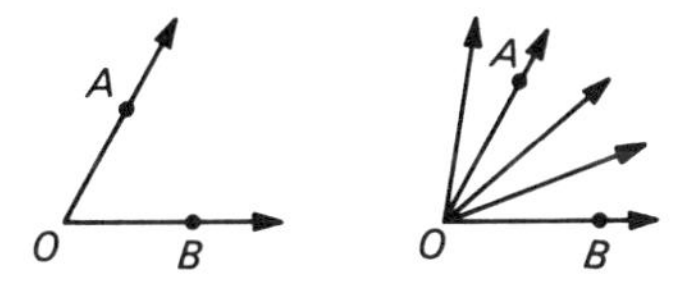

Figure 3.16. Using scale to measure an angle

shows $\angle AOB$, whose measurement is to be found. Then, the angle is pictured with a superimposed scale consisting of rays that have been drawn to represent angles with a measurement of 20, 40, 60, and 80 degrees. Since side $\overrightarrow{OB}$ coincides with the 0 ray and since side $\overrightarrow{OA}$ coincides with the 60-degree ray in the scale, $\angle AOB$ and its interior can be partitioned into 60 congruent angles, each with a measurement of 1 degree; hence the measurement of $\angle AOB$ is 60 degrees. The measurement of an angle is not always a whole number of degrees; angles could have a measurement of $62\frac{1}{2}$ degrees or $75\frac{3}{4}$ degrees, for example.

The use of fractions in writing degree measurements is often avoided in practice by the use of measurements called minutes and seconds. The measure of an angle of 1 minute is $\frac{1}{60}$ that of the measure of an angle of 1 degree; the measure of an angle of 1 second is $\frac{1}{60}$ that of an angle of 1 minute. Thus a measurement of $62\frac{1}{2}$ degrees can be written as 62 degrees 30 minutes (62° 30′); a measurement of $75\frac{3}{4}$ degrees could be written as 75 degrees 45 minutes (75° 45′). It is becoming increasingly common to use decimals with angle measurement. Thus an angle measurement might be given as 65.3°, for example.

Example

Express the angle measurement 65.3° in degrees and minutes.

$$\frac{3}{10} = \frac{18}{60}, \quad \text{so} \quad .3^\circ = 18'$$

$$\text{Then } 65.3^\circ = 65^\circ\ 18'$$

The theory of angle measurement discussed so far makes use of the following two axioms:

Axiom 12

There exists a correspondence that associates with each angle one real number n such that $0 < n < 180$.

Axiom 13

For every point P and every closed half-plane whose edge contains P, there is a one-to-one correspondence between the real numbers n, where $0 \leq n \leq 180$, and the set of all rays in the closed half-plane and having P as their endpoint.

The theory of measuring angles as sets of points in the geometric model, based on congruent angles, may be used to explain measurement of representations of angles in the real world. A *protractor,* shown in Figure 3.17a, is a representation of the system of rays with a common endpoint,

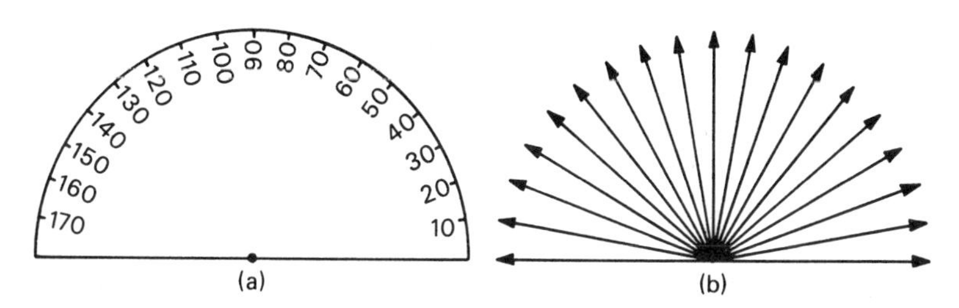

Figure 3.17. Protractor and system of rays

illustrated in Figure 3.17b. To find the measure in degrees of a representation of an angle, place the protractor over the picture so that the vertex of the angle is under the midpoint of the bottom of the protractor. Position one side of the angle along the bottom of the protractor, and then name the numerical along the curved part of the protractor that is over the second side.

Examples

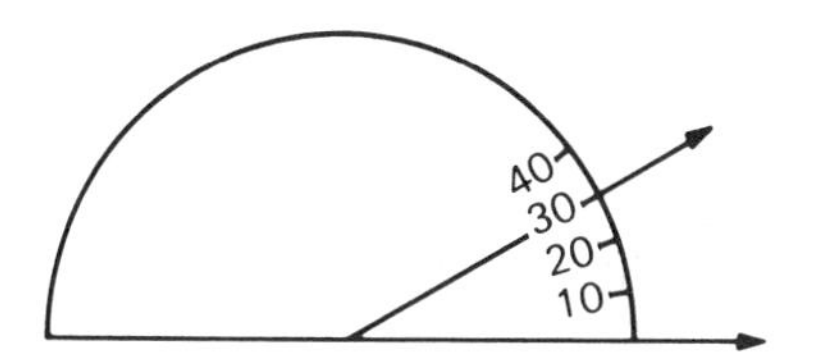

Figure 3.18. Angle with measurement of 30°

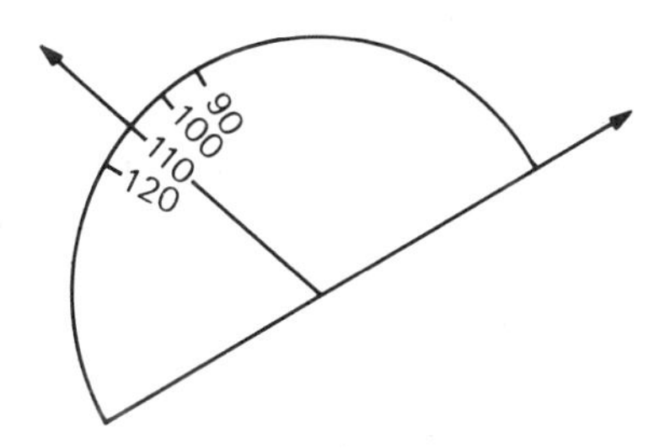

Figure 3.19. Angle with measurement of 110°

The theory of measuring angles with a protractor depends on the following basic assumption, stated as the final axiom in the chapter.

Axiom 14

For any angle ABC there exists exactly one ray-coordinate system with $\overrightarrow{BA}$ as a zero-ray such that for any point X in the C-side of $\overrightarrow{BA}$, $\overrightarrow{BX}$ corresponds to a unique real number n such that $0 < n < 180$.

The shape of the ordinary protractor may cause a person to fail to grasp its significance as a pattern or scale of numerals positioned in such a way as to correspond to a set of rays used in defining measurement of angles. It may

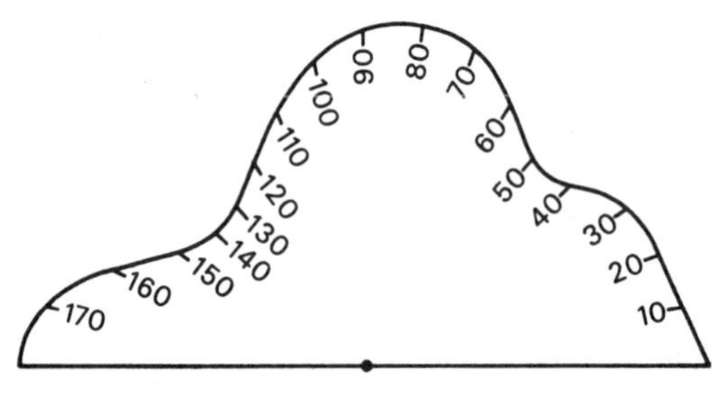

Figure 3.20. Irregular protractor

seem that you are measuring around a circle, when actually this is not necessary. To understand this point thoroughly, study the picture of the protractor in Figure 3.20. Such an instrument can actually be used for measurement of angles, yet certainly it is not in the usual shape.

Angles may be classified according to their measures in degrees. An angle with a measurement of 90 degrees is a *right* angle. Angles whose measures are less than 90 in degrees are called *acute* angles, and angles whose measures are more than 90 but less than 180 in degrees are called *obtuse* angles. It should be pointed out that the measure of angles has been limited, so far, to angles whose measures are between 0 and 180 in degrees. Extending the measure to 180 and beyond would require a change in the interpretation of an angle—but one that is actually necessary in more advanced mathematics courses. Figure 3.21 shows acute angles AOB, BOC, and COD, obtuse angles AOD and BOD, and right angle AOC. The sides of a right angle are called perpendicular rays. Also, lines or segments intersecting at the vertex of a 90-degree angle and lying on the sides of the angle are said to be perpendicular.

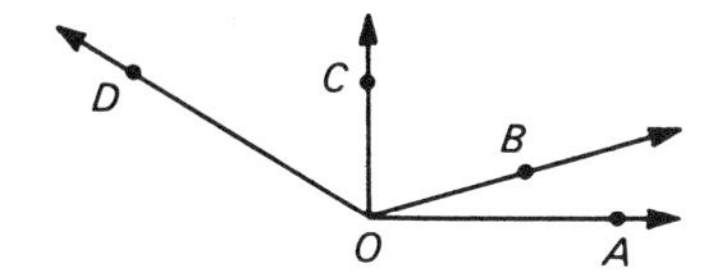

Figure 3.21. Acute, right and obtuse angles

Vertical angles and adjacent angles are pairs of angles identified without regard to metric considerations. Other special pairs of angles are named because of the measurement of the angles. Two examples are *complementary* and *supplementary angles.* Two angles are complementary if the sum of

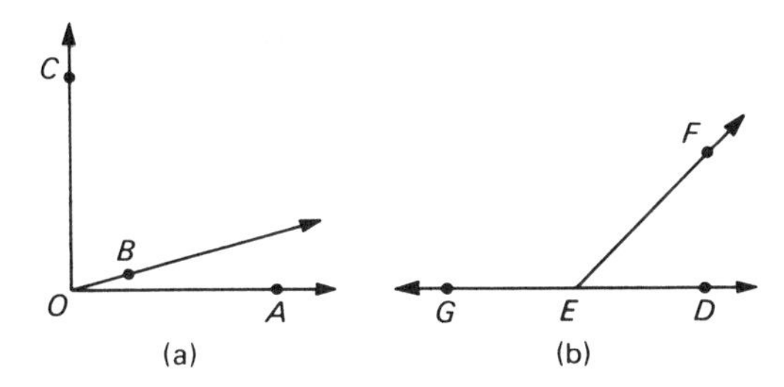

Figure 3.22. Complementary and supplementary angles

their measures in degrees is 90. Figure 3.22a shows angles *AOB* and *BOC*, which are complementary and adjacent. In this case, $\overrightarrow{OA}$ and $\overrightarrow{OC}$ are perpendicular. Angles may be complementary without being adjacent, of course, as long as the sum of their measures in degrees is 90. Two angles are supplementary if the sum of their measures in degrees is 180. Figure 3.22b shows angles *DEF* and *FEG*, which are supplementary and adjacent. Rays $\overrightarrow{ED}$ and $\overrightarrow{EG}$ lie on the same line. It is assumed that a *straight angle* such as angle *DEG* has a measure of 180. Angles do not need to be adjacent to be supplementary. For example, angles *ABC* and *DEF* in Figure 3.23 are supplementary, since the measure in degrees of the first is 32 and the measure of the second is 148.

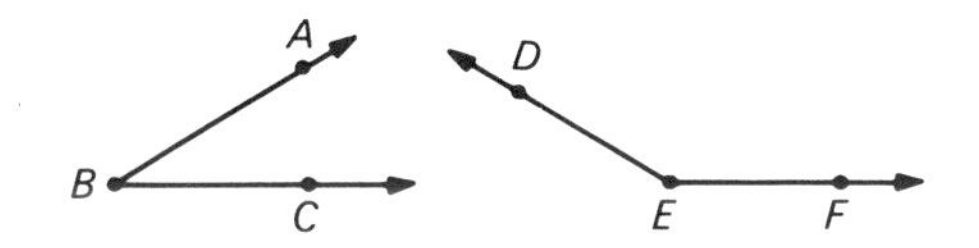

Figure 3.23. Supplementary non-adjacent angles

In Figure 3.24 angles AOB and COD are vertical angles. Both are supplementary to the same angle, AOC (or BOD). The measure of both is 180

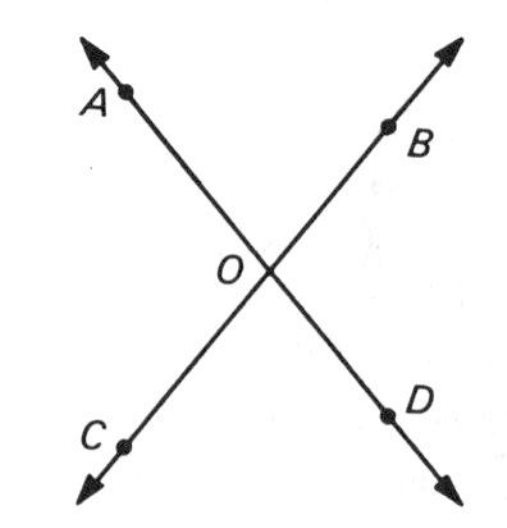

Figure 3.24. Vertical angles

minus the measure of $\angle AOC$. This informal proof leads to the following theorem:

Theorem

Two vertical angles have the same measure and are congruent to each other.

Other pairs of angles are associated with parallel lines. Accepting the assumptions that Euclid (300 B.C.) made about parallel lines (assumptions that have formed the basis for the ordinary geometry taught in high school) leads to the conclusion that, if one line intersects two parallel lines, the corresponding angles are congruent. In Figure 3.25, assume that lines AG and DE are parallel and that $\overleftrightarrow{OC}$ is a line cutting both. This line OC is a *transversal.* The angles AOB and ECF are a pair of *corresponding angles* and are assumed to be congruent. Another pair of corresponding angles are ECB and AOH. Since angles ECF and DCB are vertical angles, angles AOB and DCB are also congruent, because two angles congruent to the

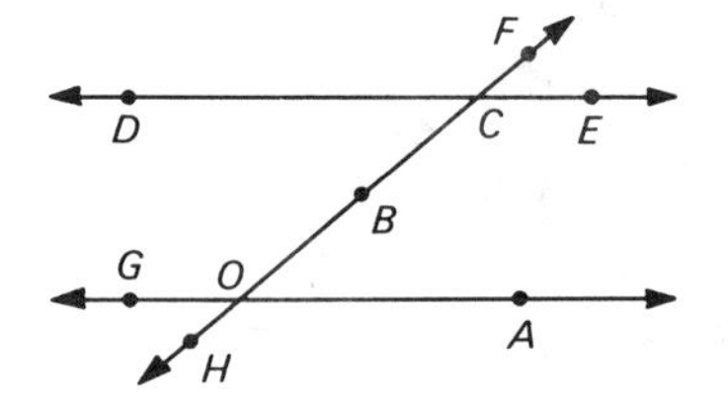

Figure 3.25. Angles associated with parallel lines

same angle are congruent to each other. These two angles, $\angle AOB$ and $\angle DCB$, are *alternate interior angles*; they lie on different sides of the transversal and are between the parallel lines.

Theorem

Alternate interior angles are congruent.

You should be able to find, in Figure 3.25, other pairs of corresponding angles and other pairs of congruent angles that are not corresponding angles.

Exercise 3.3

1. Find the measure of the three angles shown in Figure 3.26, correct to the nearest whole number, using unit angle A.

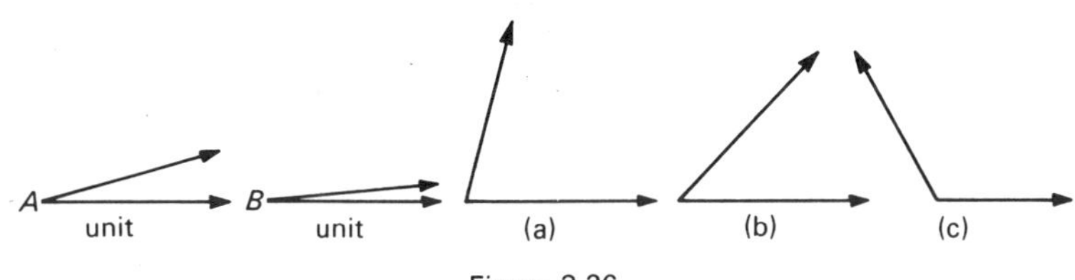

Figure 3.26

2. Find the measure of each angle shown in Figure 3.26, correct to the nearest whole number, using unit angle B.
3. Make a protractor similar to the one in Figure 3.20 (on p. 58) and use it to find the measurement of the three angles in Figure 3.26, correct to the nearest 5 degrees. Check by using an ordinary protractor.

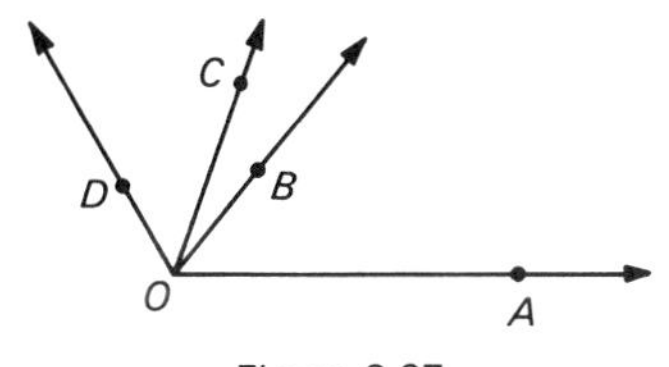

Figure 3.27

4. Use the protractor you made for Exercise 3 to find the measurement of each angle in Figure 3.27, correct to the nearest 10 degrees. Use an ordinary protractor to check.
5. Classify each of the five angles (including the unit angles) in Figure 3.26 as acute, right, or obtuse.
6. Classify each angle in Figure 3.27 as acute, right, or obtuse.
7. What is the measure of the complement of each angle whose measure is given below?
   (a) 35 (b) 63 (c) 15.4 (d) $49\frac{3}{4}$ (e) 89.99
8. What is the measure of the supplement of each angle whose measure is given below?
   (a) 95 (b) 32 (c) $16\frac{1}{2}$ (d) 138.4 (e) 179.93
9. Write the measurement of each angle in degrees and minutes.
   (a) 63.5° (b) 75.1° (c) 12.9°
10. List all the angles pictured in Figure 3.25 (on p. 60) that are congruent to angle *AOB*.
11. List all the angles pictured in Figure 3.25 that are congruent to angle *ECB*.
12. State and prove a theorem about the sum of the measures of angles *AOB* and *BCE* (pictured in Fig. 3.25).
13. State and prove a theorem about the sum of the measures of angles *GOB* and *DCB* (pictured in Fig. 3.25).
14. Without using a protractor, experiment to see how close you can come to drawing angles with these measurements:
   (a) 90° (b) 60° (c) 45° (d) 30°
15. A corner of a piece of paper can be used as a guide to draw a 90° angle. Name three other angles that could also be drawn easily if the paper corner could be folded over repeatedly.

This chapter extends the study of the non-metric geometry of the plane by considering more general sets of points, called curves, and then defining interesting and significant subsets. The vocabulary of union and intersection of sets of points studied in Chapter 2 can be employed to help you understand what is meant by a simple closed curve, a triangle, a quadrilateral, and other polygons. The chapter also uses the ideas of measurement introduced in Chapter 3, since it discusses the measurement of angles associated with triangles, some special properties of a right triangle involving measurement, and the perimeter of polygons.

Basic Concepts

The word *curve* will be used in this text to mean plane curve, just as the word *line* is used to mean straight line. That is, all the points of a curve lie in the same plane. Actually, *curve* will be considered an undefined term in this informal approach to geometry. Instead of a definition, a description can be given for illustrating a curve.

A curve is a set of points. A representation of a curve can be drawn by a single continuous motion of a pencil point on a piece of paper, provided that the point is not lifted from the paper and no portion of the drawing—other than single points—is retraced. Figure 4.1 shows four different

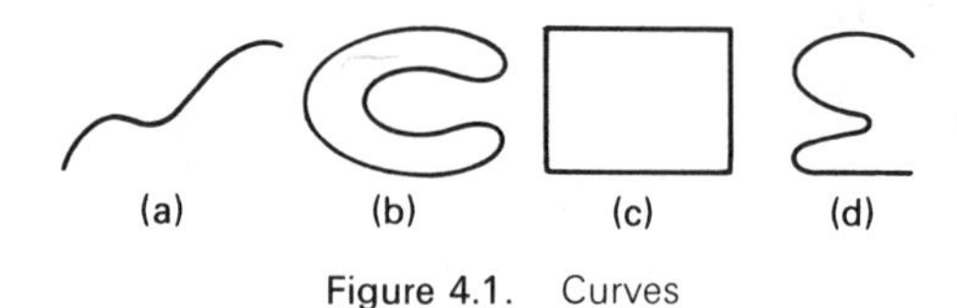

Figure 4.1. Curves

curves; notice that curves may be composed of segments, as in Figure 4.1c. Except for the trivial case of a single point being considered a curve, all curves are infinite sets of points.

Although many commonly used geometric figures are examples of curves, not all sets of points in a plane are curves. In Figure 4.2 are four sets of points that are not curves. It is interesting to notice, however, that figures that are not themselves curves may often be described as the union of two or more curves.

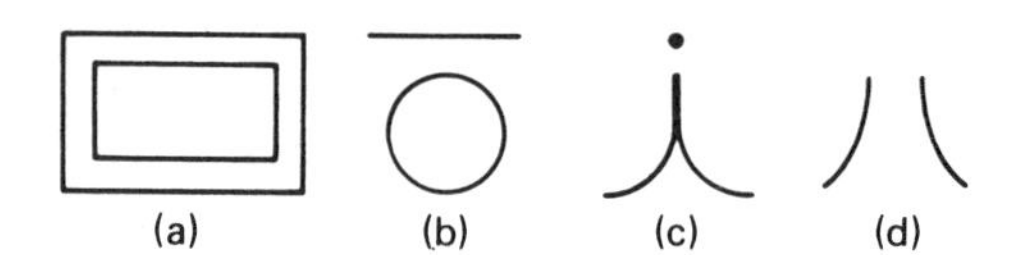

Figure 4.2. Sets of points that are not curves

One of the most noteworthy things about the concept of a curve is that it includes straight lines, segments, and rays. To think of a line as a very special case of a curve may seem unnatural at first, but it is consistent with the way *curve* is used in various branches of mathematics, including geometry.

The subset of the set of curves that we will study in more detail is the set of *simple closed curves*. To draw a picture of a closed curve, you must start and stop at the same place. A simple curve is one that does not cross itself. Thus a curve that is both simple and closed can be drawn by starting and stopping at the same point, with the curve never crossing itself. Figures 4.1b and 4.1c show simple closed curves. Figure 4.3 shows two curves that are not simple closed curves.

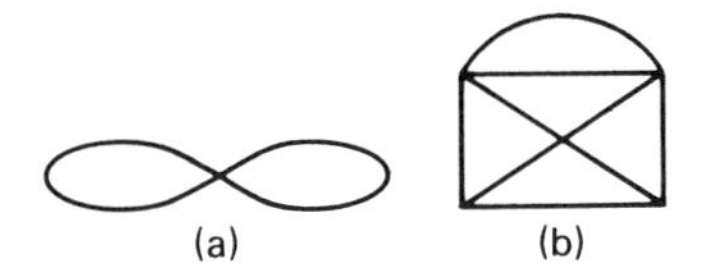

Figure 4.3. Curves without single interiors

An important statement about a simple closed curve that is proved in higher geometry includes part of the *Jordan Curve Theorem.* The statement is that any simple closed curve partitions the plane into three disjoint sets. These three sets are the set of points on the curve, the set of points in the interior, and the set of points in the exterior. A simple closed curve is seen to be a curve with a single interior. Note that the curves in Figure 4.3 cannot be said to have a single interior.

Tracing puzzles are closely related to the concept of curve and closed curve. A curve can be traced without lifting the pencil or retracing a portion of the path. A closed curve can be traced in the same way by starting and stopping at the same point.

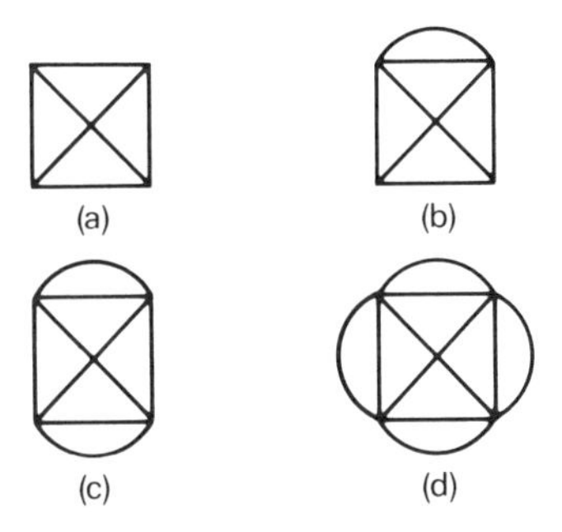

Figure 4.4. Tracing puzzles

Check each tracing puzzle in Figure 4.4 to see whether it represents a set of points that is not a curve, that is a curve, or that is a closed curve. Now try to decide what the basic differences are that would make it possible to discover by a quick inspection whether a tracing puzzle can be traced (whether it represents a curve). Do so before reading further.

The points at which the various segments and arcs meet in Figure 4.4 may be called *vertices.* Each segment or part of a curve leading to or from a vertex is called a *path,* so that the number of paths to each vertex may be counted. If there is an odd number of paths, the vertex is called an *odd vertex*. If there is an even number of paths, the vertex is called an *even vertex*. The following table summarizes the number of odd and even vertices for the sets of points in Figure 4.4.

	number of odd vertices	*number of even vertices*	*can it be traced?*	*by starting and stopping at same point?*
(a)	4	1	no	
(b)	2	3	yes	no
(c)	0	5	yes	yes
(d)	4	1	no	

Since an even number of paths makes it possible to go to a point and leave each time, it is only the odd vertices that cause trouble. In tracing, it is necessary when there are odd vertices to either start or stop the tracing at one of them so that the last odd path will be used up. For this reason, there can be no more than two odd vertices if the figure is to be traced. Even then, it cannot always be traced by starting and stopping at the same point, unless there are no odd vertices. These results are summarized by the statement that a tracing puzzle represents a curve when it has no more than two odd vertices and that it represents a closed curve when it has no odd vertices.

Although the concept of simple closed curves has been introduced as a rather special subset of the more general set of points called curves, the set of simple closed curves is still quite extensive. Of particular significance is the fact that some simple closed curves are the union of line segments. In this case, the name given to the set of points is a *simple polygon*. The word *polygon* is sometimes used to refer to figures that are not simple; however, the word *polygon* in this text implies simple polygon, with one interior. Examples of polygons are drawn in Figure 4.5. Polygons are themselves general sets of points and may be classified further according to the number of line segments in the union. They are discussed in later sections of this

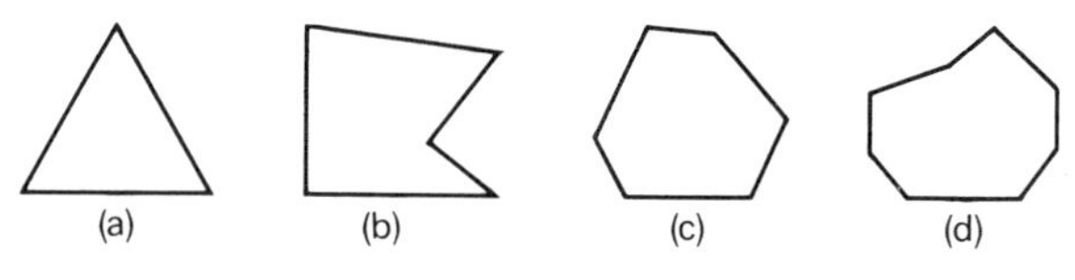

Figure 4.5. Polygons

chapter. The infinite set of points in the union of a simple closed curve and its interior is called a *plane region* or just a *region.* What is meant by interior and exterior of a simple closed curve is intuitively evident. Also, a straight line could lie entirely in the exterior but not entirely in the interior of a simple closed curve.

A simple closed curve and its interior may or may not be a convex set. A test for convexity can be discovered by studying Figure 4.6. If the region is

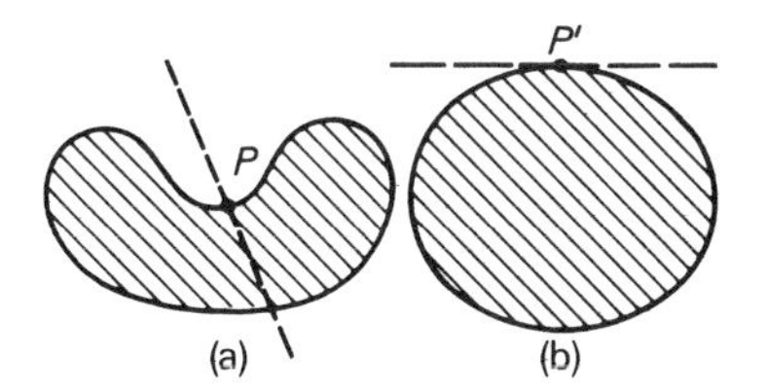

Figure 4.6. Test for convexity

not convex, then there must exist points like P such that no line can be drawn through P with all the interior points on one side of the line. For a convex region, every point on the boundary has at least one line through it such that the interior lies entirely on one side of this line.

Another test of the convexity of a simple closed curve and its interior is shown in Figure 4.7. If it is convex, every line through a point in the interior

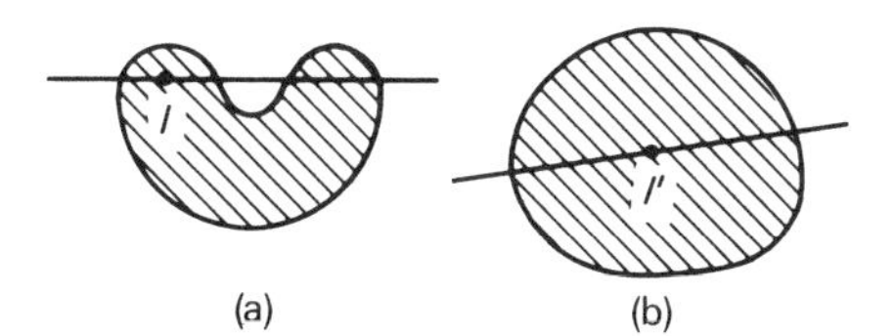

Figure 4.7. Second test for convexity

intersects the curve in exactly two points. This statement is not always true if the region is non-convex. The line through I, for example, intersects the curve in four points.

Exercise 4.1

In Exercises 1–4, refer to Figure 4.8.

1. Which sets of points are curves?
2. Which sets of points are closed curves?

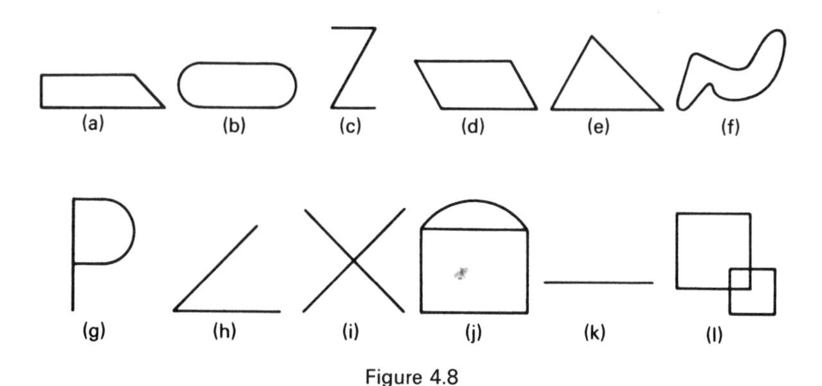

Figure 4.8

3. Which sets of points are simple closed curves?
4. Which sets of points are polygons?

Answer yes or no for Exercises 5–12.

5. All straight lines are curves.
6. Some straight lines are simple closed curves.
7. Some curves are straight lines.
8. All curves have at least one interior.
9. If a point in the interior and a point in the exterior of a simple closed curve are endpoints of a line segment, then the intersection of the segment and the simple closed curve is the empty set.
10. A plane region is a simple closed curve.
11. Every line through an interior point of a simple closed curve has exactly two points in common with the curve.
12. If a tracing puzzle represents a curve, then it can be traced without lifting the pencil or retracing a portion of the path.

For Exercises 13–17, tell whether each tracing puzzle represents a curve, a closed curve, or not a curve.

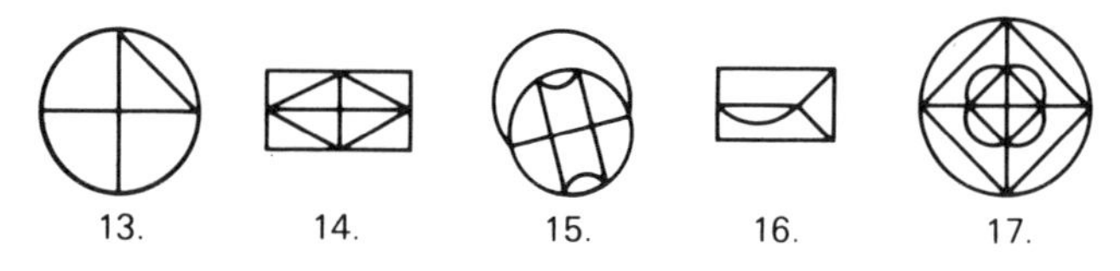

Figure 4.9

Classification of Triangles

A *triangle* is a polygon with three sides. A triangle is one of the important subsets of the set of simple closed curves; it is a simple closed curve composed of three line segments. A second way to define a triangle is "the union of three non-collinear points in a plane and the three segments connecting them." Non-collinear points are points that do not all lie on the same line. The three points, which are the endpoints of the segments, are the *vertices* of the triangle, and the three segments are the *sides.* A triangle in the geometric model reminds a person of physical objects, or the shapes of certain physical objects, such as the sails of a boat, certain pieces of land, a banner, or a wedge.

A triangle is named by stating the three vertices. Figure 4.10 shows triangle ABC, written $\triangle ABC$. In a triangle, each vertex is the intersection of two

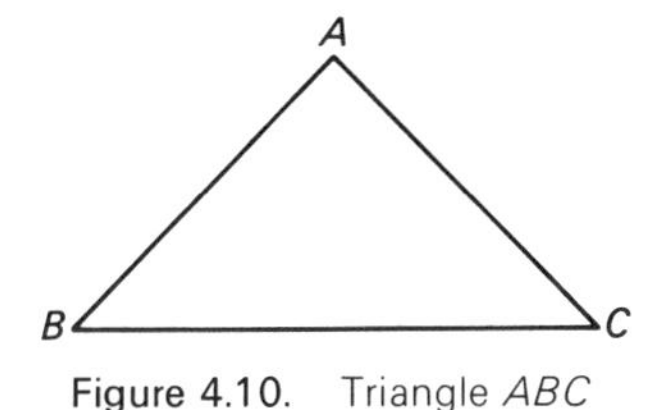

Figure 4.10. Triangle *ABC*

sides and is said to be opposite the third side. In Figure 4.10, vertex A is the intersection of $\overline{AB}$ and $\overline{CA}$; hence it is opposite $\overline{BC}$.

The possibilities for union and intersection of a triangle with many of the other sets of points previously studied are numerous; a few are shown in Figure 4.11. What set of points is each of these intersections: $\triangle ABC \cap \overleftrightarrow{DE}$? $\triangle ABC \cap \overline{BE}$? $\triangle ABC \cap \angle ABC$?

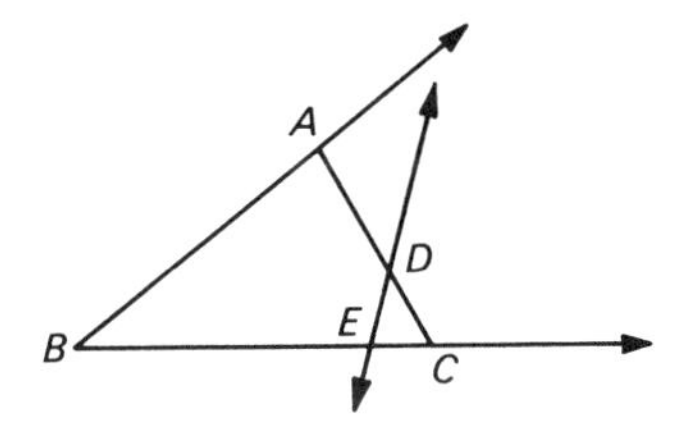

Figure 4.11. Union and intersection of triangle and another set

One way to classify triangles depends on the lengths of the sides—that is, on the measurements of the line segments. If a triangle has no two sides congruent, then it is a *scalene* triangle. Figure 4.12a illustrates a scalene triangle. If at least two sides are congruent, the triangle is *isosceles* (Fig. 4.12b). If all three sides are congruent, the triangle is *equilateral* (Fig. 4.12c).

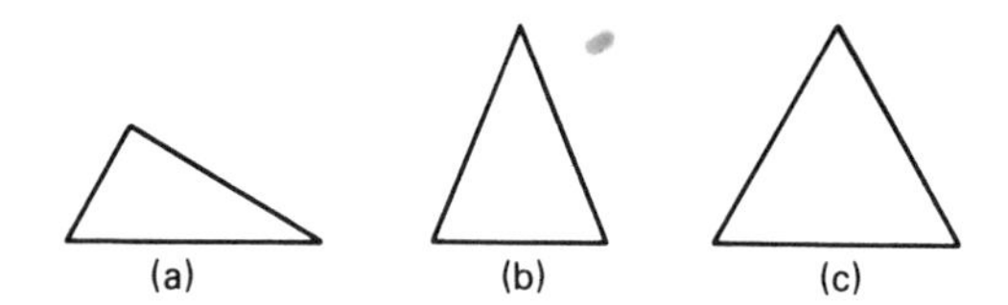

Figure 4.12. Triangles classified according to length of sides

From a very technical point of view, the answer to the question "How many angles are in a triangle?" is "None!" By definition, an angle is composed of two rays, but the sides of a triangle are segments. However, there are angles connected or associated with a triangle. In Figure 4.13, the

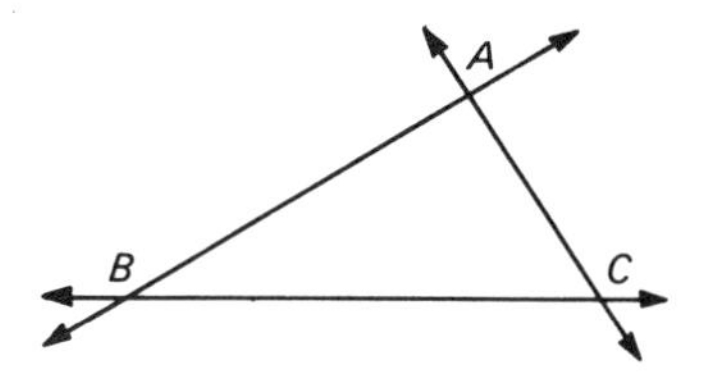

Figure 4.13. Angles of a triangle

angles ABC, BCA, and CAB are called the three angles *in* the triangle, or the three angles *of* the triangle. Each has a vertex of the triangle for its vertex. The sides of the triangle lie on the sides of the angles. Each of the points in the interior of the triangle is in the interior of each of three angles.

The three angles in a triangle have measures that are not all independent. In Figure 4.14, $\overleftrightarrow{DE}$ is the line through point A that is parallel to $\overleftrightarrow{BC}$. Then angles BCA and CAE are alternate interior angles of the two parallel lines and hence congruent. Similarly, $\angle CBA$ and $\angle BAD$ are also alternate interior angles and congruent to each other. The sum of the measures of the three angles at A($\angle BAD$, $\angle BAC$, and $\angle CAE$) is 180 in degrees, which means that the sum of the measures of the three angles in triangle ABC is 180 in degrees. The argument leads to the statement of an important theorem in Euclidean plane geometry.

Theorem

The sum of the measures of the three angles in a triangle is 180 in degrees.

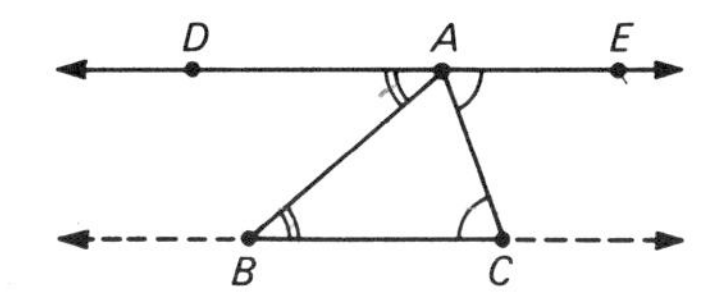

Figure 4.14. Total measure of angles in a triangle

Triangles may also be classified by considering the measurements of the three angles. They may be classified as *acute, right,* or *obtuse.* In Figure 4.15a, $\triangle ABC$ has three acute angles and is an acute triangle. In Figure

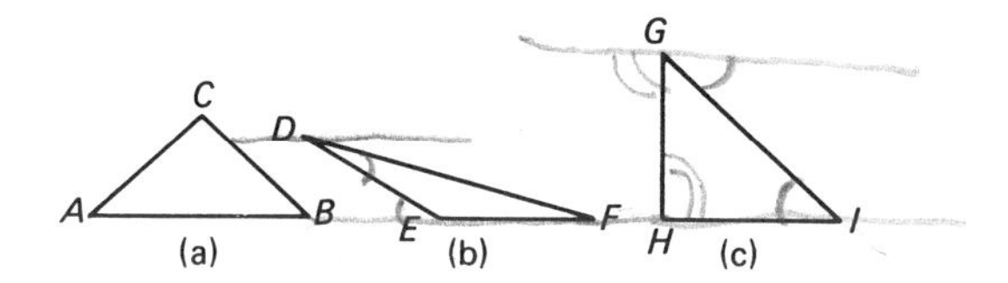

Figure 4.15. Triangles classified by measurement of angles

4.15b, $\triangle DEF$ has one obtuse angle at E and is an obtuse triangle. In Figure 4.15c, $\triangle GHI$ is a right triangle, since it has one right angle at H. The sides $\overline{GH}$ and $\overline{HI}$ of the right triangle are perpendicular. At least two angles of a triangle must be acute, because otherwise the sum of the measures of the angles would be greater than 180. In summary, an acute triangle has three acute angles, a right triangle has one right angle, and an obtuse triangle has one obtuse angle.

Theorem

The longest side of a triangle is opposite the greatest angle.

Figure 4.16 shows how an indirect proof can be used for this theorem. Assume that $\angle B$ is the greatest angle but that some other side, say $\overline{BC}$, is the longest side.

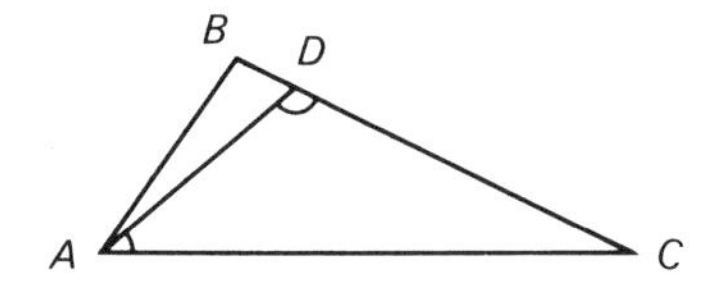

Figure 4.16. Use of indirect proof

If $\overline{BC}$ is the longest side, there must be a point D of $\overline{BC}$ such that $\overline{CD} \cong \overline{AC}$. The base angles $\angle DAC$ and $\angle ADC$ of isosceles triangle ADC are congruent. An *exterior angle* of a triangle is formed by one side and the extension of another side. In Figure 4.16, $\angle ADC$ is an exterior angle for triangle ADB. The exterior angle and the interior angle at the same vertex are supplementary. Because the sum of the measures of the three angles of a triangle is 180, an exterior angle has a measure equal to the sum of the interior angles at the other two vertices. In Figure 4.16, the measure of $\angle ADC$ is the sum of the measures of $\angle ABD$ and $\angle BAD$. For this reason, the measure of an exterior angle is greater than the measure of either of the interior angles at the other two vertices. Then the measure of $\angle ADC$ is greater than the measure of $\angle ABD$. This statement is a contradiction, because it makes $\angle BAC$ greater than $\angle ABC$, contrary to the assumption. Thus $\overline{BC}$ cannot be the longest side. In the same way, it can be shown that $\overline{AB}$ cannot be the longest side, so $\overline{AC}$ is proved to be the longest side.

Congruent triangles have exactly the same size and the same shape; all the corresponding sides and angles of the two triangles are congruent. The important concept of a *transformation* is useful in describing the relationship between two congruent triangles. In Figure 4.17, $\triangle ABC$ is transformed into congruent $\triangle DEF$ by a mapping that preserves size and shape. Point A is mapped into its *image,* point D; point B is mapped into its image, point E; and so on. Each point on $\triangle ABC$ has another point as its image under the isometry, which changes the focus of attention from the first triangle to the second.

When two triangles are congruent, it is possible to identify the three corresponding vertices, the three pairs of corresponding sides, and the three pairs of corresponding angles. If $\triangle ABC$, in Figure 4.17, is congruent to $\triangle DEF$, written $\triangle ABC \cong \triangle DEF$, then you know the vertices that correspond are A and D, B and E, and C and F. There are three pairs of congruent segments, $\overline{AB} \cong \overline{DE}$, $\overline{AC} \cong \overline{DF}$, and $\overline{BC} \cong \overline{EF}$. Also, there are three pairs of congruent angles, $\angle ABC \cong \angle DEF$, $\angle BAC \cong \angle EDF$, and $\angle BCA \cong \angle EFD$.

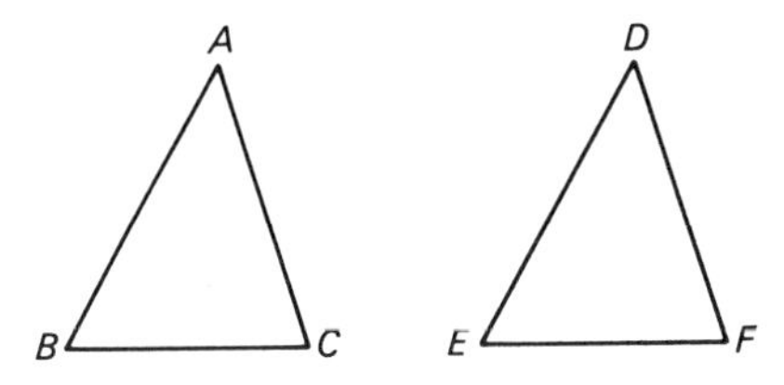

Figure 4.17. Congruent triangles

Do you think you would have to know that all six pairs of corresponding sides and angles were congruent to conclude that the two triangles were congruent? Suppose, for example, that the three pairs of corresponding sides and the two pairs of corresponding angles were congruent. Would the remaining pair of angles also have to be congruent? What is the minimum amount of information you would have to know about two triangles to be able to conclude that they were congruent? Try some examples to see if you can reach any conclusions before reading the next paragraph.

At this point it is necessary to extend the structure of Euclidean geometry by stating three additional axioms about congruence of triangles. These axioms give three sets of conditions for which it may be assumed that triangles are congruent.

Axiom 15

If two sides and the included angle of one triangle are congruent to the corresponding two sides and the included angle of the other triangle, then the correspondence is a congruence. (The *included angle* of two sides is the interior angle of the triangle whose vertex is the vertex of the triangle and whose rays contain the two sides.)

Axiom 16

If two angles and the included side of one triangle are congruent to the corresponding two angles and the included side of the other, then the correspondence is a congruence.

Axiom 17

If the three sides of one triangle are congruent to the corresponding three sides of the other triangle, then the correspondence is a congruence.

To illustrate, using the notation in Figure 4.18, the two triangles are

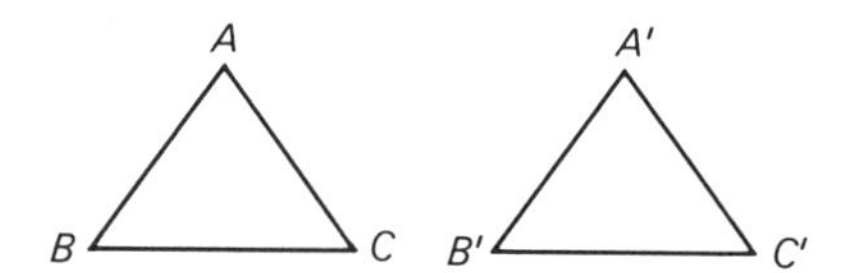

Figure 4.18. Conditions for congruence of triangles

congruent if:

(a)
$$\overline{AB} \cong \overline{A'B'}$$
$$\overline{AC} \cong \overline{A'C'}$$
$$\angle BAC \cong \angle B'A'C'$$

(b)
$$\angle BAC \cong \angle B'A'C'$$
$$\overline{AB} \cong \overline{A'B'}$$
$$\angle ABC \cong \angle A'B'C'$$

(c)
$$\overline{AB} \cong \overline{A'B'}$$
$$\overline{BC} \cong \overline{B'C'}$$
$$\overline{AC} \cong \overline{A'C'}$$

An additional set of minimum conditions can be proved to establish congruence.

Theorem

Two triangles are congruent if in the first triangle two angles and the side opposite one of them are congruent to the corresponding angles and sides in the second triangle.

In a triangle, the *side opposite* an angle is the side not lying on the rays of the angle. The proof of the theorem includes the fact that the third angle must also be congruent; hence Axiom 16 applies.

The property of congruence is preserved if a triangle is rotated or translated in the plane. This statement can be verified intuitively by moving a paper model of a triangular region to illustrate that the size and shape do not change.

If two triangles are right triangles, then you know to begin with that the two right angles are congruent; you need only two other independent facts to show that the two are congruent. You could modify the previous sets of data to apply to right triangles.

Two right triangles, ABC and DEF (Fig. 4.19), are congruent if:

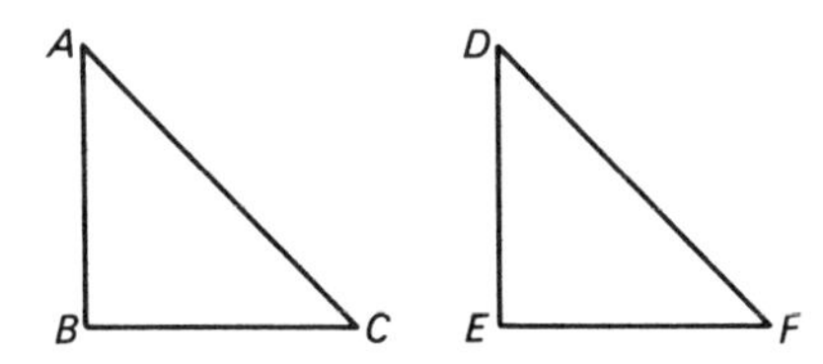

Figure 4.19. Congruent right triangles

(a) $\overline{AB} \cong \overline{DE}$ and $\angle A \cong \angle D$,

(b) $\overline{AB} \cong \overline{DE}$ and $\overline{BC} \cong \overline{EF}$, or

(c) $\overline{AC} \cong \overline{DF}$ and $\angle C \cong \angle F$.

If two triangles are congruent, their corresponding segments and angles are congruent. For example, you may wish to show that two angles are congruent. If you can show that they are corresponding angles in congruent triangles, you can conclude that they are congruent. In the rest of the text, congruent triangles will be used to prove some previously stated theorems and to explain some constructions.

Exercise 4.2

1. Which of these endings make correct statements (although possibly not good definitions) for a triangle? "A triangle is... (a) a curve." (b) a closed curve." (c) a simple closed curve." (d) a polygon." (e) the union of three segments." (f) the union of three vertices and three sides." (g) a convex set of points."
2. For a triangle PQR, name each vertex and the side opposite it.
3. List all possibilities for the intersection of two sets of points in the same plane if one of the sets of points is a triangle and the second set of points is a... (a) line. (b) ray. (c) angle. (d) triangle.

4. Are all isosceles triangles equilateral?
5. Are all equilateral triangles isosceles?
6. If the measures in degrees of two angles in a triangle are the numbers given, find the measure of the third angle.
   (a) 40, 90 (b) 33, 75 (c) $15\frac{1}{2}$, $37\frac{3}{4}$
7. If the measures in degrees of one angle in a triangle are the numbers given, and if the second angle is obtuse, use the symbol for "less than" to state how small the measure of the third angle must be.
   (a) 50 (b) 27 (c) 14.3
8. List all the possibilities for the numbers of these kinds of angles a triangle may have.
   (a) acute (b) right (c) obtuse
9. Experiment to see what seems to be true about the measures of the angles of an equilateral triangle.
10. Experiment to see what seems to be true about the measures of the two angles at the vertices opposite the congruent sides of an isosceles triangle.
11. Experiment to see which of these statements are true about the relative lengths of the sides of a triangle, using the same units of measurement.
    (a) The sum of the measures of any two sides must be greater than the measure of the third side.
    (b) The difference of the measures of any two sides must be less than the measure of the third side.
12. Are two triangles congruent if each angle of one is congruent to the corresponding angle of the other?
13. If you know that two triangles are equilateral, what is the minimum additional information you would need to conclude that they are congruent?
14. In Figure 4.20, assume that triangle ABC is isosceles, with $\overline{AB}$ congruent to $\overline{AC}$.

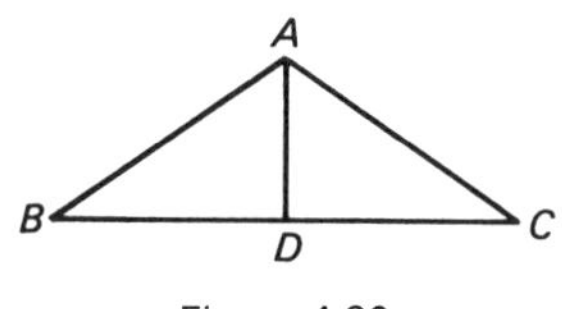

Figure 4.20

    (a) If $\overline{AD}$ is the *bisector* of angle BAC so that $\angle BAD \cong \angle CAD$, how do you know that triangle ABD is congruent to triangle ACD?
    (b) Why can you conclude that the angles at B and C are congruent?
    (c) Put the statement in Exercise 10 in the form of a theorem.
15. Use Exercise 14c to prove the theorem implied in Exercise 9.

Quadrilaterals

A *quadrilateral* is a polygon with four sides. The set of quadrilaterals is a subset of the set of simple closed curves. Quadrilaterals may also be considered as the union of four segments. Many common objects are representations of quadrilaterals or have shapes that remind you of a quadrilateral; examples are picture frames, luggage, paper money, and books.

Figure 4.21a shows a convex quadrilateral. Any segment connecting two points of the region lies entirely in the region. Non-convex quadrilaterals, such as the one shown in Figure 4.21b, will be excluded from more detailed

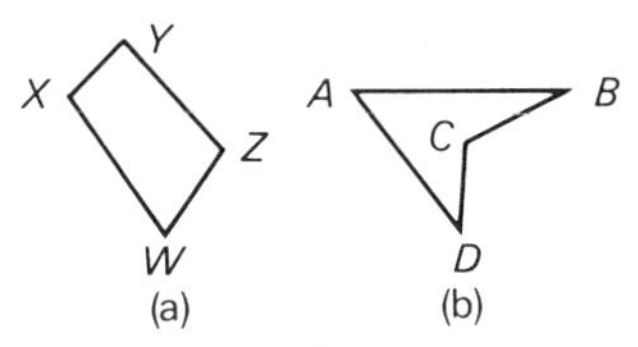

Figure 4.21. Convex and non-convex quadrilaterals

discussion later. A quadrilateral can be said to have four interior angles, one at each vertex, analogous to the angles of a triangle. Technically, these angles consist of the rays on which the sides of the quadrilateral lie. In Figure 4.21b, angle C is greater than 180°. The expression "greater than 180 degrees" is used here for simplicity, rather than the less familiar "greater than 180 in degrees." Similar simplifications will often be used throughout the rest of the text, although the reader should not lose sight of the distinction between measure and measurement.

The discussion of angles in previous chapters was limited to those less than 180 degrees. However, it is possible to extend the concept of angle to angles of any measure, positive or negative, and in particular to those between 180 and 360 degrees. These angles are called *reflex* angles. In Figure 4.22, you can identify the angle whose interior is to the left of C

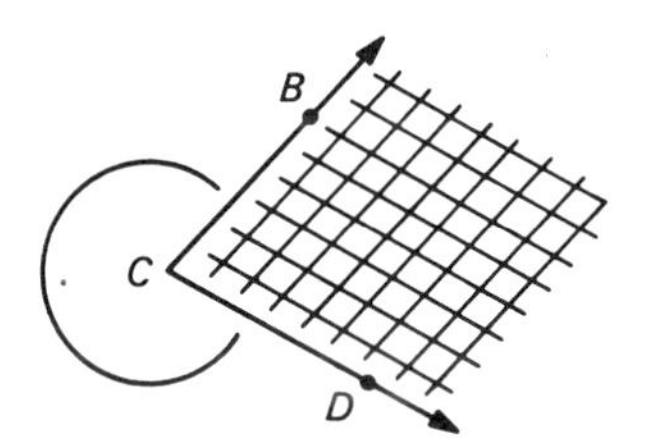

Figure 4.22. Reflex angle

simply by reversing the definitions of interior and exterior of angle *BCD*. If the shaded portion of the plane is the *exterior* of the angle, then angle *BCD* is a reflex angle. When each of the interior angles of a polygon is less than 180 degrees, the polygonal region is *convex*. Polygons with one or more reflex angles are called concave or non-convex. It is with sets of convex quadrilaterals that this section deals.

Quadrilaterals may be classified into several subsets, not all disjoint, according to special properties. A quadrilateral with no two segments lying on parallel lines is called a *trapezium*. If exactly two segments lie on parallel lines (if exactly two sides are parallel), the quadrilateral is a *trapezoid*. Figure 4.23a shows trapezoid *ABDC*. A quadrilateral is named by beginning at any vertex and naming the vertices in order, going around the picture either in a clockwise or a counterclockwise direction. For example, another name for Figure 4.23a is trapezoid *DCAB*.

If both pairs of opposite sides of a quadrilateral are parallel, as in Figure 4.23b, the quadrilateral is a *parallelogram*. Some parallelograms may have further specifications imposed on them, resulting in sets of points shown in Figure 4.23c, d, and e. If each of the angles of a parallelogram is a right angle, as in Figure 4.23c, the set of points is a *rectangle*. If the four segments of a rectangle are congruent, the rectangle is a *square* (Fig. 4.23d). Figure

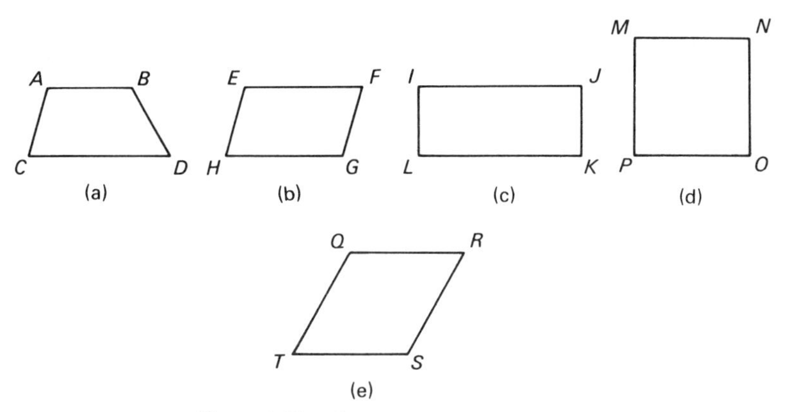

Figure 4.23. Examples of quadrilaterals

4.23e shows a *rhombus*, which is a parallelogram with all four segments—but not necessarily all four angles—congruent. A square is an example of a special kind of rhombus, but a rhombus does not have to be a square.

A trapezium, a trapezoid, a parallelogram, a rectangle, a square, and a rhombus have been listed as special quadrilaterals. The relationship among the various sets of points is shown in the form of a diagram in Figure 4.24.

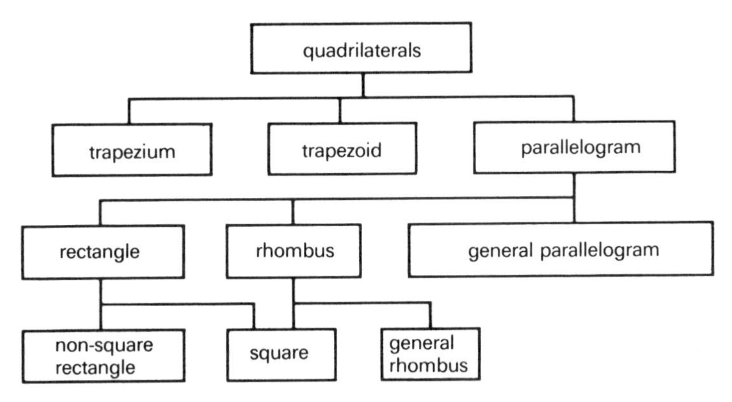

Figure 4.24. Classification of quadrilaterals

Next, consider some of the special properties of quadrilaterals, which may be stated in the form of theorems. In Figure 4.25, $\overline{AC}$ is called a *diagonal.*

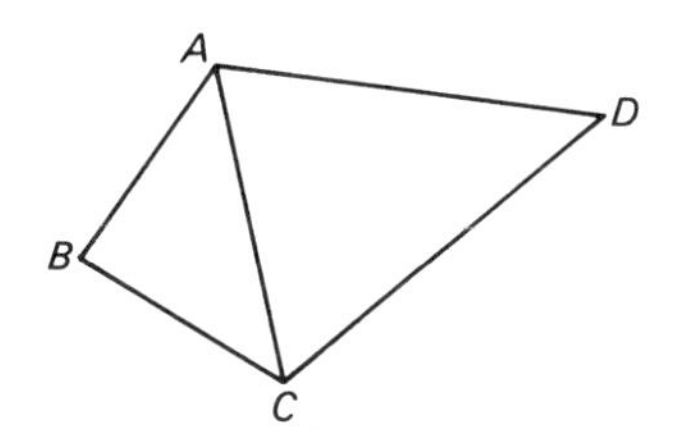

Figure 4.25. Quadrilateral with diagonal

Since the sums of the measures of the angles in triangles ADC and ACB are both 180 degrees, the sum of the measures of the angles in quadrilateral $ABCD$ is 360 degrees.

Theorem

The sum of the measures of the angles in any quadrilateral is 360 degrees.

Incidentally, notice that Figure 4.25 could actually be described as the union of the triangular regions ACD and ACB.

Parallelograms have interesting properties that can be investigated in intuitive geometry. One basic theorem is that the opposite sides of a parallelogram are congruent segments. This theorem can be proved by using congruent triangles. In Figure 4.26, for example, $\overline{BC}$ and $\overline{AD}$ will

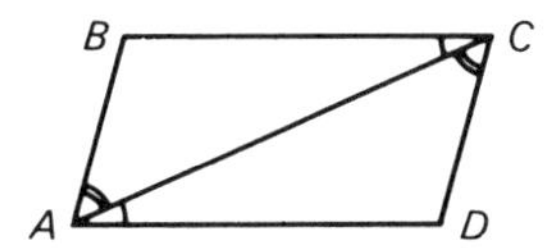

Figure 4.26. Congruent triangles in parallelogram

be congruent if they can be shown to be corresponding sides of congruent triangles. But $\triangle ABC \cong \triangle CDA$. Why? The fact that the opposite sides of a parallelogram are parallel leads to conclusions about relationships of angles in the set of points.

In Figure 4.27, consecutive angles *CDA* and *DAB* are supplementary, but angles *ADC* and *DCB* are also supplementary, which means that opposite angles *A* and *C* are congruent.

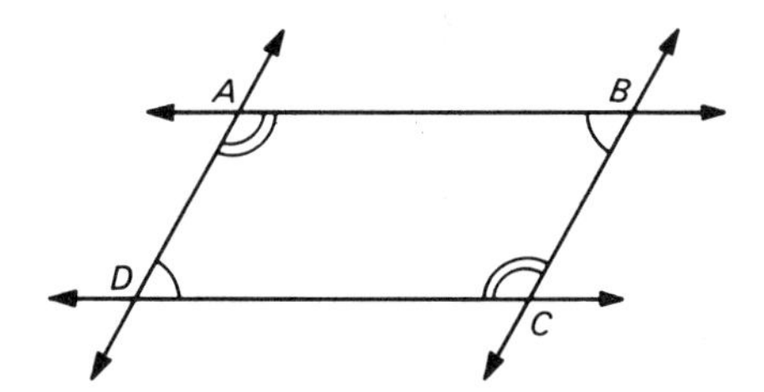

Figure 4.27. Angles of a parallelogram

Theorem

The opposite angles of any parallelogram are congruent, and the consecutive angles are supplementary.

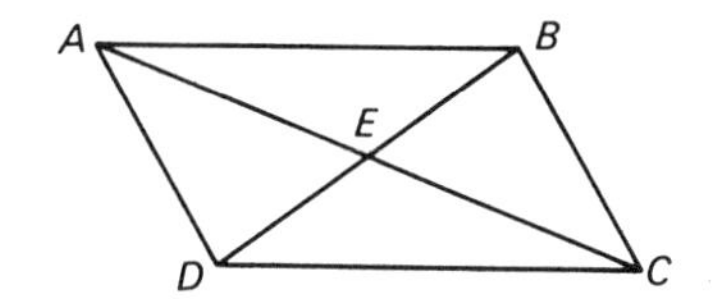

Figure 4.28. Diagonals of a parallelogram

Figure 4.28 shows quadrilateral $ABCD$, which is a parallelogram with its two diagonals intersecting at point E.

Theorem

For a parallelogram, the point of intersection of the two diagonals is the midpoint of each of the diagonals.

The notation of Figure 4.28 may be used in proving the theorem.

$$\overline{AB} \cong \overline{CD} \qquad \text{Why?}$$

$$\angle BAE \cong \angle DCE \qquad \text{Why?}$$

$$\angle ABE \cong \angle CDE \qquad \text{Why?}$$

Therefore

$$\triangle ABE \cong \triangle CDE \qquad \text{Why?}$$

Thus $\overline{BE} \cong \overline{DE}$ and $\overline{AE} \cong \overline{CE}$, since these are corresponding sides of congruent triangles.

It is intuitively obvious that other important sets of points besides segments, angles, and triangles may also be congruent. Two squares are congruent if they have the same measure of one edge (since that would make all the sides congruent) and all the angles are known to be right angles. You would need more information to conclude that two rectangles are congruent. Would the knowledge that two consecutive sides were congruent to the corresponding sides of the other rectangle be sufficient?

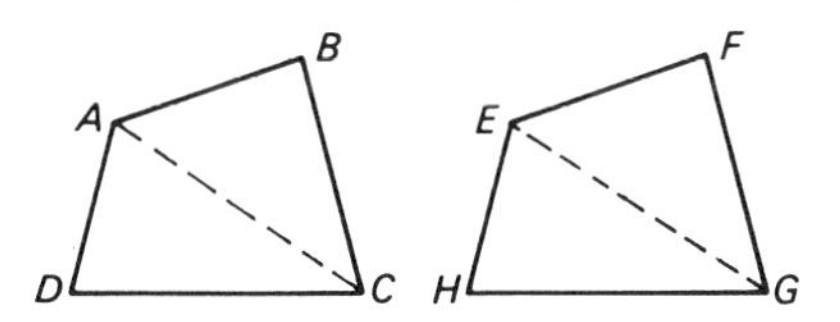

Figure 4.29. Congruent quadrilaterals

For a parallelogram, even more information would be needed. You would also need to know, in addition to the two consecutive sides, that a pair of corresponding angles were congruent. For a quadrilateral, still more information is needed. Often the congruence of more general polygons is approached through stating that the two polygons can be partitioned in the same way into sets of congruent triangles. For example, in Figure 4.29, if you knew that $\triangle ABC \cong \triangle EFG$ and $\triangle ADC \cong \triangle EHG$, you could conclude that the two quadrilaterals were congruent.

The ancient Greeks often used a rectangle of a particular shape in their architecture. For example, the Parthenon is shaped like the rectangle in Figure 4.30. The ratio of the measures of the sides of this rectangle is called the *golden ratio,* and a rectangle of this shape is called a *golden rectangle.*

Figure 4.30. Golden rectangle

The exact value of the golden ratio is $(\sqrt{5} + 1)/2$, and the approximate decimal value is 1.618. Closer and closer approximations to the golden ratio can be found by taking ratios of consecutive numbers in the famous sequence called *Fibonacci numbers*: 1, 1, 2, 3, 5, 8, 13, 21, ... A number in the set after the first two is the sum of the two previous numbers. Some ratios are:

$$\frac{1}{1} = 1, \quad \frac{2}{1} = 2, \quad \frac{3}{2} = 1.5, \quad \frac{5}{3} \approx 1.67,$$

$$\frac{8}{5} = 1.6, \quad \frac{13}{8} = 1.625, \quad \frac{21}{13} \approx 1.615$$

Exercise 4.3

Answer yes or no for Exercises 1–4.

1. The set of all parallelograms is a subset of the set of all quadrilaterals.
2. The set of squares and the set of rectangles are disjoint sets.
3. The set of all parallelograms is a subset of the set of all trapezoids.
4. Some rectangles are rhombuses.
5. Each vertex of a quadrilateral is an endpoint for how many diagonals?
6. How many names are possible for any quadrilateral, using the four vertices in any order permitted?
7. Draw several quadrilaterals in which the two diagonals intersect at a point that is the midpoint of each diagonal. What kind of quadrilateral must this be?
8. Draw several parallelograms with the two diagonals congruent segments. What kind of parallelogram must this be?

9. Compare the measure of the segment joining the midpoints of the two non-parallel sides of a trapezoid with the sum of the measures of the two parallel sides. What seems to be the relationship between these two numbers?

For Exercises 10–16, tell whether or not the two sets of points are always congruent.

10. Two squares.
11. Two squares with exactly one side in common.
12. Two rectangles, both of which have a diagonal with the same measure.
13. Two rectangles, if the measures of their perimeters are the same.
14. Two parallelograms, if each angle of one is congruent to the corresponding angle of the other.
15. Two parallelograms, if they can be partitioned into two pairs of corresponding congruent triangles by corresponding diagonals.
16. Two quadrilaterals, each partitioned by a diagonal into two isosceles triangles.

For Exercises 17–18, find the approximate length of the longer side of a golden rectangle if the shorter side is:

17. 3.1 m
18. .24 in.
19. Prove that a diagonal partitions a rectangle into two congruent triangles.
20. Prove that a diagonal partitions a rhombus into two congruent triangles.
21. Experiment with examples and then state which of the sets of quadrilaterals shown in Figure 4.24 (on p. 79) always have diagonals that are perpendicular to each other.
22. Does a diagonal partition every quadrilateral into two congruent triangles? Which quadrilaterals?

Polygons with More than Four Sides

In previous sections you have read about polygons with three sides (triangles) and with four sides (quadrilaterals). A polygon may be an n-sided figure, where n is a counting number 3 or greater. In this section polygons that are the union of five or more segments will be studied. Unless otherwise stated, the word *polygon* is used to mean a polygon that is the boundary of a convex region. At each vertex is one interior angle, consisting of rays on which the two adjacent sides lie, and each of these angles has a measure less than 180 degrees.

The names for polygons with more than four sides are sometimes used in everyday language: a polygon with five sides is a *pentagon*; a polygon with six sides is a *hexagon*; one with seven sides is a *heptagon*; and one with eight sides is an *octagon*. Some examples of actual objects that have characteristics of polygons with more than four sides are the Pentagon building; cells of a beehive (six sides); hexagonal crystals; a stop sign (eight sides); and some buildings, columns of buildings, and tables (six or eight sides).

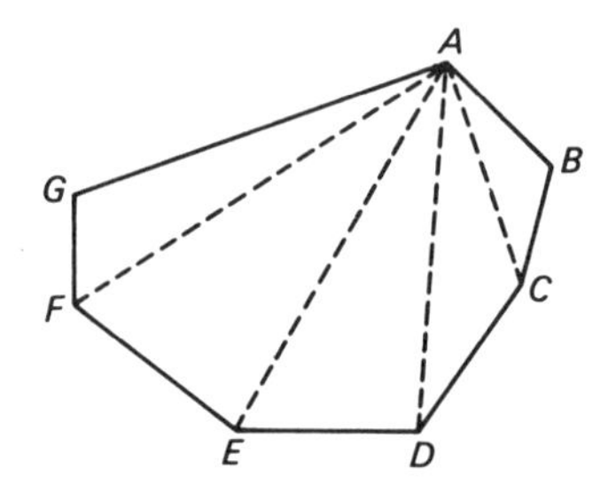

Figure 4.31. Heptagon

Figure 4.31 illustrates a heptagon. Segments with vertices as endpoints, but not lying on the sides of the polygon, such as $\overline{AC}$, are called diagonals. For convex polygons, diagonals are in the interior. From any vertex of this polygon, such as A, four diagonals can be drawn. A diagonal cannot be drawn from A to itself, nor to the two next vertices, G and B, but a diagonal can be drawn from A to each of the other four vertices. In this case the number of diagonals that can be drawn from a vertex is three fewer than the number of vertices. Do you think this statement is true for all polygons? Draw an octagon, and determine how many diagonals can be drawn from each vertex of it.

In Figure 4.31, the four diagonals from A partition the polygon and its interior into five triangles and their interiors. Each triangle has A as a common vertex. The number of triangles is two less than the number of vertices. Is this statement also true for the octagon you drew? This method of partitioning into triangles makes it possible to investigate the sum of the measure of the angles in any convex polygon. In Figure 4.31, the sum of the measures of the angles in the five triangles is 5×180 degrees. Since the polygon has seven vertices, $(7 - 2) \times 180$ degrees also indicates this measure. Can you write a similar statement for the octagon you drew? A polygon with n sides has $(n - 2) \times 180$ for the sum of the measures of its angles.

Example

Find the sum of the measures of the angles of a polygon of 20 sides.

$$(20 - 2) \times 180 = 18 \times 180$$
$$= 3240$$

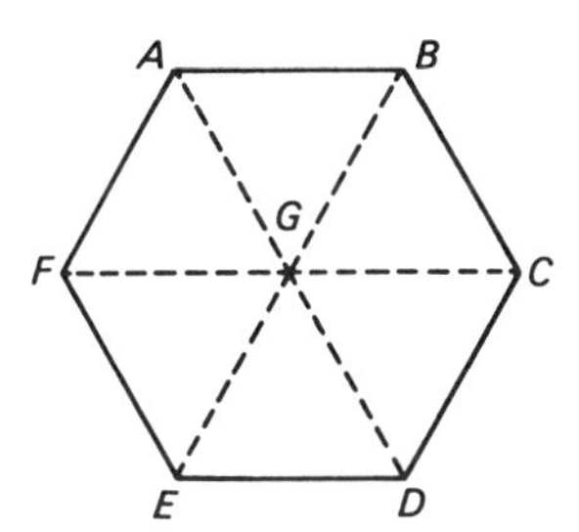

Figure 4.32. Regular hexagon

Many common examples of polygons have all sides congruent and all angles congruent. Convex polygons of this type are called *regular* polygons. Figure 4.32 illustrates a regular hexagon. Point G may be called the center of the polygon, since it is the same distance from each vertex. The fact that a regular polygon has a center is established from the discussion of circles. That is, all of the segments, $\overline{AG}$, $\overline{BG}$, and so on, are congruent. Vertices A and D, B and E, and C and F may be designated as opposite vertices. The diagonals connecting opposite vertices intersect at the center of the polygon. For any regular polygon with an even number of vertices, such as 4, 6, 8, 10, 12, . . . , a similar statement about opposite vertices may be made.

Since a formula has been given for the sum of the measures of angles in a polygon, and since all of the angles of a regular polygon are congruent, it is easy to find the measure of each angle of a regular polygon if you know the number of sides.

Example

For a regular pentagon, the measure of each angle in degrees is

$$\frac{3 \times 180}{5} = \frac{540}{5} = 108.$$

Do you think the measure of each angle of a regular hexagon is more or less than 108? The general expression for the measure of each angle of a regular polygon with n sides is

$$\frac{(n - 2) \times 180}{n}$$

Example

Find the measure of each angle of a regular polygon with ten sides.

$$\frac{(10 - 2) \times 180}{10} = \frac{8 \times 180}{10}$$
$$= 144$$

The concept of congruence of polygons is an extension of the ideas discussed for quadrilaterals.

Two polygons are congruent if they can be partitioned in the same way into corresponding congruent triangles.

The concept of congruence will be left undefined for simple closed curves that are not polygons (except for circles). Two sets of congruent points differ only in location. Intuitively, you can see that the pictures of two congruent curves would look the same. If you traced one and placed it over the other, it would seem to match exactly.

Exercise 4.4

1. Name additional examples of objects that have characteristics of polygons with more than four sides.
2. From each vertex of a polygon with n sides, where n is a counting number 4 or greater, how many diagonals can be drawn?

3. In Figure 4.31 (on p. 84), how many diagonals can be drawn from each vertex named, not including those diagonals already drawn from all of the preceding vertices? Vertex:
   (a) B (b) C (c) D (d) E (e) F (f) G
4. What is the total number of diagonals for a heptagon?
5. Complete the table of the total number of diagonals possible for various polygons.

number of sides in polygon	*total number of diagonals*
4	___
5	___
6	___
7	___
8	___

6. Develop a formula for the total number of diagonals in a polygon with n sides.
7. Use the formula in Exercise 6 to find the total number of diagonals in a polygon with 16 sides.
8. Find the sum of the measures of the angles in a hexagon.
9. Find the sum of the measures of the angles in a decagon (ten sides).
10. What special name is given to a regular polygon that is a triangle?
11. What special name is given to a regular polygon that is a quadrilateral?
12. Complete the table for the measure in degrees of each angle of a regular polygon.

number of sides	*measure in degrees of each angle*
4	___
5	___
6	___
7	___
8	___
16	___

13. Could a pentagon be congruent to a hexagon?
14. Are two regular polygons with the same number of sides always congruent?
15. What is the smallest possible measure for an interior angle of a regular polygon?

Perimeter

The *perimeter* of a simple closed curve is the length of the curve—that is, the distance around it. In this section the perimeter of polygons, a subset of the set of simple closed curves, will be discussed. The measure of the perimeter of a polygon is the sum of the measures of the sides—assuming that the same unit of measurement is used.

Example

If the sides of a triangle are 8, 10, and 16 inches, then the perimeter is 34 inches.

Example

If the measures of the sides of a polygon are 6, 5, 8, 7, 14, 20, 12, and 3, then the measure of the perimeter is

$$6 + 5 + 8 + 7 + 14 + 20 + 12 + 3 = 75.$$

In general, you add to find the measure of perimeter for a polygon. In the case of regular polygons, however, you may multiply the number of sides by the measure of each. This useful idea may be expressed as a formula for P, the measure of perimeter, where n represents the number of sides and e represents the measure of each side.

$$P = ne$$

Example

The measure of perimeter of a regular octagon, if the measure of each side is 5, is $P = 8 \times 5 = 40$.

If any two of the three numbers indicated in the formula $P = ne$ are known, the third may be found.

Example

If the perimeter of a regular polygon is 54 inches, and the polygon is a hexagon, then $54 = 6 \times e$, and $e = 9$. In this case each side is 9 inches.

It is also interesting to investigate the effect of varying one of the numbers represented in the formula for perimeter and holding a second one constant. For example, the effect of doubling the number of sides and keeping the measure of each the same is to double the measure of perimeter for the polygon.

Do you think it is possible to have one polygon inside another and for the inside one to have the greater perimeter? This is not possible for convex polygonal regions but is easily possible otherwise, as is shown in Figure 4.33.

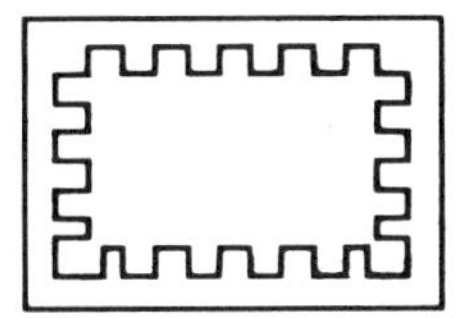

Figure 4.33. Comparing perimeters

The Pythagorean Theorem

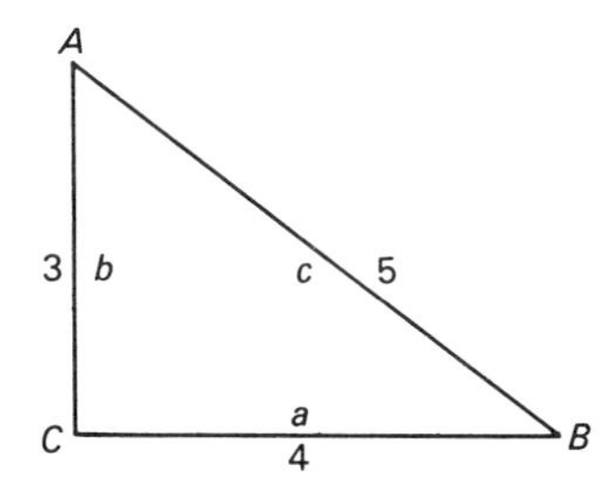

Figure 4.34. Lengths of sides of a right triangle

The ancient Egyptians and Babylonians knew and used a significant relationship among the lengths of the three sides of a right triangle; it is called the *Pythagorean Relationship,* or Pythagorean Theorem. They found, for example, that a piece of rope 12 units long, if held by three people to represent a triangle (see Fig. 4.34) with sides 3 units, 4 units, and 5 units, would give them a representation of a right triangle with the right angle opposite the 5-unit side.

In Figure 4.34, the measures of the sides opposite the vertices are indicated by small letters, so that the side with the measure a is opposite vertex A, and so on. Then the *Pythagorean Theorem states that*

$$c^2 = a^2 + b^2$$

The side opposite the right angle of a right triangle is called the *hypotenuse.* In words, the theorem states that the square of the measure of the hypotenuse is equal to the sum of the squares of the measures of the other two sides of a right triangle. An alternative form of the formula, using the symbol for square root, is

$$c = \sqrt{a^2 + b^2}$$

Example

If the two shorter sides of a right triangle have measures of 2 and 3, find the measure of the hypotenuse.

$$c = \sqrt{2^2 + 3^2} = \sqrt{4 + 9} = \sqrt{13}$$

In practice, the square root of 13 is often named approximately by a rational number, using the symbol for approximately equal. For example, you might write

$$c = \sqrt{13} \approx 3.6$$

The method of divide and average can be used to find a square root correct to any desired number of decimal places. The method is illustrated for $\sqrt{13}$.

First estimate (guess): 4

Divide: $13 \div 4 = 3.25$

Since $\sqrt{13}$ must be between 4 and 3.25, take the average as the next estimate and repeat the process until an estimate with the desired precision is obtained.

Second estimate: $\dfrac{4 + 3.25}{2} = 3.625$

Divide: $13 \div 3.625 \approx 3.586$

Third estimate: $\dfrac{3.625 + 3.586}{2} = 3.6055$

The Greek mathematician Pythagoras, who lived about 500 B.C. and was head of a famous mystical brotherhood, is given credit for proving the theorem that bears his name. Since that time, several hundred distinct proofs have been offered, including one by President Garfield. One proof is given in Chapter 7. It is significant that Pythagoras probably arrived at a method of proving his theorem by purely intuitive means, possibly by observing a pattern of tiles on a floor, similar to the one pictured in Figure 4.35. Count to verify that the number of small triangles in the large square on the hypotenuse is equal to the sum of the numbers of small triangles in the squares on the other two sides.

In general, a right triangle will not have three counting numbers for the measures of its three sides. What is the measure of the hypotenuse in Figure 4.35, for example? The "3, 4, 5" triangle mentioned earlier is an exception in this regard, as is a triangle with measures 5, 12, and 13. These triples of counting numbers that may serve as measures for the sides of a

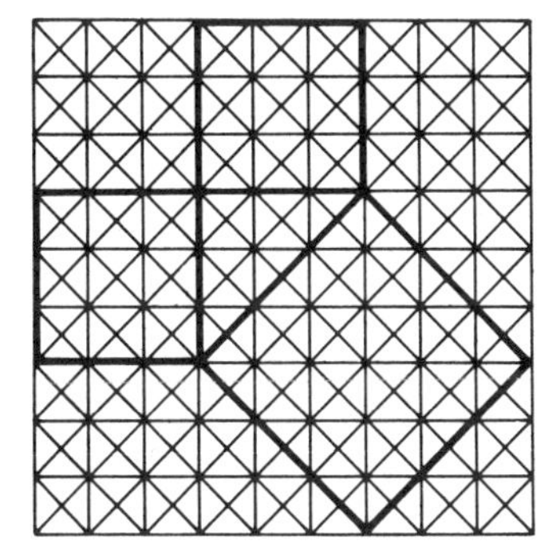

Figure 4.35. Tile pattern illustrating Pythagorean Theorem

right triangle are called *Pythagorean triples,* although they were investigated by the Babylonians even before the time of Pythagoras. Given a Pythagorean triple, such as 3, 4, 5, you can multiply each of the numbers by a constant to find another Pythagorean triple. For example, 6, 8, 10 is also a Pythagorean triple.

Pythagoras was especially interested in the length of the hypotenuse of a right triangle with two sides 1 unit long. Applying the Pythagorean Theorem leads to the conclusion that the hypotenuse of this isosceles right triangle has a measure of $\sqrt{2}$, an irrational number. The Pythagoreans were greatly disturbed by the discovery of this number, which could not be expressed as a ratio of integers. To them it represented a flaw in the universe.

Using the Pythagorean Theorem, it is possible to find the measure of the third side, and hence the measure of the perimeter, of a right triangle, knowing the measures of just two sides. Since this process involves finding the square root of a number, results are usually approximate. An example illustrates the type of problem for which the Pythagorean Theorem can be used.

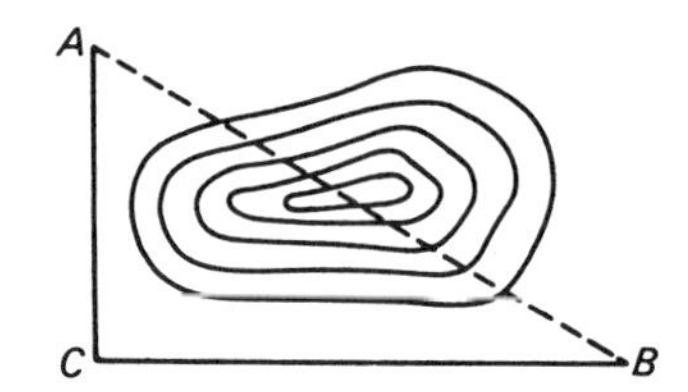

Figure 4.36. Use of Pythagorean Theorem

Example

In Figure 4.36, A and B are on opposite sides of a mountain. How many miles apart are these two locations, if it is 2.1 miles from A to C and 3.7 miles from C to B?

$$c = \sqrt{(2.1)^2 + (3.7)^2} = \sqrt{4.41 + 13.69}$$

$$= \sqrt{18.1} \approx 4.3$$

The approximate distance is 4.3 miles.

Exercise 4.5

1. Find the perimeter of a triangle if the sides are 3.5, 2.1, and 4.4.
2. Do two congruent triangles always have the same measure of perimeter?
3. Are two triangles that have the same measure of perimeter always congruent?
4. Find the perimeter of a parallelogram if one side has a measure of 12 and the next side has a measure of 16.
5. Experiment to see if this statement seems true or not: The sum of the lengths of the segments drawn from a vertex to the midpoint of the opposite side of a triangle is greater than the perimeter of the triangle.
6. Complete this table for regular polygons, using the relationship $P = ne$.

measure of perimeter	*number of sides*	*measure of each side*
____	8	3.2
126	10	____
144	____	6
____	15	3.8

7. In the formula $P = ne$, what is the effect on:
   (a) P if n and e are both doubled?
   (b) P if n is doubled and e is halved?
   (c) n if e is doubled and P is left unchanged?
   (d) n if P is halved and e is doubled?
   (e) e if P and n are both doubled?

In Exercises 8–9, find the third number so that the set of three numbers will be a Pythagorean triple if the first two numbers are:

8. 9, 12
9. 5, 12

In Exercises 10–13, find the length of the hypotenuse of a right triangle if the lengths of the other sides are the measurements given. Leave your answer in the form of a square root.

10. 4 in., 9 in.
11. 7 m, 10 m
12. 14 in., 5 in.
13. 20 m, 19 m

In Exercises 14–20, express answers to the nearest whole number.

14. If a ladder 12 ft long is placed against a house so that the base of the ladder is 4 ft from the house, how high will it reach?
15. How long must a wire be to reach from a point 20 ft up a pole to a point on the ground 10 ft from the base of the pole?
16. How long a piece of canvas is required for the roof of the shelter shown in Figure 4.37a?

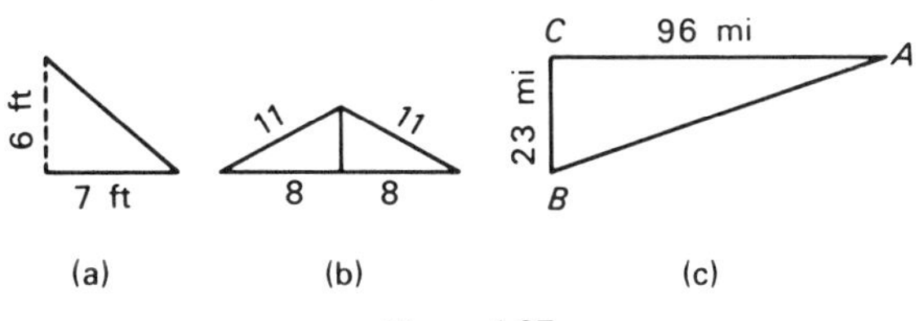

Figure 4.37

17. Find the length of the diagonal of a rectangle if the sides are 7 in. and 3 in. long.
18. Find the length of the altitude of an equilateral triangle if each of the sides is 4 in. long.
19. In Figure 4.37b, if a rafter 11 ft long is used for a roof 16 ft wide, how much higher is the roof in the center of the building than at the sides?
20. In Figure 4.37c, how much shorter is the direct distance from A to B than the distance by way of point C?

One set of simple closed curves of particular importance is the set of *circles.* Many examples of circular objects can be given, such as a tire, a clock face, a dinner plate, a ring, a doorknob, a coin, or a jar lid. In this chapter the circle as a set of points will be investigated. Following an analysis of some basic definitions and practice in thinking of a circle in connection with other geometric sets, the topics of central angle, arc, and circumference will be introduced. The study of circles leads to an analysis of the useful topic of symmetry and a look at some applications.

Basic Concepts

A circle is the set of all points in a plane at a fixed distance from a given point in the plane. *Distance* is used here as an undefined term and implies the shortest possible distance measured along a segment. A circle is an example of a curve, a closed curve, and a simple closed curve. Figure 5.1 shows circle A, named by stating the point that is the center (the given point of the definition).

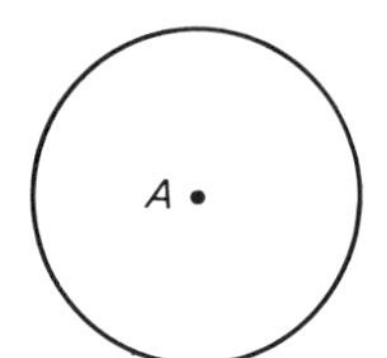

Figure 5.1. Circle *A*

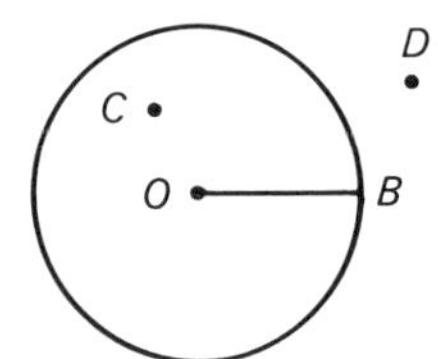

Figure 5.2. Radius, interior, exterior

Each point on a circle, such as *B* in Figure 5.2, may be considered as one endpoint of a segment called a *radius*, with the center as the other endpoint. All of these segments are congruent for any given circle. The word *radius* is also used for the common length of all the segments as well as to designate the actual set of points. Thus two circles with the same radius are congruent. The idea of congruence rather than distance may be used to explain what a circle is. In Figure 5.2, think of all the segments in the plane with *O* as one endpoint and congruent to $\overline{OB}$. The other endpoints constitute a circle that contains *B*, and with *O* as the center.

Since a circle is a simple closed curve, it has a single *interior*. A circular region is another example of a convex set of points. A circle partitions the points of a plane into three disjoint subsets: the points in the interior, the points on the circle, and the points in the exterior. Any point in the interior, such as *C*, is closer to the center than is a point on the circle; a point in the exterior, such as *D*, is farther from *O*. In other words, the measure of $\overline{CO}$ is less than the measure of $\overline{BO}$, which in turn is less than the measure of $\overline{DO}$.

In Figure 5.3, $\overline{FE}$ is a *diameter*. A diameter is a line segment that has its two endpoints on the circle and contains the center of the circle. Each circle has an infinite number of diameters. Notice that a diameter can be thought of as the union of two radii that lie on the same line. The word *diameter* is also often used in a second way to indicate the length of the segment rather than the segment itself. A segment such as $\overline{GH}$, which

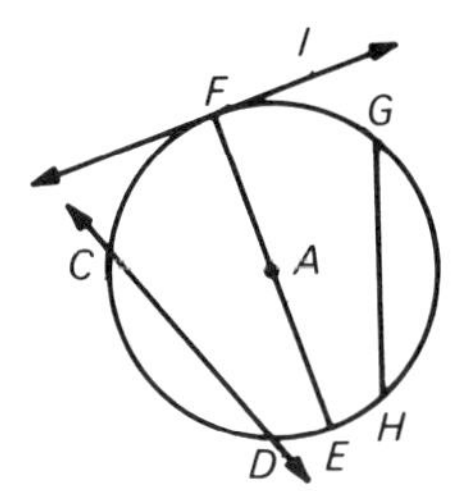

Figure 5.3. Intersection of a line and circle

has endpoints on the circle but which does not necessarily pass through the center, is called a *chord.* A diameter may be thought of as a special chord. The line on which a chord lies, such as $\overleftrightarrow{CD}$, is called a *secant* of the circle. A secant intersects a circle in two distinct points. A line such as $\overleftrightarrow{FI}$, which intersects a circle in only one point, is called a *tangent* to the circle. The point at which a tangent intersects the circle— in this case, point F— is the *point of tangency.* A tangent is perpendicular to the radius at the point of tangency.

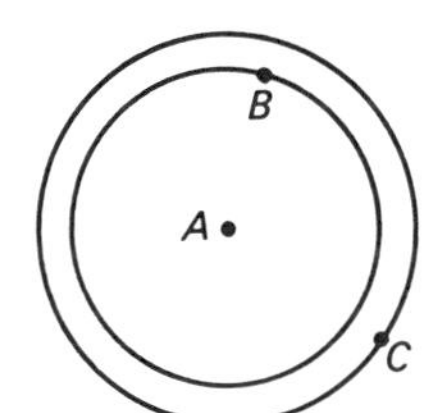

Figure 5.4. Concentric circles

Figure 5.4 shows two circles that are *concentric.* They have the same center but not the same radius. In a case such as this, when naming the center would not distinguish between them, circles can be named by stating one point of each circle, such as "circles B and C." Points in the interior of circle B, and points on circle B itself, are all in the interior of circle C. The radius of circle C is longer than the radius of circle B. The smaller circular region is a subset of the larger.

Some of the interesting examples of intersection of a circle and other sets of points have already been shown in Figures 5.1 through 5.3. A line can intersect a circle in two, one, or no points, as can a ray, a segment, and a half-line. In Figure 5.5, the intersection of $\triangle ABC$ and circle D consists of the three vertices, A, B, and C. In this case the triangle is said to be *inscribed* in the circle, or the circle is *circumscribed* about the triangle.

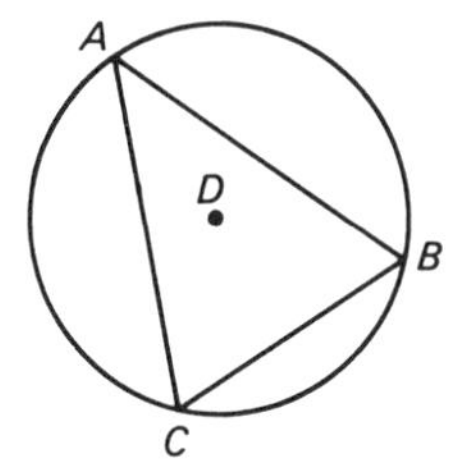

Figure 5.5. Triangles inscribed in circle

Polygons inscribed in a circle have special properties, which will be investigated later in this chapter.

In Figure 5.6, the circle is inscribed in the polygon. Circle E is inscribed in quadrilateral $ABCD$. Here each of the sides of the quadrilateral has exactly

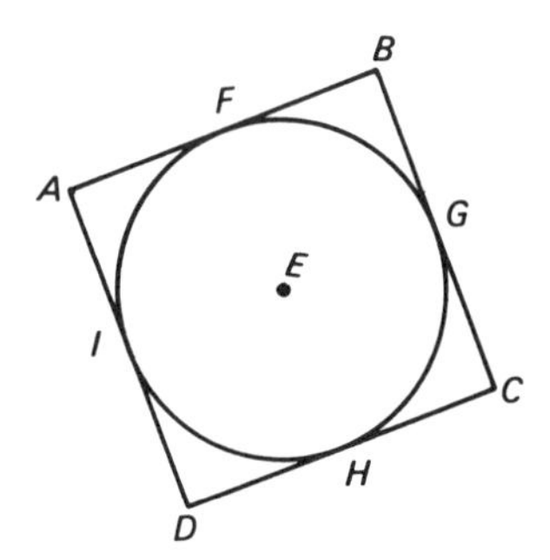

Figure 5.6. Circle inscribed in polygon

one point in common with the circle. Each side is tangent to the circle; the four points in the intersection of the polygon and the circle are the four points of tangency, F, G, H, and I. Another way of explaining the relationship is by saying that the polygon is circumscribed about the circle.

Figure 5.7 shows circle O with two angles in special positions. Angle BAC is called an *inscribed* angle. Its vertex, point A, is a point on the circle. The intersection of the angle and the circle consists of points A, B, and C. The second angle shown, $\angle DOE$, has its vertex at the center of the circle and intersects the circle in points D and E. It is called a *central* angle. Both an inscribed angle and a central angle will be used in the next section.

The diameter of a circle is sometimes called its width. For any set of points in a plane, the width in any particular direction is the distance between

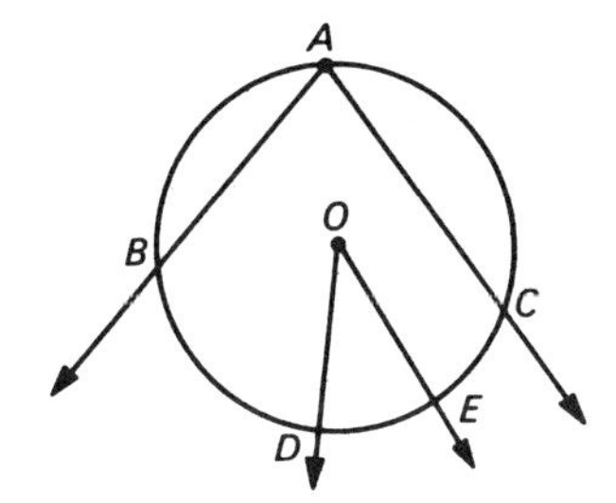

Figure 5.7. Central and inscribed angles

two parallel lines, as shown in Figure 5.8. A circle is an example of a *set of constant width,* because its width is the same regardless of the direction in which it is measured. Do you think the circle is the only set of constant width?

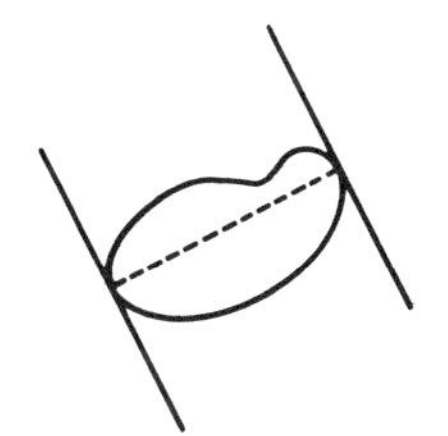

Figure 5.8. Width of a set

Sets of constant width have been examined from a general point of view only during the past 25 years. Figure 5.9 shows a set of constant width

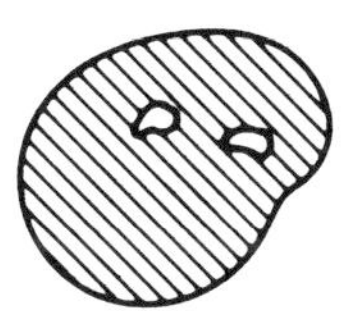

Figure 5.9. Non-convex set of constant width

that is not convex. Figure 5.10 shows a special set of constant width called a Reuleaux triangle. Note that A, B, and C are the vertices of an equilateral triangle, whereas arcs AB, BC, and CA are parts of circles with the opposite vertices as centers. This shape is used for a part in a Wankel engine, a snowmobile engine, and for the drive gear on some film projectors.

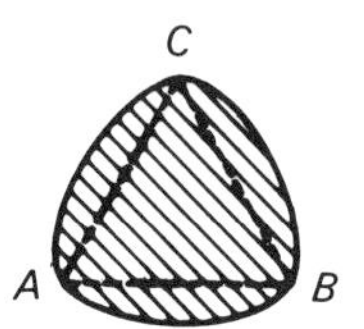

Figure 5.10. Reuleaux triangle

Exercise 5.1

1. (a) Is a circlc a polygon? (b) Why?
2. In Figure 5.2, on p. 96, a segment with C and D as endpoints intersects the circle in how many points?
3. What is the name given to the longest possible chord that can be drawn in a circle?
4. In Figure 5.3, on p. 97, what is the measure of the angle between radius AF and tangent FI?
5. What are the possibilities for the intersection of a line in the plane with the set of points on two concentric circles in the plane?
6. What are the possibilities for the intersection of two different circles in the same plane?
7. What is the greatest number of points possible for the intersection of two triangles if one is inscribed in a circle and another is circumscribed about the same circle?
8. Could the center of a circle be a point on both rays of an angle inscribed in that circle?
9. If an angle is a central angle for one of two concentric circles, then is it also a central angle for the other circle?
10. Describe the possible sets of points in the intersection of a circular region and:
    (a) a line in the same plane.
    (b) an angle in the same plane.
11. Is a triangular region an example of a set of constant width?
12. Draw another example of a non-convex set of constant width, other than the one pictured in Figure 5.9.
13. Answer yes or no for a Reuleaux triangle.
    (a) It is a convex set.
    (b) It is a triangle.
    (c) It is a triangular region.

Central Angles and Arcs

An *arc* is defined as the set of points on a circle consisting of two points and all the points on the circle between these two points. An arc is a part of a circle or the entire circle; it is a subset of the set of points on a circle. The use of the word *between* for points on a circle seems intuitively clear, but, in Figure 5.11, either point C or point D could be considered as on the arc

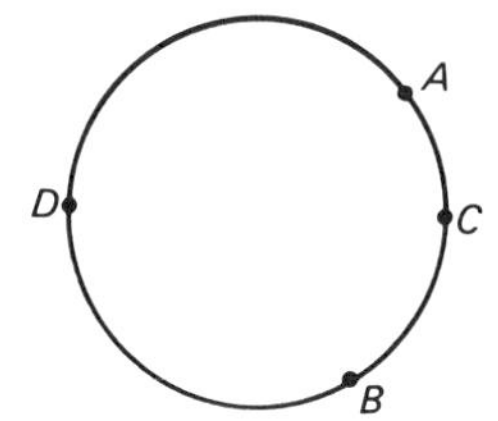

Figure 5.11. Points determining arcs

between A and B. That is, points A and B actually determine two distinct arcs, since the meaning of the word *between* is not definite enough to allow you to tell which of the two sets of points is meant. To avoid this ambiguity, an arc is ordinarily identified by three letters, such as arc ACB, written $\overparen{ACB}$. In this notation, the first and third letters indicate the endpoints of the arc; the second letter, by naming one point on the correct part of the circle, indicates which of the two possible arcs is meant.

In Figure 5.11, if points C and D are endpoints of a diameter of the circle, then $\overparen{DAC}$ and $\overparen{DBC}$ are called *semicircles.* The prefix *semi* indicates that a semicircle is half of a circle.

Comparing arcs gives rise to the need for a system of measuring arcs. What does it mean to measure an arc? The most obvious answer is that measuring an arc could mean finding its length in terms of linear units, and this idea is explored in the next section. A first approach to the measurement of arcs, however, is concerned with determining how the length of the arc compares with the length of the circle of which it is a subset.

The unit of measurement for arcs is called an *arc degree.* An arc degree can be explained by considering a central angle. In Figure 5.12, central angle AOB is pictured for circle O. If the measure of this angle is 15 in degrees, then arc ACB is said to have a measure of 15 in arc degrees. There is a one-to-one correspondence between the arcs and central angles, so that each

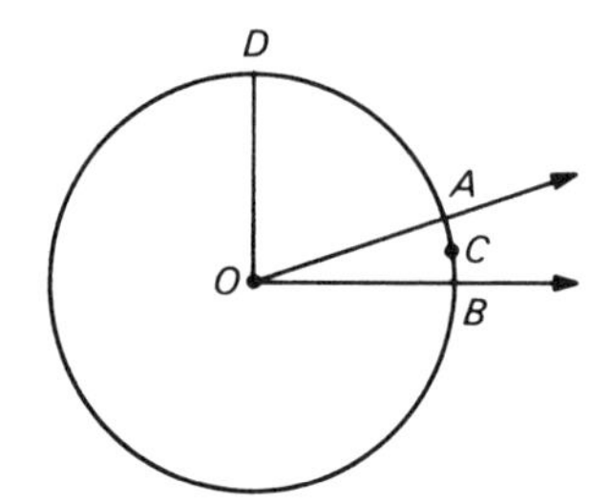

Figure 5.12. Central angles

arc has a corresponding central angle and each central angle has a corresponding arc. In Figure 5.12, angle AOB is said to *subtend* or to intercept arc ACB. Central angle BOD with a measure of 90 in degrees subtends an arc with a measure of 90 in arc degrees. It should be pointed out that this is one of the central ideas in the construction of circle graphs. An arc with a measure of 90 in arc degrees has a measure $\frac{1}{4}$ as great as the measure of the entire circle in arc degrees. The measure of the circle in arc degrees is 360, and the measure of a semicircle in arc degrees is 180. The statement that a central angle is measured by its intercepted arc is another way of saying that the measure of the central angle in degrees is the same as the measure of the arc in arc degrees. To distinguish again between degree and arc degree, remember that the first is a measurement used for angles and the second is a measurement used for arcs. For the same circle or congruent circles, congruent arcs are arcs having the same measurement in arc degrees.

Suppose that central angles AOB and COD in Figure 5.13 are congruent. As a consequence of the last paragraph, what can you conclude about arcs AEB and DFC? What seems to be true about the lengths of segments AB and CD? Of $\overline{AD}$ and $\overline{CB}$? Chords $\overline{AB}$ and $\overline{CD}$ are said to subtend congruent arcs. Chords $\overline{AD}$ and $\overline{CB}$ also subtend congruent arcs. If, in the same figure, the measure in arc degrees of $\widehat{CGB}$ is twice the measure in arc degrees of

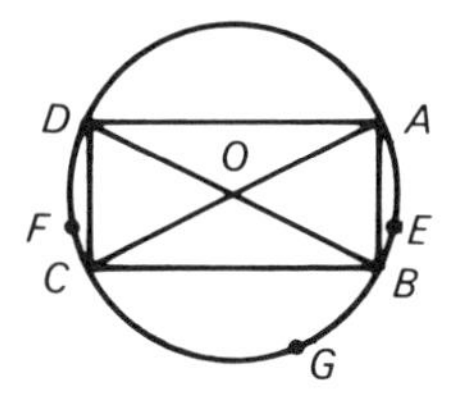

Figure 5.13. Central angles and their arcs

$\widehat{AEB}$, then you could conclude that central angle COB has a measure twice the measure of central angle AOB (written m $\angle COB = 2\text{m} \angle AOB$).

One additional axiom needs to be stated to formalize the structure necessary for the concept of measurement of arcs.

Axiom 18

If $\widehat{AP}$ and $\widehat{PB}$ are arcs of the same circle and have exactly one point P in common, then $m\ \widehat{AP} + \text{m}\ \widehat{PB} = \text{m}\ \widehat{AB}$ (See Fig. 5.14.)

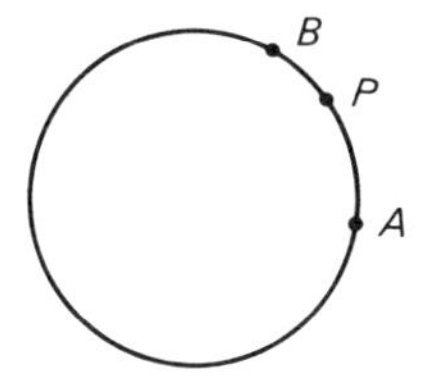

Figure 5.14. Addition of arc measure

Additional properties of circles and sets of points connected with circles can be proved as theorems.

A chord of a circle, as you have learned, is a segment with its two endpoints on the circle. In Figure 5.15, consider chord AB. Let C be the midpoint of this chord. What seems to be true about the relationship of $\overline{OC}$ and $\overline{AB}$? Draw other circles and chords and then consider the relationship of the chord and the segment whose endpoints are the center of the circle and the midpoint of the chord.

Theorem

A line through the center of a circle and the midpoint of a chord (not a diameter) is perpendicular to the chord.

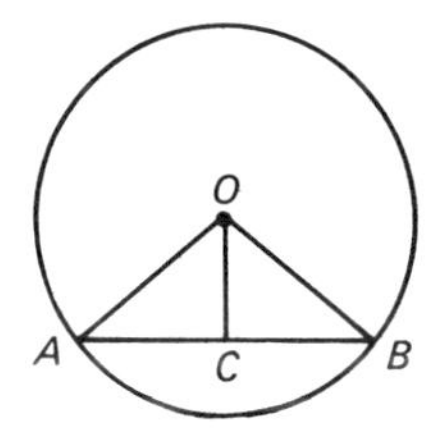

Figure 5.15. Line through center perpendicular to chord

In Figure 5.15, $\triangle AOC \cong \triangle BOC$. Why?

It follows that angles OCA and OCB are right angles, since they are congruent and supplementary.

In Figure 5.16, imagine other chords with the same length as $\overline{AB}$. One of them, chord CD, is drawn with its midpoint F. Now compare the measures of segments OE and OF. Make similar comparisons in the other circles you have drawn.

Theorem

In a circle, congruent chords are the same distance from the center of the circle.

In Figure 5.16, $\triangle OEB \cong \triangle OFD$. Why? Then $\overline{OE} \cong \overline{OF}$, since they are corresponding sides of congruent triangles.

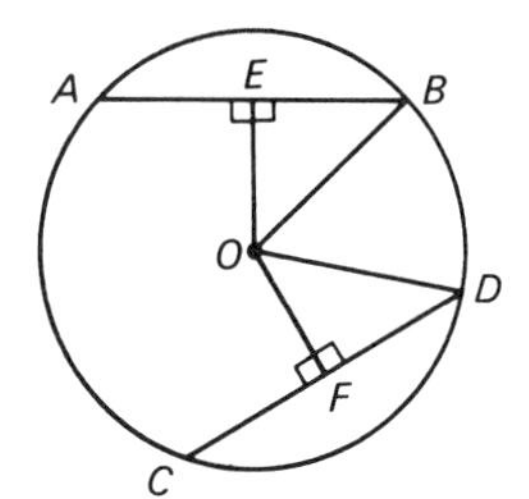

Figure 5.16. Congruent chords

You have found that points F and E in Figure 5.16 are the same distance from the center of the circle. If other chords congruent to $\overline{AB}$ and $\overline{CE}$ are drawn, and if their midpoints are determined, then each of these midpoints will also be the same distance from the center of the circle. Midpoints of chords of a constant length in a given circle lie on another circle concentric to the given circle. The radius of this new circle is the perpendicular distance from the given chords to the center of the given circle.

Theorem

The line of centers for two intersecting circles is the perpendicular bisector of the common chord.

In Figure 5.17, $\triangle OO'A \cong \triangle OO'B$. Why? Then $\angle AO'C \cong \angle BO'C$. $\triangle CAO' \cong \triangle CBO'$. Why? The conclusion is that $\overline{AC} \cong \overline{BC}$ and that angles ACO' and BCO' are right angles. Why?

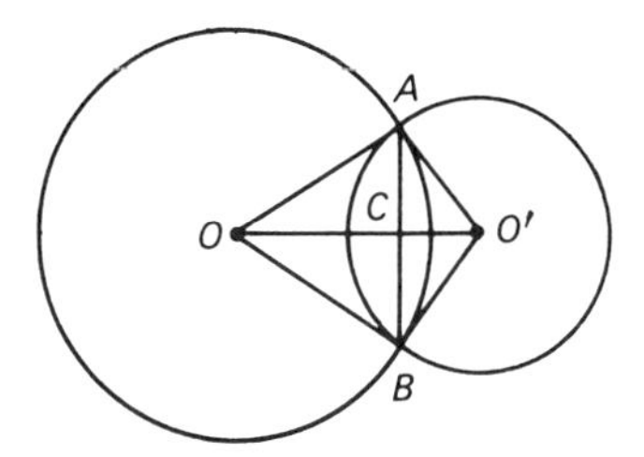

Figure 5.17. Line of centers and common chord

Theorem

The center of a regular polygon is the same distance from all the vertices.

In Figure 5.18, let A, B, and C be three vertices with D as one of the next vertices. If you can show that D lies on circle O (that $\overline{OD}$ is a radius), you

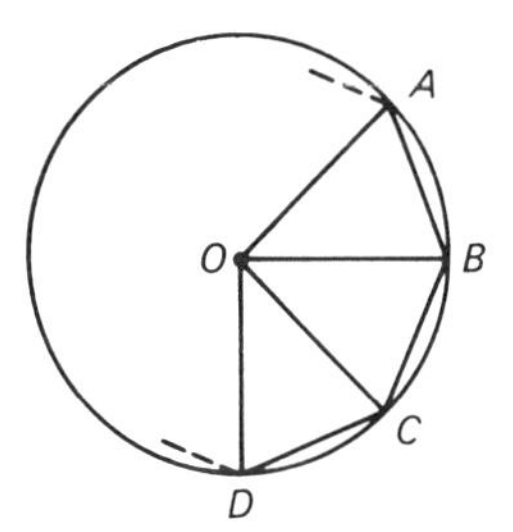

Figure 5.18. Inscribed regular polygon

can repeat the procedure for each vertex and prove the theorem. $\angle OBC \cong \angle BCO$. $\angle ABC \cong \angle BCD$, $\angle ABO \cong \angle DCO$. $\triangle AOB \cong \triangle DOC$. Why? $\overline{OD} \cong \overline{OA}$, since they are corresponding sides of congruent triangles.

Figure 5.19 illustrates that, from a point outside a circle, two tangents to the circle can be drawn. Segments AB and AC are congruent, as you can establish. Triangles OBA and OCA are right triangles, with the right angles at B and C, since tangents are perpendicular to the radii at the point of tangency. Furthermore, $\overline{OB}$ and $\overline{OC}$ are congruent, because they are radii.

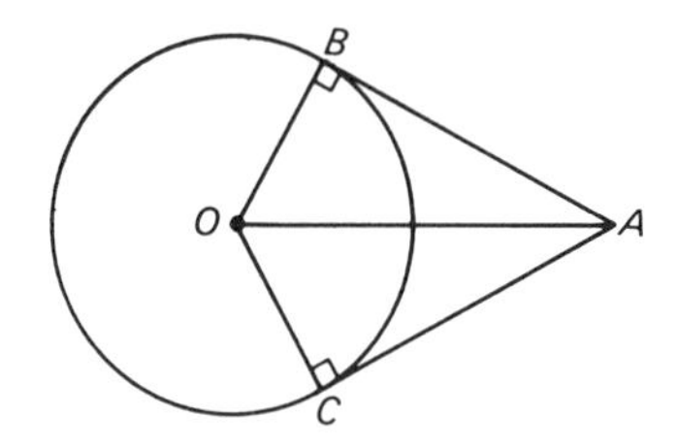

Figure 5.19. Tangents from point to circle

Then you may use congruent triangles to find that the measures of $\overline{AB}$ and $\overline{AC}$ are equal.

You have found that a central angle of a circle is measured by its intercepted arc and that an inscribed angle is measured by one-half of its intercepted arc. The sides of an inscribed angle lie on two chords whose point of intersection is a point on the circle. In Figure 5.20, one chord is drawn with B as an endpoint, but the other side of $\angle ABC$ lies on the tangent to the circle at point B. Since $\angle ABC$ might be considered as just a special case of an inscribed angle (or more technically, a limiting case), it seems reasonable that $\angle ABC$ (the angle between a chord and a tangent) can also be measured by one-half of its intercepted arc. Verify that this relationship is reasonable by drawing other circles. For example, if $\widehat{ADB}$ is 120 arc degrees, then $\angle ABC$ is 60 degrees.

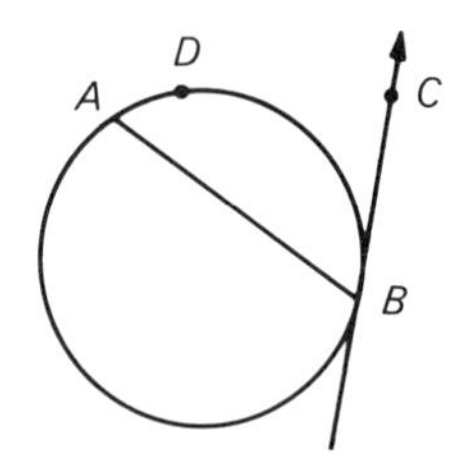

Figure 5.20. Tangent and chord

Exercise 5.2

1. Is an arc always a: (a) curve? (b) closed curve? (c) simple closed curve?
2. Is an arc sometimes a: (a) polygon? (b) circle? (c) semicircle?

In Exercises 3–4, name all labeled points in Figure 5.21 that are points on the indicated arcs.

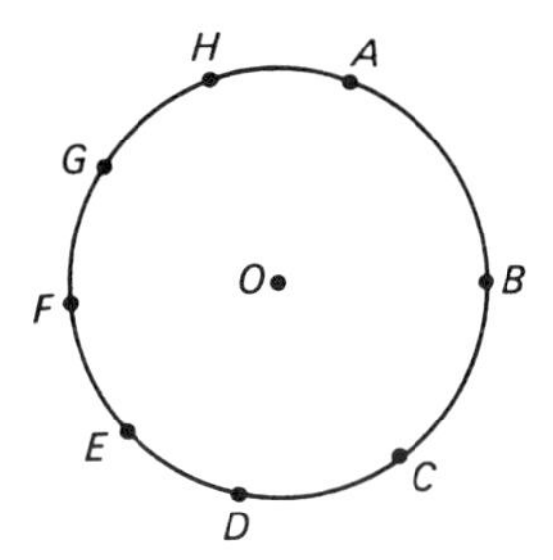

Figure 5.21

3. $\widehat{AEF}$
4. $\widehat{CFH}$
5. In Figure 5.21, what can you conclude about $\overline{BF}$ if $\widehat{BCF}$ is a semicircle?
6. In Figure 5.21, is $\widehat{DEG} \cong \widehat{DFG}$ a true statement?
7. Find the approximate measure in arc degrees of the arc with a measure less than 180 subtended by each central angle indicated in Figure 5.21.
   (a) $\angle DOC$ (b) $\angle COH$ (c) $\angle EOG$ (d) $\angle BOE$
8. What is the sum in arc degrees of the measures of $\widehat{BAG}$ and $\widehat{GDB}$?
9. Suppose that $\overline{GE} \cong \overline{EC}$. What can you conclude about:
   (a) $\angle GOE$ and $\angle EOC$? (b) $\widehat{GFE}$ and $\widehat{EDC}$?
10. If two chords of a circle are not congruent, which do you think is closer to the center of the circle?
11. The radii from the center of a circle to the vertices of an inscribed regular hexagon, along with the sides of the hexagon, form how many congruent triangles?

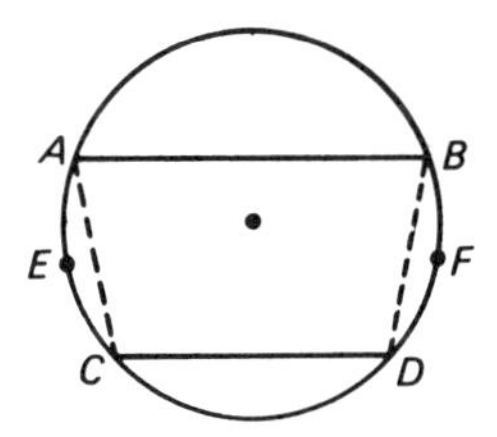

Figure 5.22

12. Chords AB and CD in Figure 5.22 are parallel but not necessarily congruent.
    (a) Do you see two congruent arcs in the picture? Try drawing parallel chords in other circles.
    (b) State a theorem about parallel chords and the arcs they intercept.

13. In Figure 5.23, if $\overleftrightarrow{CD}$ and $\overleftrightarrow{EF}$ are tangents to the smaller of the two concentric circles, how do you know that $\overline{CD} \cong \overline{DF}$?

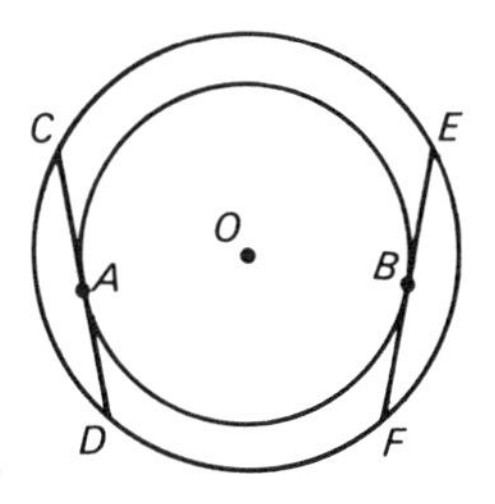

Figure 5.23

14. In Figure 5.23, how do you know that $\overline{CA} \cong \overline{AD}$?
15. If two circles are tangent to each other, what seems to be the relative position of the line of centers and the point of tangency?
16. In Figure 5.24, $\overleftrightarrow{BC}$ is tangent to the circle at point B.

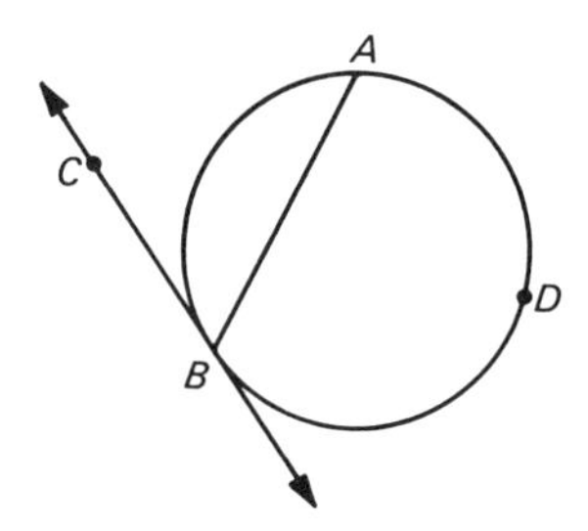

Figure 5.24

If you know that the measure in arc degrees of $\widehat{ADB}$ is 280, what is the measure in degrees of angle ABC?

17. In Figure 5.25a, compare the measure in arc degrees of $\widehat{ADC}$ with the measure in degrees of inscribed $\angle ABC$. Draw other inscribed

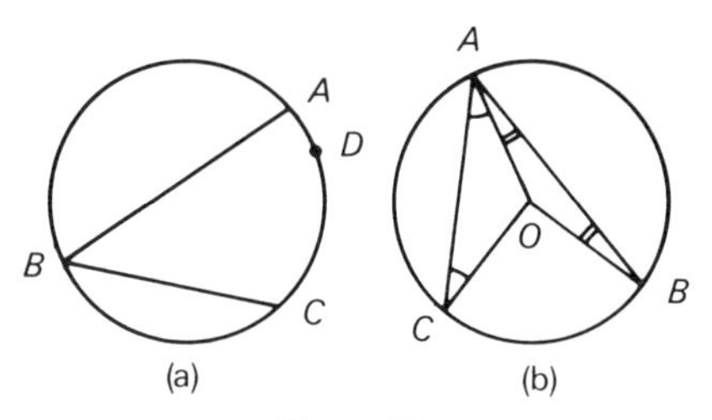

Figure 5.25

angles and experiment to see what relationship seems to exist between the measures of an inscribed angle and its intercepted arc for any circle.

18. Refer to Figure 5.25b. Give the reasons in each set and then write a theorem showing the relationship existing between the measure of an inscribed angle and its intercepted arc. Let O be the center of the circle.
    (a) $m\,\angle OAC = m\,\angle OCA$ Why?
    (b) $m\,\angle OAB = m\,\angle OBA$ Why?
    (c) $m\,\angle COA = 180 - 2m\,\angle OAC$ Why?
    (d) $m\,\angle BOA = 180 - 2m\,\angle OAB$ Why?
    (e) $m\,\angle COB = 360 - (180 - 2m\,\angle OAC) - (180 - 2m\,\angle OAB)$ Why?
    (f) $m\,\angle COB = 2m\,\angle OAC + 2m\,\angle OAB$ Why?
    (g) $m\,\angle COB = 2m\,\angle CAB$ Why?
    (h) Write the general theorem.
19. Prove that the opposite angles of a quadrilateral inscribed in a circle are supplementary.

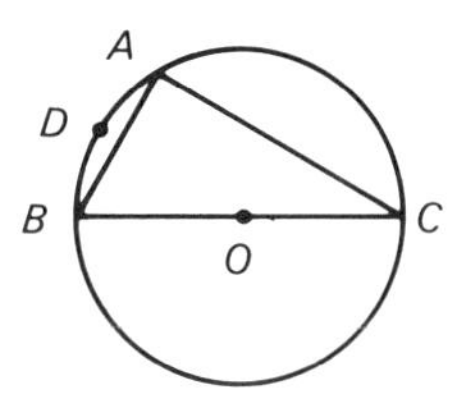

Figure 5.26

Figure 5.26 shows $\angle BAC$ inscribed in a semicircle. Using the figure, answer yes or no for Exercises 20–22.

20. The intercepted arc of $\angle BAC$ is also a semicircle.
21. The measure of $\angle BAC$ is 90 degrees.
22. The measure in arc degrees of $\widehat{BDA}$ is half the measure in degrees of $\angle BCA$.

Circumference

The *circumference* of a circle is its measurement of length in linear units rather than in arc degrees. Recall that this type of measurement of arcs was mentioned in the previous section. Intuitively, it is easy to explain that the concept of length of a line segment can be expanded to apply to measuring the length of a curve. For example, you might think of placing a string along the picture of a curve and then stretching the string along a ruler to find its length. In Figure 5.27, it seems clear that the perimeter of an

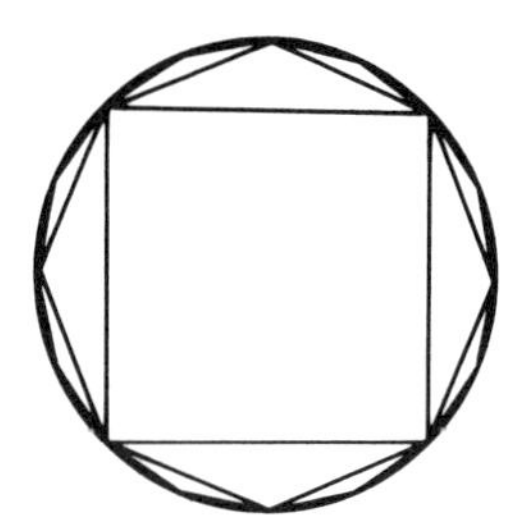

Figure 5.27. Inscribed polygons

inscribed polygon is always less than the circumference of the circle. As the number of sides of the polygon increases, the perimeter gets closer and closer to the circumference. Assume that the circumference of a circle can be measured in linear units.

The ratio of the measure of the circumference (C) to the measure of the diameter (d) of a circle is a constant for any given circle. By careful measurement of the diameter and circumference of a circular object, you can probably estimate that the constant is between 3.1 and 3.2. The exact number that is the quotient C/d is named *pi* and is written π. Then $C/d = \pi$, or $C = \pi d$. The number pi is an irrational number; it cannot be expressed as the ratio of two counting numbers. On the other hand, there is a point on the number line of real numbers that corresponds to π. Irrational numbers, when expressed as infinite decimals, do not repeat. The decimal expression for pi, naming the first five digits to the right of the decimal point, is 3.14159.... Displays of the decimal expression for pi correct to thousands of places past the decimal point are possible with computers. Such expressions may seem of little practical value to most people, but one seemingly odd use is that they are sometimes assumed to be a random set of digits when random digits are needed by statisticians.

Here are two of the curious expressions that mathematicians have discovered for π.

$$\frac{\pi}{2} = \frac{2 \cdot 2 \cdot 4 \cdot 4 \cdot 6 \ldots}{1 \cdot 3 \cdot 3 \cdot 5 \cdot 5 \ldots}$$

$$\frac{\pi}{4} = 1 - \frac{1}{3} + \frac{1}{5} - \frac{1}{7} + \cdots$$

Comte de Buffon, in 1760, stated a way of using probability to find π. This method is called the needle problem and is illustrated in Figure 5.28.

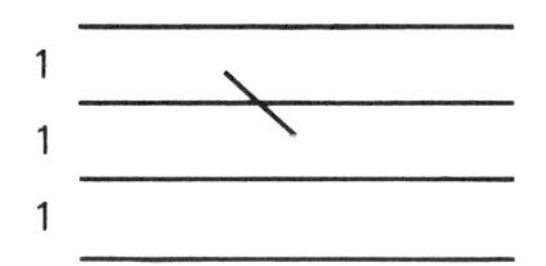

Figure 5.28. Needle problem

Suppose a needle of length l (less than one unit) is dropped at random onto a surface marked with parallel segments one unit apart. The probability that the needle will touch one of the lines is $p = 2l/\pi$. If the experiment is repeated a large number of times, an approximation to the value of π is $2l/p$.

For practical purposes of measurement, pi is often approximated by a rational number. Some common approximations are 3.14, 3.1416, and 22/7. Remember that none of these is equal to pi. Whenever any one of these numbers is used, the symbol for "approximately equal" should be written.

Example

Find the circumference of a circle if the radius is 3/4 inch, using 22/7 for pi.

$$
\begin{aligned}
C &= 2\pi r \\
&= 2\pi\left(\frac{3}{4}\right) \\
&\approx 2\left(\frac{22}{7}\right)\left(\frac{3}{4}\right) \\
&\approx \frac{66}{14}, \text{ or } 4\frac{5}{7}
\end{aligned}
$$

The circumference is approximately $4\frac{5}{7}$ inches.

The use of the formula $C = \pi d$ or $C = 2\pi r$ has a significance that might not be immediately apparent. The formulas make it possible to find the length of a curve (part of the circumference) if you know the length of a line segment (the diameter or radius). Finding the measurement of a circle is essentially reduced to finding the length of a line segment.

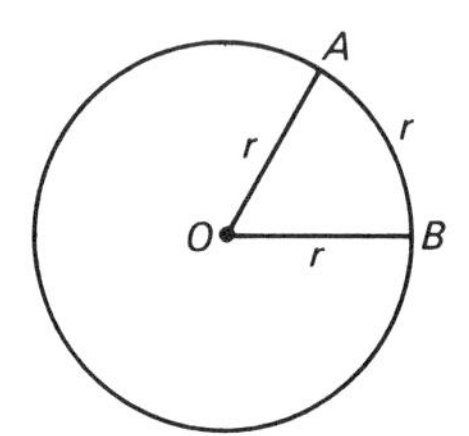

Figure 5.29. Radian measure

The discussion about central angles and measurements of arcs in circles makes it possible to introduce an important standard unit angle widely used in higher mathematics. Study Figure 5.29. Central angle AOB subtends an arc that has the same length as the radius of the circle. The measurement of $\angle AOB$ is 1 *radian.* Note that the concept of radian measure is easy to extend to angles whose measure is greater than 180 in degrees. The *radian measure* of an angle is the ratio of its intercepted arc to the radius of the circle in which the angle is a central angle.

Since the length of a semicircle is πr, it is π times as long as the radius. An angle with a measurement of 180° has a measurement of π radians. From the relationship $180° = \pi$ radians, other statements of equivalent measurements can be derived, such as $60° = \pi/3$ radians and $45° = \pi/4$ radians. Since $C = 2\pi r$ and $C = 360°$, then $2\pi r = 360°$. Hence 1 radian $= 180°/\pi$ and $1° = \pi/180$ radians. By attempting to draw a picture of an angle with a measurement of 1 radian, you could find that the angle would have a measurement of about 57°.

Example

Express 7° in radian measure.
Since $1° = \pi/180$ radians
$7° = 7\pi/180$ radians

Example

Express $\pi/5$ radians in degree measure.
Since π radians $= 180°$
$\pi/5$ radians $= 180°/5$
$= 36°$

The arrangement of points on a circle is the intuitive basis for what is called modular arithmetic. You may have studied modular arithmetic in other books, but it is reviewed or introduced here as an application of the concept of circle and also to show the fundamental difference between a regular number line and a "circular number line." As an illustration, Figure 5.30 shows a "clock" with 0 replacing 12. Addition in the set of

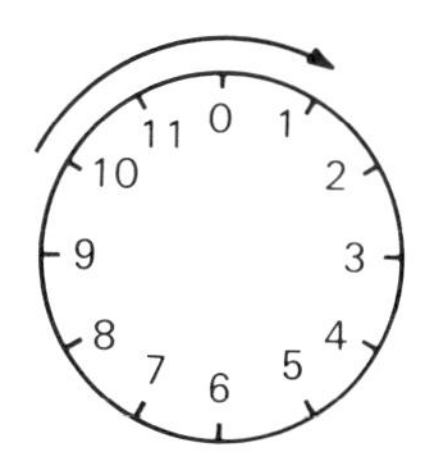

Figure 5.30. Modular arithmetic, modulus 12

numbers shown is taken to mean movement in a clockwise direction; subtraction is indicated by movement in the opposite direction. Thus, $10 + 3 = 1$ in the modular arithmetic of Figure 5.30.

Examples

In the modular arithmetic of Figure 5.30,

$$8 + 6 = 2 \qquad 2 - 3 = 11$$

$$4 + 5 = 9 \qquad 5 - 9 = 8$$

$$6 + 6 = 0 \qquad 0 - 2 = 10$$

For the modular arithmetic above, 12 is called the modulus. Another illustration is given, using a modulus of 5, as in Figure 5.31.

Examples

Arithmetic using a modulus of 5.

$$2 + 3 = 0 \qquad 2 - 3 = 4$$

$$3 + 3 = 1 \qquad 1 - 4 = 2$$

$$4 + 3 = 2 \qquad 0 - 2 = 3$$

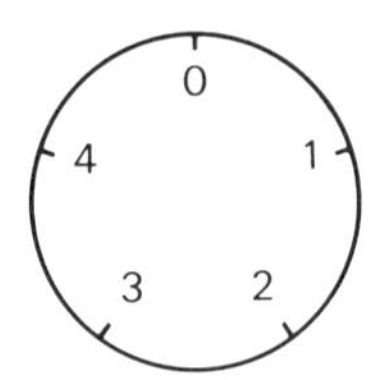

Figure 5.31. Modular arithmetic, modulus 5

Several important differences between modular arithmetic and ordinary arithmetic should be noticed.

1. In modular arithmetic, concepts of order do not apply. Decisions cannot be made about which of two numbers is greater.
2. The answer to a problem in modular arithmetic is one of a finite set of numbers.
3. The sum of two non-zero numbers may be zero in modular arithmetic.
4. Subtraction in modular arithmetic is always possible, without the need for negative numbers.

Modular arithmetic is an example of a finite mathematics. Some finite geometries are discussed in Chapter 14.

Exercise 5.3

In Exercises 1–2, can you tell if the two sets of points are congruent or not?

1. Two circles, each of which has a circumference of 7 ft.
2. Two circles, both of which have regular hexagons inscribed in them.

In Exercises 3–6, do not use an approximation for pi. Leave the answer in terms of π.

3. If the circumference of a circle is 6π in., what is the diameter?
4. If the circumference of a circle is 7π m, what is the radius?
5. If the measure of the radius of one circle is twice the measure of the radius of another, how do the measures of the circumferences compare?

6. The measure of length of a semicircle is how many times the measure of length of the diameter of that same circle?
7. Complete the table below, using 22/7 as an approximation for pi.

measure of radius	*measure of diameter*	*measure of circumference*
$1/2$	____	____
____	$4/3$	____
____	____	$57/4$
$5/2$	____	____

8. Complete the table below, using 3.14 for pi.

measure of radius	*measure of diameter*	*measure of circumference*
4.5	____	____
____	16.3	____
____	____	18.4
2.03	____	____

9. Complete the table of measurements in degrees and radians.

degrees	*radians*
30	____
____	$\pi/2$
120	____
____	$5\pi/4$
9	____
____	$\pi/12$

In Exercises 10–15, find the answers in modular arithmetic, using a modulus of 12.

10. $8 + 8 = ?$
11. $9 + 9 = ?$
12. $10 + 11 = ?$
13. $3 - 8 = ?$
14. $2 - 11 = ?$
15. $5 - 11 = ?$

In Exercises 16–21, find the answers in modular arithmetic, using a modulus of 7.

16. $4 + 4 = ?$
17. $5 + 5 = ?$
18. $6 + 6 = ?$
19. $2 - 6 = ?$
20. $3 - 5 = ?$
21. $1 - 6 = ?$

In Exercises 22–27, find the answers in modular arithmetic, using a modulus of 10.

22. $3 + 7 = ?$
23. $7 + 6 = ?$
24. $9 + 9 = ?$
25. $2 - 9 = ?$
26. $3 - 6 = ?$
27. $0 - 3 = ?$

Symmetry

The word *symmetry*, in its everyday usage, means to have corresponding parts, to be in harmony, or to be proportioned in some balanced and pleasing way. Examples of symmetry may be observed in nature in a butterfly, a shrub, or a flower. Symmetry is used in homes, art, and industrial design to create pleasing patterns that seem balanced because one part of the pattern is the same as the other. In this section you will study some of the mathematics underlying these intuitive ideas of symmetry.

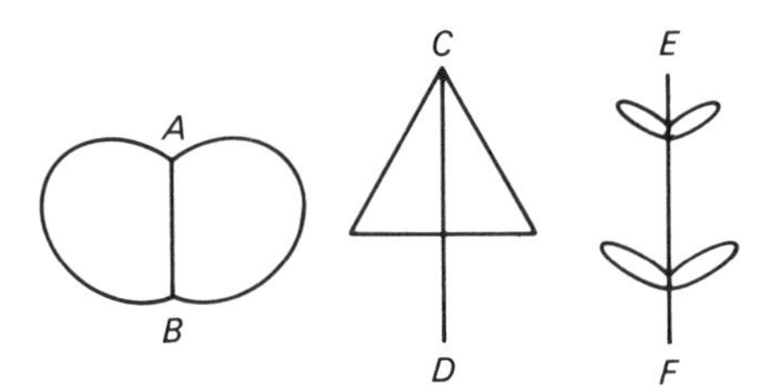

Figure 5.32. Symmetry about a line

Two kinds of symmetry that will be investigated within the mathematical model are symmetry *with respect to a line* and symmetry *with respect to a point.* Figure 5.32 shows three examples (which may remind you of a tree, the wings of a butterfly, and a plant) of sets of points that are symmetric with respect to a line. In these three examples, the lines, called *axes of symmetry,* are $\overleftrightarrow{AB}$, $\overleftrightarrow{CD}$, and $\overleftrightarrow{EF}$. Intuitively, you realize that the set of points on one side of the line has the same shape as the set of points on the other side of the line. The picture seems balanced, with the line of symmetry in the middle. If you traced the figures and cut them out, you could fold them down the middle on the axis of symmetry, and the points of one half would fit directly over the points of the second half. To explain this idea further, you could say that the set of points on one side of the axis of symmetry is congruent to the set on the other side of the axis.

Symmetry with respect to a line can be tested by seeing if the set of points meets the specific requirement that it have an axis of symmetry. This approach is illustrated in Figure 5.33. $\overleftrightarrow{AB}$ is the axis of symmetry in this picture. For any points on one side of this axis, such as points C and E, there must be corresponding points on the other side of the axis, such as D and F. In each case, the axis of symmetry must be the perpendicular bisector of the segments with the corresponding points as endpoints. This means that $\overline{AB}$ is perpendicular to both $\overline{CD}$ and $\overline{EF}$ and that $\overline{CG} \cong \overline{GD}$ and $\overline{EH} \cong \overline{HF}$. If a corresponding relationship holds for every point in a set of points, then the set of points has an axis of symmetry.

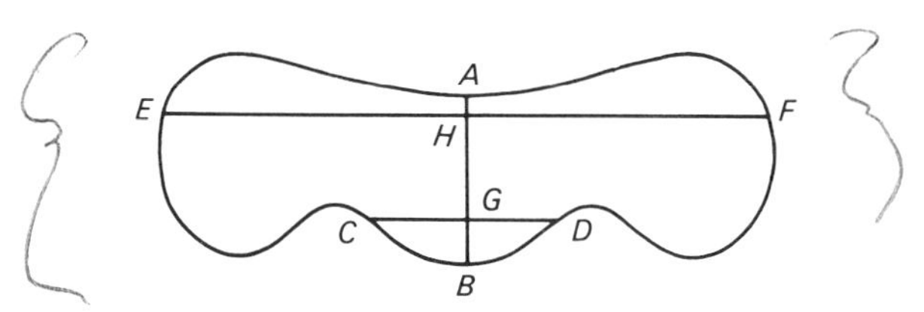

Figure 5.33. Test for symmetry about a line

Symmetry with respect to a line may be explained in terms of a mapping or transformation that is an isometry. In Figure 5.33 the points to the right of $\overline{AB}$ may be transformed into the points to the left of $\overline{AB}$. Stated another way, the entire set of points in Figure 5.33 undergoes a transformation in which each point has an image (C is the image of D, for example) in the set. Distance is preserved, and the transformation is an isometry.

Axis of symmetry is actually a new phrase for an idea you have met before. For example, the diameter of a circle is an axis of symmetry for the circle. Other examples are a diagonal of a square and an altitude of an equilateral triangle. In each of these cases, the set of points has more than one axis of symmetry. For example, each altitude of the equilateral triangle meets the criterion of being the perpendicular bisector of segments whose endpoints are pairs of corresponding points.

A second kind of symmetry, which can be observed both in nature and in man-made designs, is *symmetry about a point*. The pattern seems to have a center and to be arranged about this center in a balanced fashion. Familiar objects with this characteristic include some flowers, spider webs, electric fans, and lighting fixtures. Figure 5.34 shows three examples of centers or points of symmetry. It is obvious that symmetry with respect to a point is different from symmetry with respect to a line; yet they are alike in that each set of points presents a balanced appearance, with a pattern and harmony.

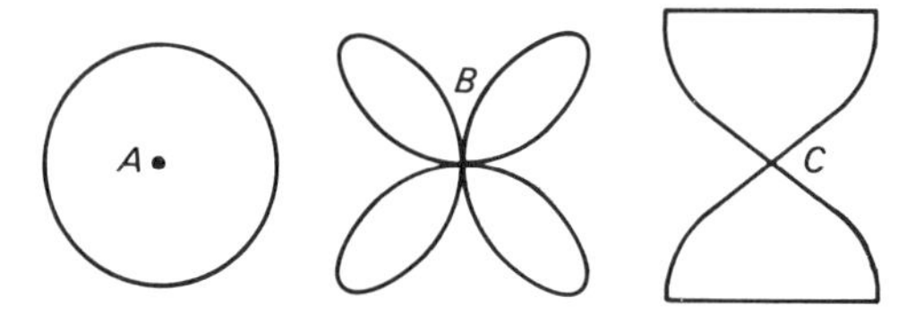

Figure 5.34. Symmetry about a point

A precise way of explaining what is meant by a "point of symmetry" can be formulated by studying Figure 5.35. Here, point A is the center of symmetry. For any points in the set of points, such as points B and E, there are

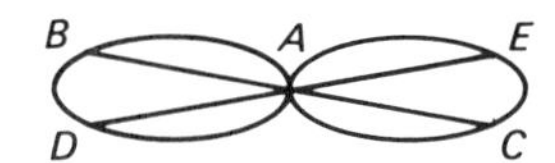

Figure 5.35. Test for symmetry about a point

corresponding points C and D. If all these pairs of corresponding points, such as B and C or D and E, are considered as the endpoints of segments, and $\overline{BA} \cong \overline{AC}$ and $\overline{DA} \cong \overline{AE}$, and so on, then A is the center of symmetry. Each point in the set is mapped into another point in the set by the transformation that designates A as the center of symmetry. Again, distance is preserved and the transformation is an isometry.

For symmetry with respect to a line or to a point, congruence has been used in the explanation. Congruence is a relationship between two distinct sets of points; in a sense, symmetry is an analogous relationship between two subsets of a single set of points. In a way, it is congruence within one set of points, rather than between two sets. In symmetry, a transformation results in the same set of points as before, rather than in a different set of points.

The discussion of symmetry contributes to a further understanding of the concept of isometries. Isometries of a plane onto itself are called *plane motions.* All plane motions preserve congruence, and this idea leads to intuitive ways to enumerate various motions. Using a cardboard model of a triangular region, for example, one can experiment to arrive at a classification of the types of motions.

The three basic types of motions of a plane are illustrated for triangles in Figures 5.36 through 5.40. A *translation,* in Figure 5.36, is a motion in which each point of a set may be thought of as being moved a certain

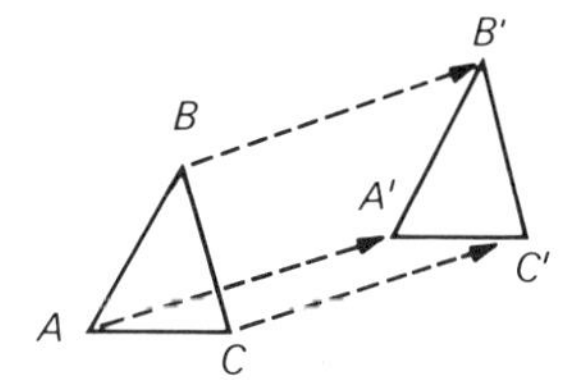

Figure 5.36. Translation

number of units in a certain direction, $\overrightarrow{BB'}$, $\overrightarrow{AA'}$, and $\overrightarrow{CC'}$ are equal vectors, having the same length and the same direction. In general, there is no symmetry in a picture of a translation, and a translation does not result in a second triangle coinciding with the first.

The second type of motion is a *rotation*, shown in Figure 5.37. Each point may be thought of as being rotated about a fixed point and through a fixed angle. For example, the image of point A is point D, found by rotating about point O through the angle AOD.

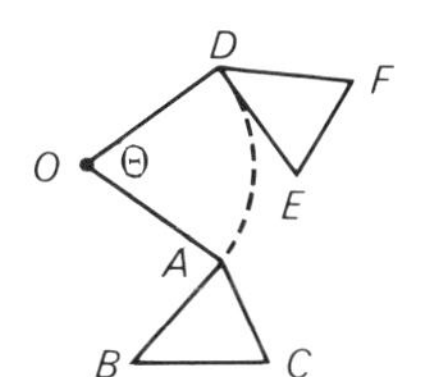

Figure 5.37. Rotation

If a set of points has a center of symmetry, a rotation of 180° about that center will result in the same set of points.

Example

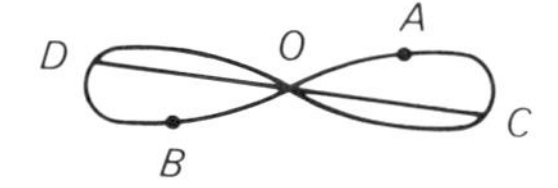

Figure 5.38. Symmetry about a point

See Figure 5.38. In a rotation of 180° about point O, B is the image of A and D is the image of C. For the entire set, however, each point in the first set has an image that is some point in the second set.

Some sets of points, even without a center of symmetry, may be made to coincide with themselves through a rotation. This concept is sometimes called rotational or turning symmetry and is illustrated by the various sets of points in Figure 5.39.

Figure 5.39. Rotational symmetry

Both of the types of motion discussed so far are called *direct* motions. For a cardboard triangle, they can be carried out by sliding or turning in the plane. The third type of motion, called a *reflection,* is an opposite motion, since it cannot be accomplished physically without flipping over the cardboard triangle.

As can be seen in Figure 5.40, a reflection results in a figure that has a line of symmetry. Triangle ABC was reflected about the line DE, resulting in $\triangle A'B'C'$. The two triangles have $\overleftrightarrow{DE}$ as an axis of symmetry.

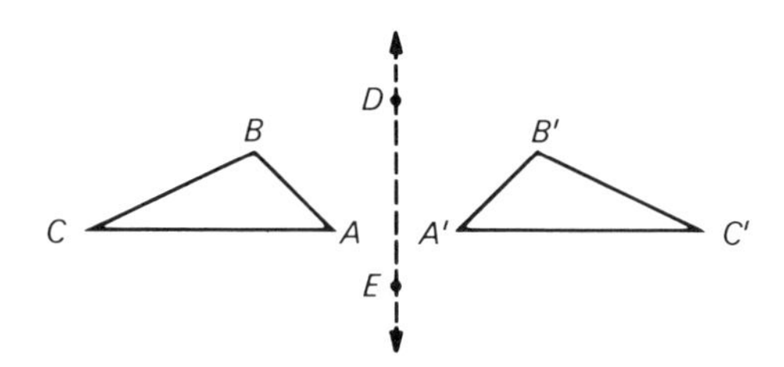

Figure 5.40. Reflection about a line

Exercise 5.4

1. Name other examples—from nature—of symmetry with respect to a line.
2. Name other examples—from man-made designs—of symmetry with respect to a line.

3. Are the following figures always examples of symmetry with respect to a line?
   (a) A parallelogram with respect to one of its diagonals.
   (b) A circle with respect to one of its diameters.
   (c) A triangle with respect to one of its medians.
   (d) An equilateral triangle with respect to one of its medians.
   (e) A circle with respect to one of its tangents.
   (f) A regular hexagon with respect to one of its diagonals joining opposite vertices.
4. In Figure 5.41, name the points that correspond to each of the given points, if $\overleftrightarrow{AB}$ is the axis of symmetry.
   (a) C (b) H (c) E (d) D

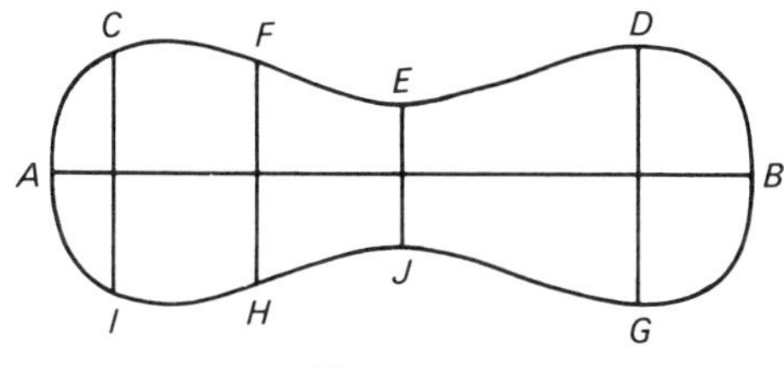

Figure 5.41

5. Tell whether or not each indicated line in Figure 5.42 seems to be an axis of symmetry for the specific set of points.

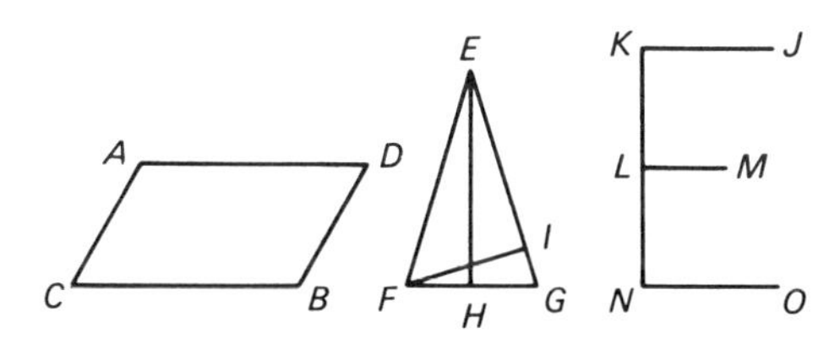

Figure 5.42

   (a) $\overleftrightarrow{AB}$ (b) $\overleftrightarrow{CD}$ (c) $\overleftrightarrow{EF}$ (d) $\overleftrightarrow{EH}$ (e) $\overleftrightarrow{FI}$ (f) $\overleftrightarrow{KJ}$
   (g) $\overleftrightarrow{KN}$ (h) $\overleftrightarrow{LM}$ (i) $\overleftrightarrow{NO}$
6. Name other examples of symmetry with respect to a point that are found in familiar objects.
7. Are the following always examples of symmetry with respect to a point?
   (a) A triangle with respect to the point of intersection of the three medians.

(b) A rectangle with respect to the point of intersection of the two diagonals.
(c) A regular hexagon with respect to the center.
(d) A circle with respect to the midpoint of a diameter.
(e) Two concentric circles with respect to the common center.
(f) A triangle with respect to the point at which the three angle bisectors meet.

8. In Figure 5.43, name the points that correspond to the given points if point A is the center of symmetry.
(a) B (b) C (c) H (d) F

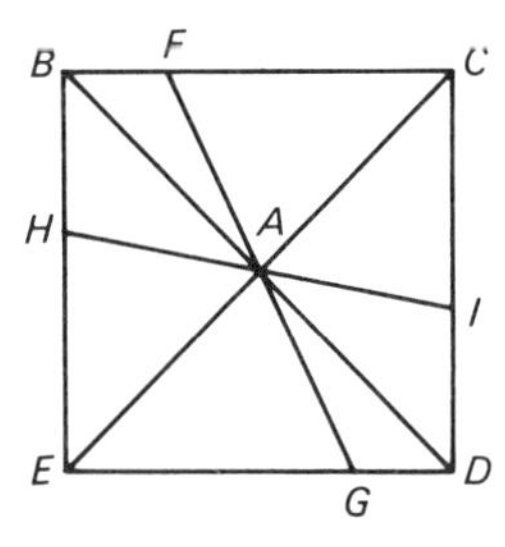

Figure 5.43

9. Use Figure 5.44. Tell whether each point indicated seems to be a center of symmetry for the specific set of points.
(a) A (b) B (c) C (d) D (e) E (f) F (g) G

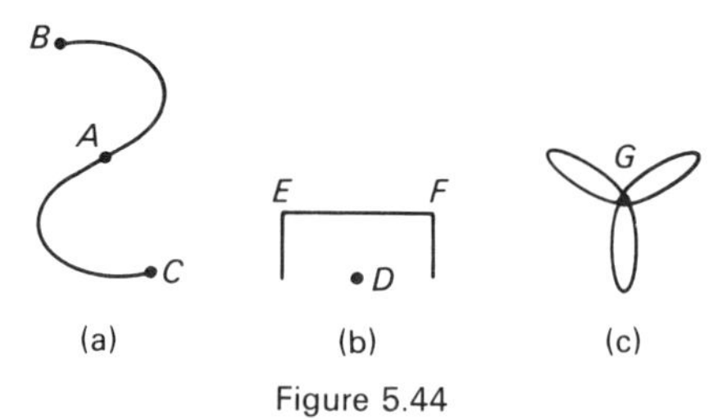

Figure 5.44

In Exercises 10–13, copy each set of points and then draw the result of the motion indicated.

10. Set of points in Figure 5.45a. Translation represented by vector shown.
11. Set of points in Figure 5.45b. Rotation of 30° about center.
12. Set of points in Figure 5.45c. Reflection about line AB.

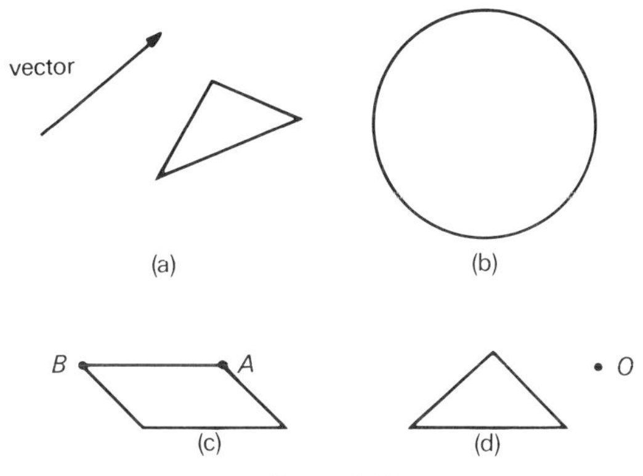

Figure 5.45

13. Triangle in Figure 5.45d. Rotation of 60° in a counterclockwise direction about point O.

In Exercises 14–21, considering the square in Figure 5.46 as the basic set of

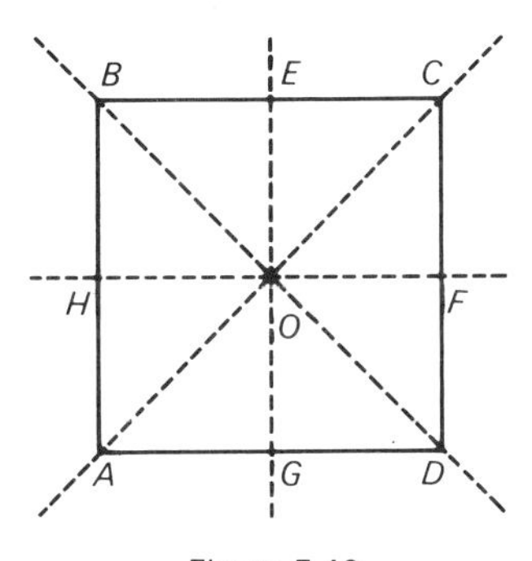

Figure 5.46

points, give the image of point A for each of these motions. (This set of motions is called the set of symmetries of a square.)

14. Reflection about $\overleftrightarrow{HF}$.
15. Reflection about $\overleftrightarrow{EG}$.
16. Reflection about $\overleftrightarrow{BD}$.
17. Reflection about $\overleftrightarrow{AC}$.
18. Rotation of 90° counterclockwise about O.
19. Rotation of 180° counterclockwise about O.
20. Rotation of 270° counterclockwise about O.
21. Rotation of 360° counterclockwise about O.

Throughout the first five chapters, you have drawn pictures of points as an aid in thinking. To draw these pictures, you have used whatever instruments you wished; in most cases, you have not been very concerned about making the drawings too carefully, since they were simply used to help you do something else. In this chapter, however, the drawings themselves become the central topic.

The subject of mathematical constructions began with the ancient Greeks; it continues to be of importance in the teaching of geometry. The meaning of *construction,* the instruments that can be used for mathematical constructions, examples of basic constructions, and some interesting construction problems will be discussed in this chapter. The subject of constructions can be used as a convenient setting for learning more about properties of sets of points studied previously.

Using Construction Instruments

To "make a construction" in geometry means to produce a carefully drawn picture of some set of points, using only certain instruments according to established rules. The Greek philosopher Plato, in about 375 B.C., is supposed to have specified that only two instruments could be used for constructions in geometry: the *straightedge* and the *dividers*. A straightedge, unlike a ruler, has no markings on it. Plato and his followers were interested in the logical problem of what kinds of pictures could be drawn, using only these two instruments. A straightedge is used only for drawing a line or some subset of points on a line, and a divider is used only for drawing an arc. Actually, the modern *compass* has replaced the dividers, but otherwise, mathematicians and novices alike have continued to be fascinated by the challenge of geometric constructions according to the ancient rules. Attempts to perform certain constructions, such as trisecting an angle, have led to extensions of knowledge in mathematics. In addition to the logical exercises encountered, students have often discovered through constructions the beauty of geometry and the wealth of geometric forms that can be produced, even under the severe restrictions imposed by Plato. For example, an angle can be trisected with a ruler; with a straightedge it cannot.

Several basic constructions in geometry will now be explained briefly. It is important to remember that you are required to produce a picture according to a fixed set of rules. Here the mathematician allows no unauthorized use of the instruments, such as using a straightedge to measure distance. You should also realize that the results of construction can ordinarily be achieved by other means much more simply, but possibly with less satisfaction. Constructions also involve approximations, since they depend on measurements. In general, proving that you have actually performed the expected construction depends on using congruent sets of points.

1. Transferring or copying a segment. See Figure 6.1. The problem is to construct a picture of a segment on $\overleftrightarrow{CD}$ with endpoint C and the same length as $\overline{AB}$. To perform this construction, draw an arc passing through

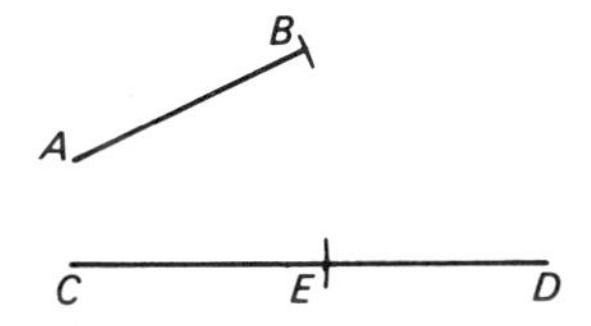

Figure 6.1. Transferring a segment

B, with A as center. With this same setting for the compass, draw an arc with C as center, cutting $\overleftrightarrow{CD}$ at the endpoint E. Then $\overline{CE}$ is the required segment.

2. *Finding the midpoint of a segment (bisecting the segment).* With each endpoint of $\overline{AB}$ in Figure 6.2 as a center, draw two arcs with the same

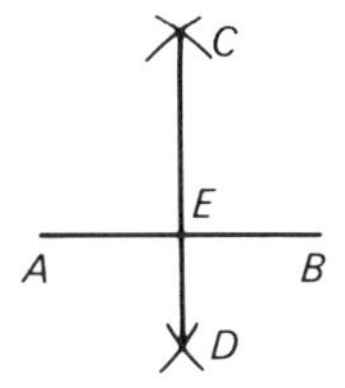

Figure 6.2. Bisecting a segment

radius that meet above and two arcs with the same radius that meet below the segment at points C and D. It is not necessary that all four arcs have the same radius. The segment joining C and D meets $\overline{AB}$ in the required midpoint, E; $\overline{CD}$ is the perpendicular bisector of $\overline{AB}$.

The proof that E is actually the midpoint depends on congruent triangles. In Figure 6.3, $\triangle CBD \cong \triangle DAC$. Why?

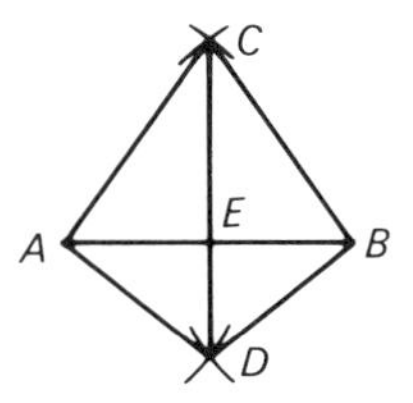

Figure 6.3. Proof of construction of midpoint

As a result, $\angle ECB \cong \angle EDA$. Then $\triangle AEC \cong \triangle BEC$ and $\overline{AE} \cong \overline{EB}$, since they are corresponding sides.

3. *Constructing a perpendicular to a line at a given point on the line.* With the given point A in Figure 6.4 as center, drawn an arc intersecting the given line in two points, B and C. With B and C as centers, draw two arcs

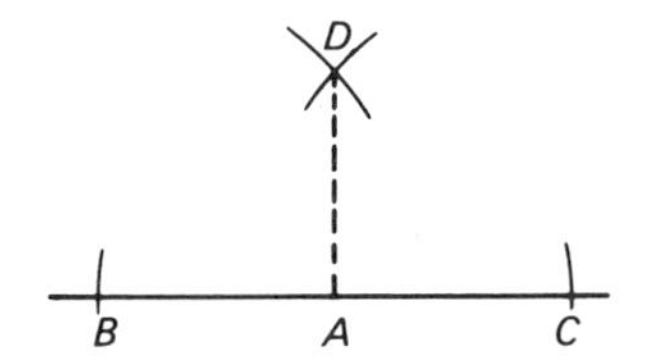

Figure 6.4. Constructing a perpendicular to line

using the same radius—which must be longer than segment BA or CA—and intersecting at a point D. Then $\overline{DA}$ is the required perpendicular.

Figure 6.5 shows how to prove that this construction results in the required perpendicular. $\triangle BAD \cong \triangle CAD$. Why? $\angle BAD$ and $\angle CAD$ are right angles, since they are congruent and supplementary.

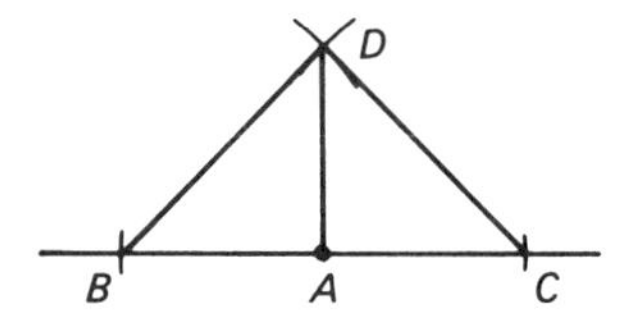

Figure 6.5. Proof of perpendicular construction

4. *Constructing the bisector of an angle.* With the vertex of the given $\angle AOB$ (Fig. 6.6) as center, draw an arc cutting the two sides of the angle at

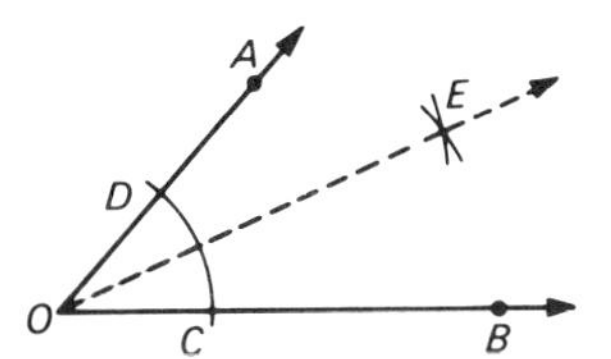

Figure 6.6. Construction of angle bisector

C and D. From these two points, and using the same radius, draw arcs that meet at point E. The segment EO, connecting this point with the vertex of the angle, lies on the bisector of the angle.

The reason why $\overrightarrow{OE}$ is actually the bisector can be seen from Figure 6.7. $\triangle ODE \cong \triangle OCE$. Why? $\angle DOE$ and $\angle COE$ are corresponding angles in these congruent triangles.

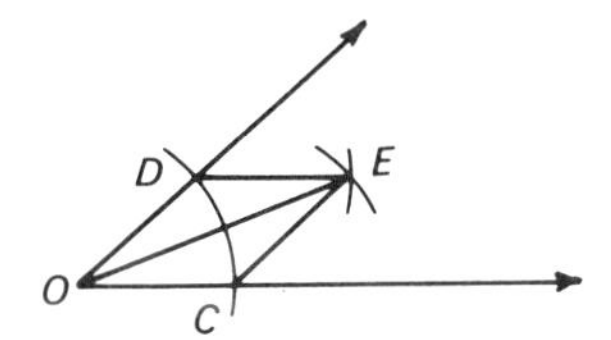

Figure 6.7. Proof of angle bisector construction

5. *Reproducing or copying an angle.* The problem is to draw an angle at E (Fig. 6.8), with $\overrightarrow{EF}$ as one side, with the same measure as given $\angle ABC$. With B as center, draw an arc intersecting both sides of the angle at G and

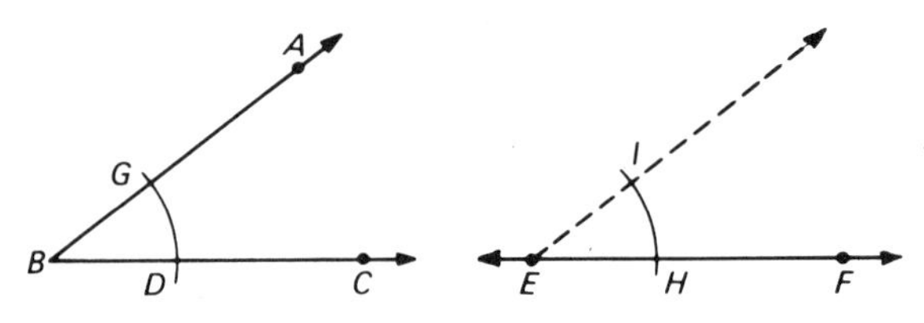

Figure 6.8. Reproducing an angle

D. With point E as center, draw an arc with the same radius intersecting $\overrightarrow{EF}$ at H. With D as center, draw an arc passing through the point G at which the original arc cut the second side of the angle. With H as center, and using the same radius as the arc with center D, draw another arc determining point I. Then $\overrightarrow{EI}$ is the second side of the required angle.

For a proof of this construction, study Figure 6.9. $\triangle ABC \cong \triangle DEF$. Why? Hence $\angle CAB \cong \angle FDE$.

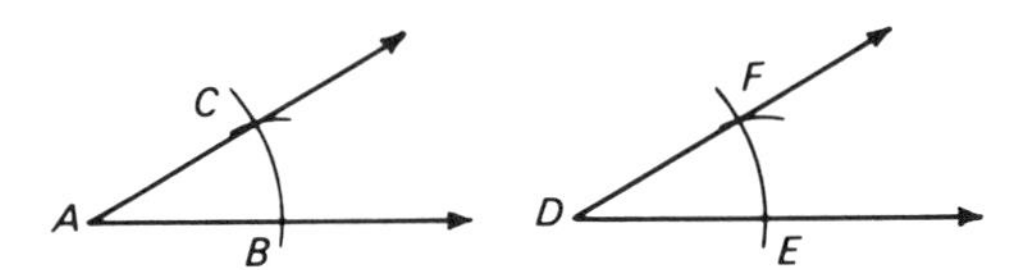

Figure 6.9. Proof of angle reproduction construction

This section concludes with examples of constructions that use several of the basic constructions to make a single drawing. Other examples will be found in the exercises.

Example

Construct a triangle, given one of the angles and the two sides lying on the sides of the angle. You are required to construct a picture of a unique triangle that has one angle with the same measure as the given angle and two sides with the same measure as the given segments. Figure 6.10 shows the given segments and angle and the completed construction with the construction marks indicated. Notice the repeated use of the basic constructions as the steps are summarized:

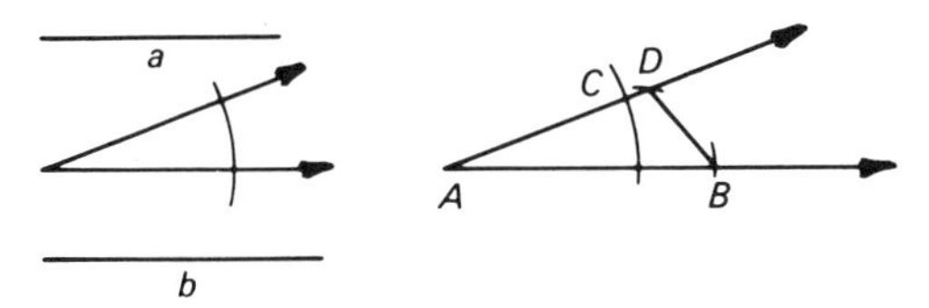

Figure 6.10. Construction of triangle

(a) Using a line through A as a beginning, locate point B so that $\overline{AB}$ has the measure b.
(b) Reproduce the given angle with A as a vertex and $\overrightarrow{AB}$ as one side, locating side $\overrightarrow{AC}$.
(c) Locate point D so that $\overline{AD}$ has the measure a; then draw segment BD.

Example

Construct a triangle, if two of the angles and the included side are given.

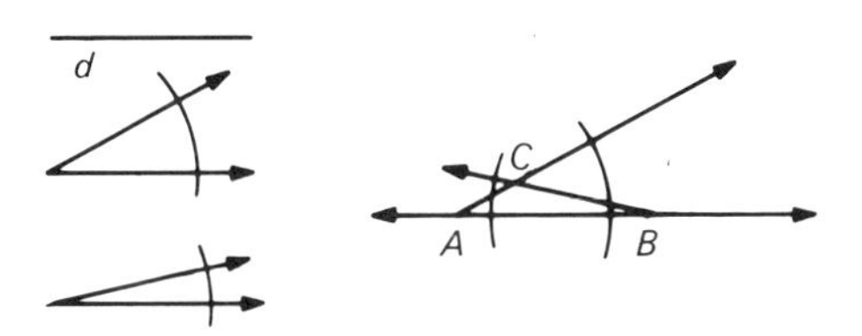

Figure 6.11. Construction of triangle

Let the given elements be those shown in Figure 6.11. Segment d is constructed ($\overline{AB}$) and the two given angles are reproduced with vertices at A and B. The intersection of the two sides of the angles determines the third vertex C.

Example

Construct a line through a point and parallel to a given line. Figure 6.12 shows the completed construction, with A the given point and $\overleftrightarrow{BC}$ the given line. The steps in the construction are briefly summarized:

(a) Draw $\overline{AD}$, which is any transversal through the given point and intersecting the given line.

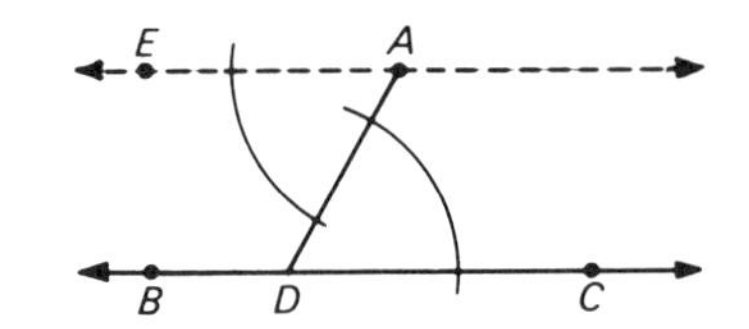

Figure 6.12. Construction of parallel line

(b) Reproduce $\angle CDA$ at A. The second side of this angle, $\overrightarrow{AE}$, lies on the required line.

Exercise 6.1

For Exercises 1–11, use the given elements in Figure 6.13. Trace the given elements and then perform the required constructions, using a straightedge and compass.

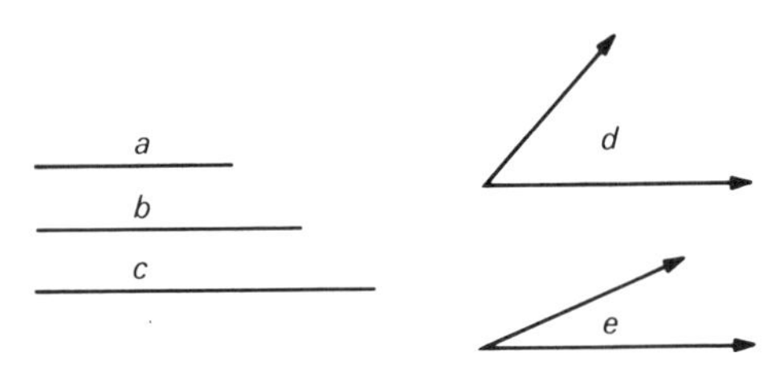

Figure 6.13

1. Construct the midpoint of segment *c*.
2. Construct the perpendicular at one end of segment *a*.
3. Construct the bisector of angle *d*.
4. Reproduce angle *e*.
5. Construct a triangle, given angle *d* and segments *b* and *c* lying on the sides of that angle.
6. Construct a triangle, given angles *d* and *e*, with segment *c* the included side.
7. Construct a triangle, given *a*, *b*, and *c* as the sides.
8. Partition segment *c* into four congruent segments.
9. Construct a right triangle, given segments *a* and *b* as the two sides on the rays of the right angle.
10. Construct a square, given segment *c* as one side.
11. Construct a parallelogram with angle *d* and with *a* and *b* as the two adjacent sides.

12. Construct the tangent to a circle at a point on the circle.
13. Technically, the difference between a compass and a divider is that, when you have drawn one arc with a divider, you cannot keep this setting so that you can draw another arc with the same radius in a different position. How can you bisect a segment, using a divider rather than a compass?

Construction Problems and Designs

The basic ideas of geometric constructions have been introduced in the previous section. Many other possible constructions exist. The following

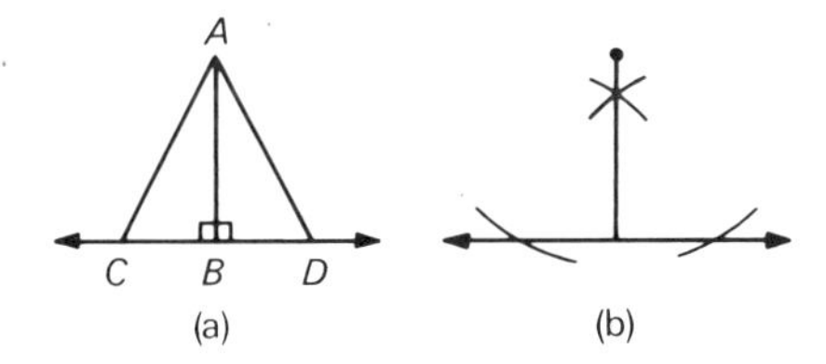

Figure 6.14. Perpendicular from point to line

example shows how it is possible to analyze a construction problem when the method of construction is not apparent.

Example

Construct the perpendicular from a point to a given line from a point not on the line.

Assume, as in Figure 6.14a, that the construction has been made, with $\overline{AB}$ the desired segment. The construction can be made if point B can be located correctly. If two points C and D can be found, then B can be located as the midpoint. Since $\triangle ABC \cong \triangle ABD$, C and D are the same distance from A.

The analysis leads to the construction shown in Figure 6.14b.

A category of construction problem that has interested many people is the construction of a regular polygon inscribed in a circle. These problems

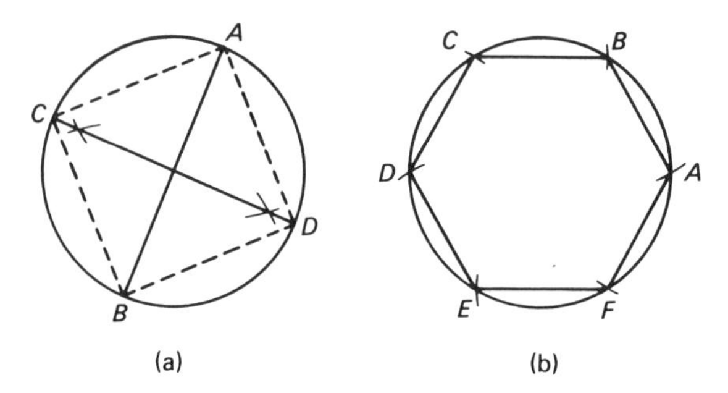

Figure 6.15. Construction of inscribed square and hexagon

assume the circle is given and the polygon is to be constructed so that all vertices lie on the circle. Two special cases are the square and the regular hexagon. The major steps in the construction of each are listed.

1. Square inscribed in a circle (see Fig. 6.15a).
   (a) Construct any diameter AB.
   (b) Construct the perpendicular bisector of this diameter, locating points C and D. Then A, B, C, D are the vertices of the required square.
2. Regular hexagon inscribed in a circle (see Fig. 6.15b).
   (a) With any point A, and a radius AO, draw an arc locating point B.
   (b) From B, following the same instructions, locate C, and so on around to F. Then the six points are the six required vertices.

The great mathematician Gauss investigated which regular polygons could be constructed inside a circle and which could not. He found, for example, that polygons with 3, 5, and 17 sides could be inscribed, but those with 7, 11, and 13 sides could not. In 1796, at the age of 19, Gauss proved that the only regular polygons having a prime number of sides that could be inscribed were those with $2^{2^n} + 1$ sides. For example, if n is 0, $2^{2^n} + 1 = 3$ and if n is 1, $2^{2^n} + 1 = 5$. It was supposedly this discovery that actually caused him to devote a lifetime of study to mathematics.

You have found how to inscribe a regular polygon with 4 and 6 sides. These constructions make it possible to construct other inscribed polygons. In Figure 6.16, a regular octagon can be inscribed simply by locating the midpoints of each of the four arcs determined by the vertices of an inscribed square. Finding the midpoints of the eight arcs of an inscribed

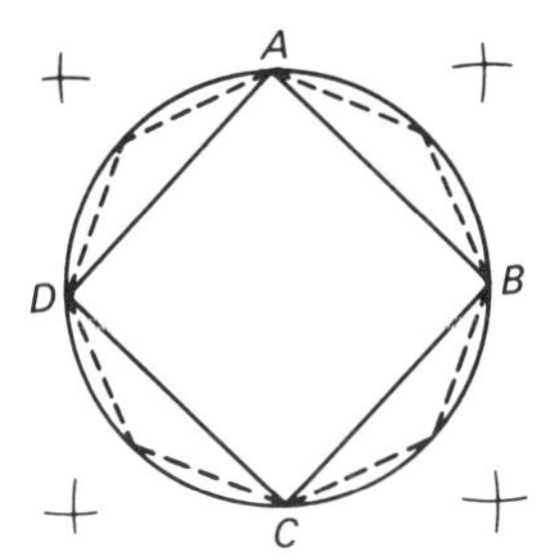

Figure 6.16. Construction of inscribed octagon

octagon makes it possible to locate the 16 points of an inscribed regular polygon with 16 sides. The midpoints of each of the six arcs determined by a regular hexagon determine a polygon with 12 vertices, which then makes possible inscribing a regular polygon with 24 sides.

Many beautiful designs can be constructed by beginning with the vertices of an inscribed regular polygon, whether these vertices can first be located by constructions or not. It has been established that given a regular polygon, the circumscribed circle can always be found. It could also be pointed out that certain constructions are impossible only because of the restriction to Euclidean tools. Two relatively simple examples are shown in Figure 6.17. Figure 6.17a shows an inscribed dodecagon—a regular polygon with 12 sides—with all its diagonals. Figure 6.17b shows an inscribed regular polygon with 18 sides and some subsets of its diagonals.

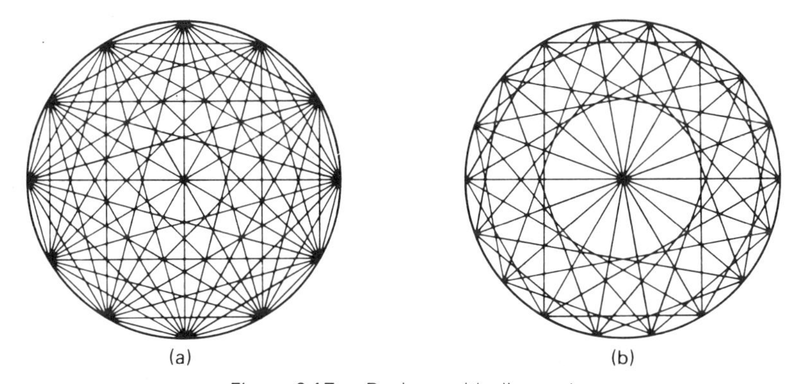

Figure 6.17. Designs with diagonals

Figure 6.18 shows other constructions involving the vertices of an inscribed polygon.

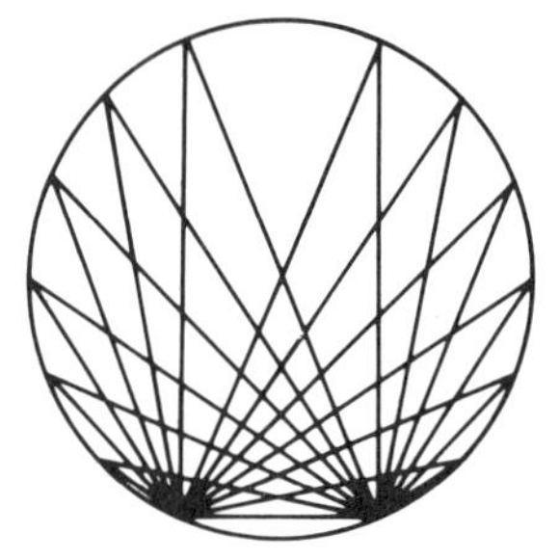
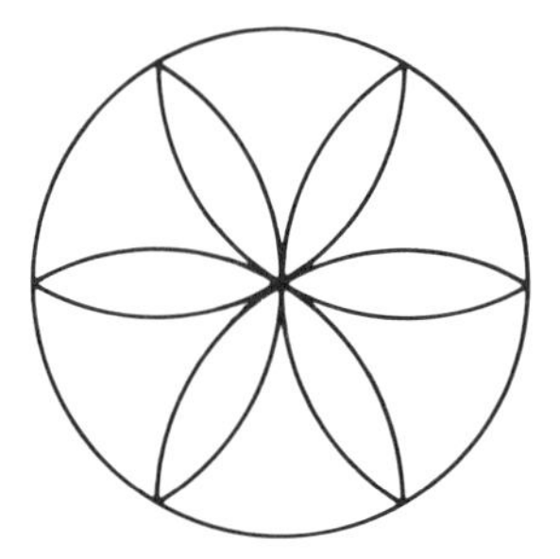

Figure 6.18. Construction designs

You have seen that Gauss showed that a regular pentagon can be constructed in a circle. This construction is not given here because it is difficult to make intuitively evident. But if it is assumed that the five vertices are located (by drawing central angles of 72°), one can construct a Christmas star very easily.

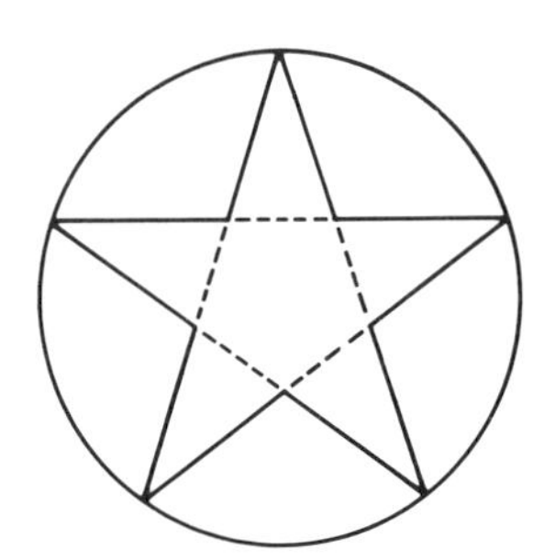

Figure 6.19. Construction of a star

Although the ancient Greeks were able to solve many construction problems by skillful methods, there were three, called the *Three Famous Greek Problems,* which they could not solve. All three have finally been proved to be impossible to construct with only the straightedge and compass. These three problems are *doubling a cube, trisecting a general angle,* and *squaring a circle.* Doubling a cube means finding the edge of a cube with twice the volume of a given cube; trisecting a general angle means finding an angle with a measure $\frac{1}{3}$ that of an arbitrary given angle; and squaring a circle means constructing a square with the same area as a given circle.

It may seem difficult enough to perform constructions with just the two prescribed instruments, but it is actually possible to solve problems with only one of the two. An Italian named Masheroni, who lived during the last half of the eighteenth century, proved that any geometric construction possible by straightedge and compass can also be made by the compass alone. To do so, you must think of a line as being determined by two points on it. Other mathematicians have investigated constructions that can be performed using only the straightedge, without the compass. In fact, if a given circle and its center are located, all of the constructions possible with both instruments may be made with the straightedge alone. This statement, called the Poncelet-Steiner Construction Theorem, assumes that the given and required elements are all points.

Exercise 6.2

1. Draw a line and a point not on it. Construct the perpendicular from the point to the given line.
2. Construct an equilateral triangle inscribed in a circle.
3. Given a square, construct a circle inscribed in it.
4. Construct a regular hexagon, without using the idea that it is inscribed in a circle. (*Hint*: Think of the hexagonal region as consisting of six triangular regions.)
5. Construct inscribed regular polygons with:
   (a) 12 sides (b) 24 sides (c) 16 sides (d) 32 sides
6. In the formula of Gauss, find $2^{2^n} + 1$ if:
   (a) $n = 2$ (b) $n = 3$
7. In Figure 6.17a (on p. 133), how many diagonals pass through the center of the circle?
8. In Figure 6.17b, how many large equilateral triangles do you see whose sides are diagonals?
9. In Figure 6.17b, how many large rectangles do you see whose sides are diagonals?

For Exercises 10–12, make up designs according to the given instructions.

10. Starting with the vertices of an inscribed regular polygon of 18 sides, make up a design using only a compass.
11. Starting with the vertices of an inscribed regular polygon of 24 sides, make up a design using only a straightedge.
12. Starting with the vertices of an inscribed regular polygon of 16 sides, make up a design using both a straightedge and a compass.

Mathematical Laboratory Experiences with Constructions

The expression *mathematical laboratory* implies a setting in which mathematical experimentation and discovery can take place. Through the use of constructions, theorems can be approached that might otherwise seem meaningless. Historically, mathematicians have often arrived at important theorems by first making drawings or constructions from which potential theorems could be guessed. The last step was the formulation of the proof itself.

In this section, constructions lead to further information about both triangles and circles.

In Figure 6.20, $\overline{AE}$ is called a *median* of the triangle. The endpoints are one vertex and the midpoint of the opposite side of a triangle. A median can be constructed by first finding the midpoint of one side. Segment AD is an *altitude*. One endpoint is a vertex, and the segment is perpendicular to the opposite side. An altitude can be constructed by the construction of a perpendicular from a point to a line. Ray AF (or sometimes just segment AF) is called the *bisector* of angle BAC in the triangle. Any point on the *bisector of an angle* is the same perpendicular distance from one side of the angle as from the other. A triangle has three medians, three altitudes, and three bisectors of the angles.

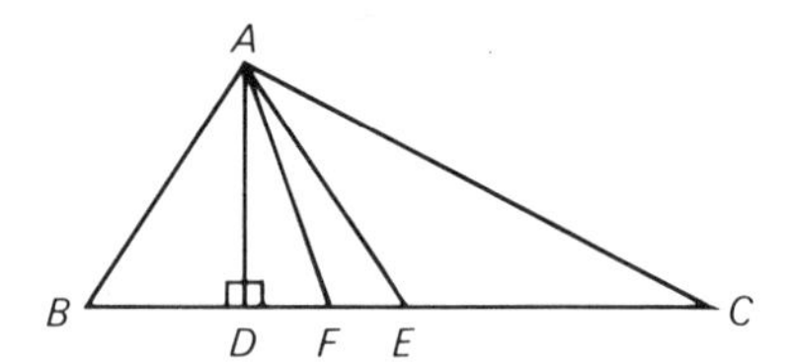

Figure 6.20. Median, altitude, angle bisector

Try constructing the medians, altitudes, and angle bisectors of various triangles before reading further. A geoboard or a model of a triangle may also be used for experimentation.

In Figure 6.21a, the medians of triangle ABC seem to be concurrent at point G. Draw other triangles and measure carefully to see if it seems reasonable that the medians of a triangle are concurrent.

Theorem

The medians of a triangle are concurrent at a point called the *centroid.*

The centroid of a triangle is the *center of gravity* for the triangle, which means that, if you cut a picture of a triangle out of cardboard, it would balance if supported at the centroid. Try this to see.

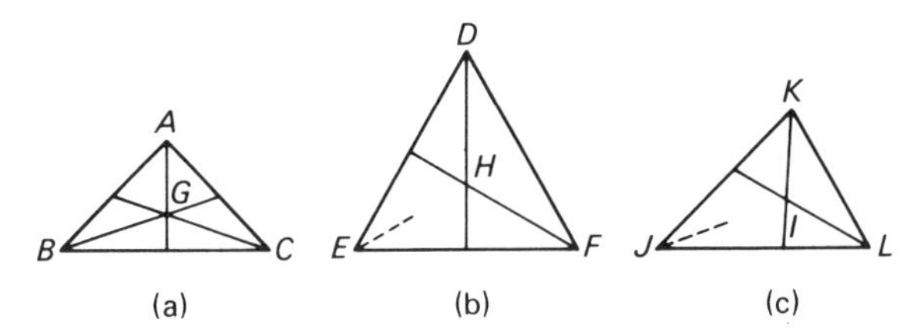

Figure 6.21. Concurrent line segments

In Figure 6.21b, two of the altitudes meet at the point labeled H. Do you believe the third altitude also passes through this same point? Draw other pictures of triangles to see if you agree with the theorem that the altitudes of a triangle are concurrent. The point of concurrency is called the *orthocenter.*

In Figure 6.21c, point I is the point of intersection of two of the bisectors of the angles. Experiment to see if you agree that the three bisectors are concurrent. The point of concurrency is called the *incenter* of the triangle.

Of course, the three theorems about concurrency are proved in a formal course in geometry. The easiest of the three proofs is for angle bisectors. As in Figure 6.22, construct two bisectors $\overline{BI}$ and $\overline{CI}$. From the point of intersection I, construct perpendiculars to the three sides of the triangle.

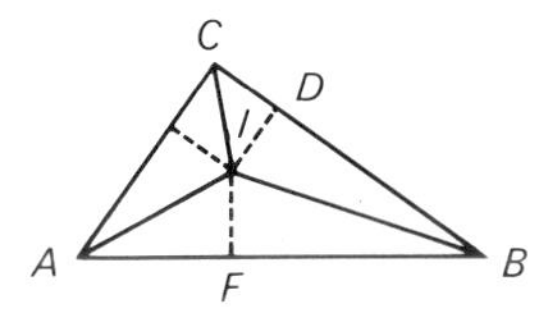

Figure 6.22. Proof of concurrency of angle bisectors

$$\triangle BID \cong \triangle BIF \qquad \text{Why?}$$

$$\triangle DIC \cong \triangle EIC \qquad \text{Why?}$$

$$\overline{FI} \cong \overline{EI} \qquad \text{Why?}$$

Since I is the same distance from $\overline{BA}$ and $\overline{AC}$, it is also on the angle bisector of angle BAC.

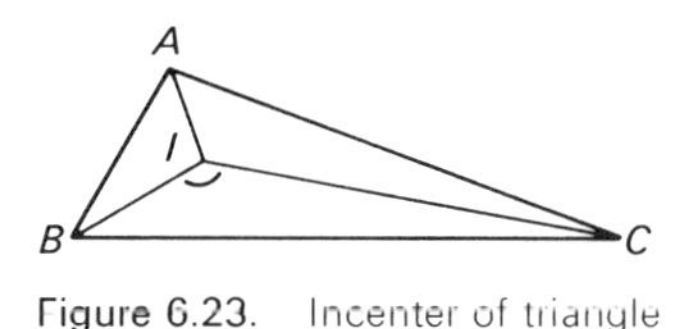

Figure 6.23. Incenter of triangle

Defining the incenter of a triangle leads to some interesting theorems involving measures of angles. In Figure 6.23, I is the incenter.

Theorem

The angle BIC has a measure in degrees that is 90 plus one-half the measure of angle A.

It is easy to verify this theorem for any particular triangle. Suppose that, in Figure 6.23, angles A, B, and C measure 70, 30, and 80 degrees, respectively. Then, in triangle BIC, $\angle IBC$ has a measure of 15 and $\angle ICB$ has a measure of 40. This means that the third angle, $\angle BIC$, has a measure of

$$180 - (15 + 40) = 180 - 55 = 125.$$

But $125 = 90 + \frac{1}{2}(70)$, which is 90 plus $\frac{1}{2}$ the measure of angle A, and this is what was to be verified.

You can prove that angle BIC has a measure in degrees of 90 plus $^1/_2$ the measure of angle A by considering the angles of triangle BIC. Let a, b, c, and i represent the measures of angles A, B, C, and BIC. Then, in $\triangle BIC$,

$$\begin{aligned} {}^1/_2 b + {}^1/_2 c + i &= 180 \\ i &= 180 - {}^1/_2 b - {}^1/_2 c \\ &= (90 + {}^1/_2 a + {}^1/_2 b + {}^1/_2 c) - {}^1/_2 b - {}^1/_2 c \\ &= 90 + {}^1/_2 a. \end{aligned}$$

Construct several triangles, construct the incenters, and verify the theorem.

Any line passing through one of the three vertices of a triangle is called a *Cevian,* after the Italian mathematician Ceva (often pronounced Shava) who lived during the seventeenth century. Medians, altitudes, and angle bisectors are all examples of Cevians. *Ceva's Theorem,* published in 1678, states a generalization for any set of three concurrent Cevians, one through each vertex of a triangle. Using the notation of Figure 6.24, the theorem states that the product of the measures of segments AM, BK, and CL is equal to the product of the measures of segments MB, KC, and LA.

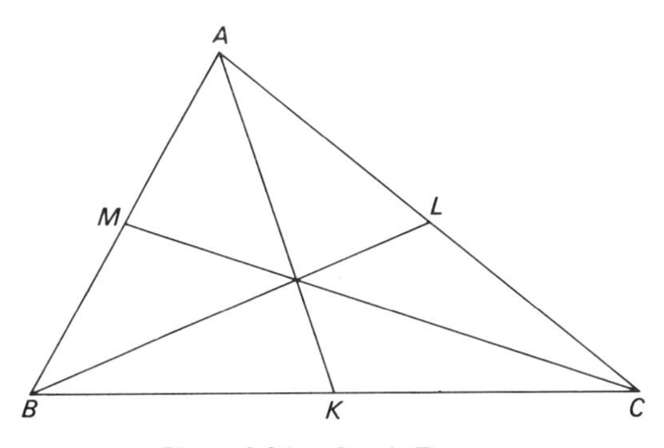

Figure 6.24. Ceva's Theorem

This means that the six segments into which points K, L, and M partition the sides of the triangle form two sets of segments. The three segments in each set have no common endpoints, and the products of the measures of the segments in the two sets are equal. Check by measuring in Figure 6.24 to see if this theorem seems reasonable. Try it for the three sets of Cevians pictured in Figure 6.21 (on p. 137). The conclusion should be rather obvious for the medians, since the endpoint of the medians is the midpoint of each of the sides. Then, if the Cevians in Figure 6.24 were medians, $\overline{AM} \cong \overline{MB}$, $\overline{BK} \cong \overline{KC}$, $\overline{CL} \cong \overline{LA}$, and the product of the measures of $\overline{AM}$, $\overline{BK}$, and $\overline{CL}$ is certainly equal to the product of the measures of $\overline{MB}$, $\overline{KC}$, and $\overline{LA}$. It is remarkable that these products are equal even in the more general cases when the three concurrent Cevians are not medians. The proof depends on the properties of similar triangles, which are introduced in a later chapter.

Here is an example of a theorem that concerns a special property of an equilateral triangle.

Theorem

For any point X within an equilateral triangle ABC, the sum of the measures of the segments $\overline{XD}$, $\overline{XE}$, and $\overline{XF}$, where these segments are perpendicular to the sides (see Fig. 6.25), is equal to the measure of the altitude of the triangle.

In Figure 6.25, measure to see that this theorem is plausible. Then try different locations for point X to see if the sum of the measures always seems the same. The proof is left as Exercise 13 of Exercise Set 7.2.

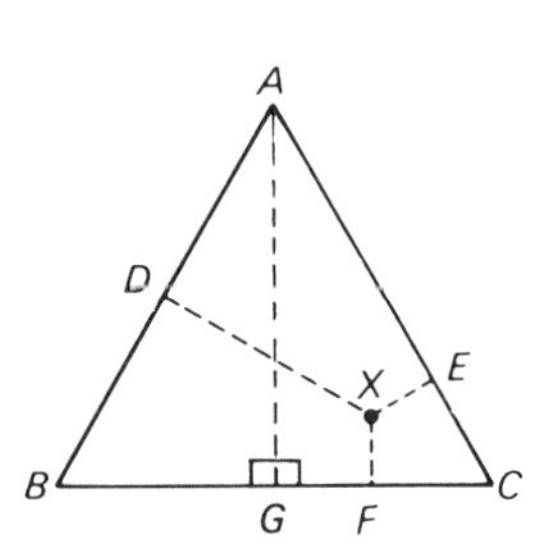

Figure 6.25. Distance from point to side of equilateral triangle

Some theorems about triangles concern inequalities rather than equalities. In Figure 6.26, $\angle ABD$ is called an *exterior* angle of $\triangle ABC$. Since it is

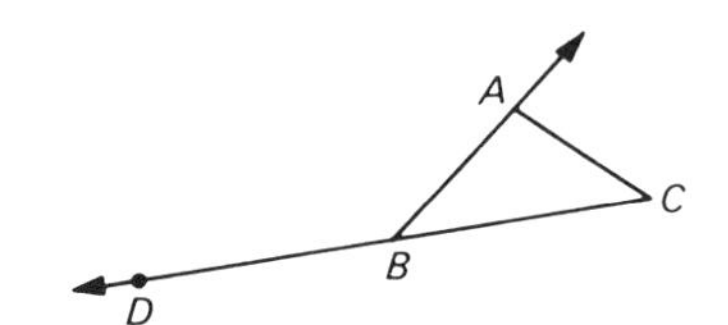

Figure 6.26. Exterior angle of a triangle

supplementary to $\angle ABC$, it has a measure equal to the sum of the measures of the interior angles at A and C. This theorem can be followed by the statement of another conclusion.

Theorem

The exterior angle at a vertex of a triangle has a measure greater than the measure of either of the two interior angles at the other two vertices.

If *all* the bisectors of both the internal and external angles of a triangle are constructed, the result is as shown in Figure 6.27. There are six bisectors,

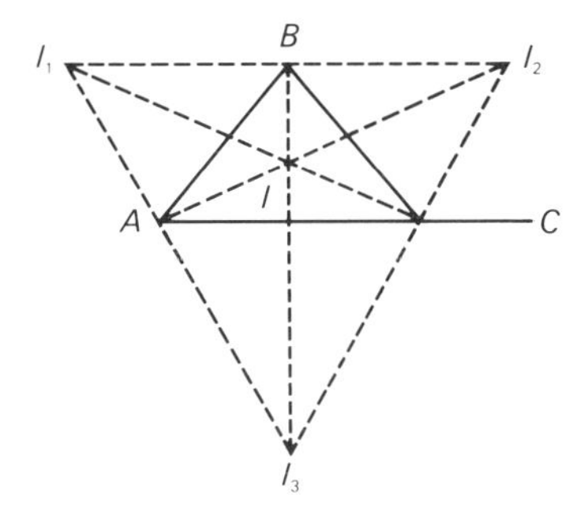

Figure 6.27. Bisectors of angles of triangle

and they intersect by threes in four points. One of these points is the incenter I. The others, I_1, I_2, and I_3, are called *excenters.* The proof that the bisectors meet at four points is very similar to the proof of the concurrency of the three internal angle bisectors.

An example of a theorem from modern geometry involving segments in a triangle comes from the word of Lemoine, who lived during the last half of the nineteenth century. In triangle ABC (Fig. 6.28), segment AD is a

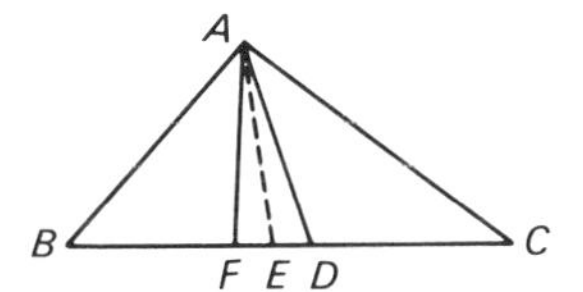

Figure 6.28. Construction of symmedian

median and $\overline{AE}$ is the angle bisector. Segment AF, constructed on the other side of the bisector from $\overline{AD}$, such that the measure of $\angle FAE$ is equal to the measure of $\angle EAD$, is called a *symmedian.* A symmedian may be thought of as the reflection of the median about the angle bisector. You should construct a triangle and its three symmedians to see if you agree with the theorem that the three symmedians of a triangle are concurrent. The point of concurrency is called the *Lemoine point.*

Many interesting theorems concern polygons inscribed in circles or circles inscribed in polygons. The incenter is the point of concurrency of the internal bisectors of the angles in a triangle. A point on the bisectors of an angle is equidistant from both sides of the angle; hence, in Figure 6.29, point I is the same perpendicular distance from side AB as from side AC. Then $\overline{IX} \cong \overline{IY}$. I is also the same distance from side AC as from side

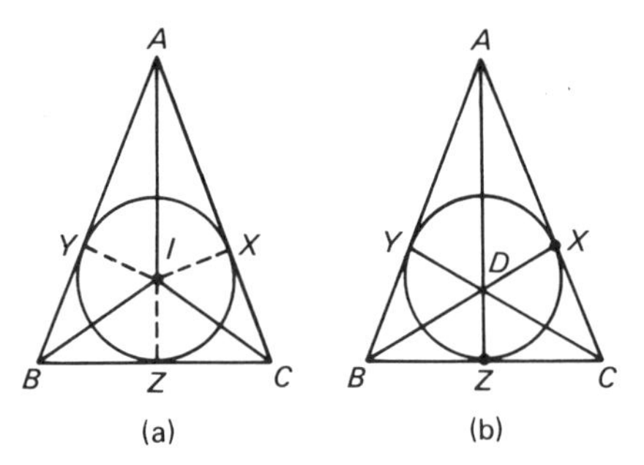

Figure 6.29. Incircle of triangle

BC; hence $\overline{IX} \cong \overline{IZ}$. Segments IX, IY, and IZ are congruent and can be considered as radii of a circle with center I. The circle with center I is inscribed in the triangle and is called the *incircle*. The incircle of a triangle can be constructed once the incenter is located.

The perpendicular bisectors of the sides of a triangle are concurrent at a point. These three bisectors are not Cevians, since they do not pass through the vertices of the triangle. A point on the perpendicular bisector of a segment is equidistant from the endpoints of the segment; hence M, in

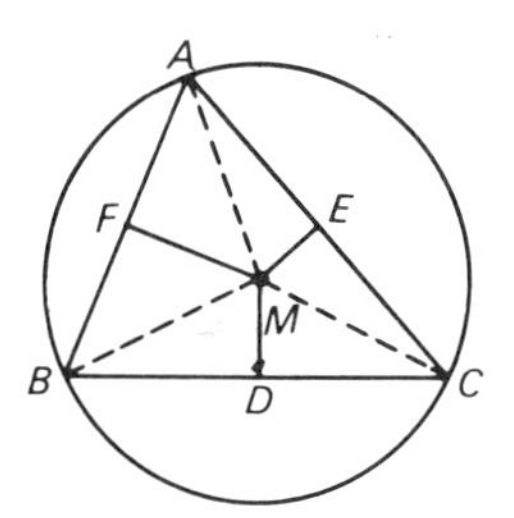

Figure 6.30. Circumcircle of triangle

Figure 6.30, is the same distance from each of the three vertices of the triangle. $\overline{AM}$, $\overline{BM}$, and $\overline{CM}$ are congruent and can be considered as radii of a circle with M as center. Since the three vertices of a triangle lie on this circle, the triangle is inscribed in the circle. The circle is called the *circumcircle* of the triangle, and point M is called the *circumcenter* of the triangle. If a triangle is given, a circle can be circumscribed about it once the circumcenter is located.

A rather unexpected result can be obtained by considering points on the circumcircle of a triangle. This theorem is credited to the English mathematician Robert Simson, who worked during the first half of the eighteenth century. Let P be any point on the circumcircle of $\triangle ABC$ in Figure 6.31. Construct $\overline{PX}$, $\overline{PY}$, and $\overline{PZ}$ perpendicular to the three sides of the triangle.

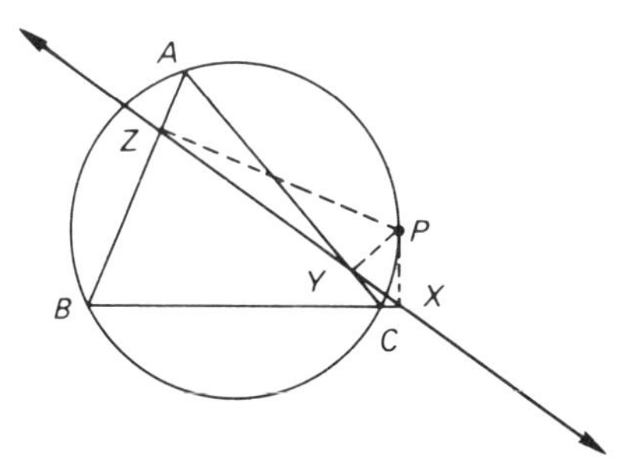

Figure 6.31. Simson line

What seems to be true about points X, Y, and Z? Choose various locations for P and see if the same statement is true. Points X, Y, and Z are collinear, and the line on which the three points lie is called the *Simson line*. The significant idea in this theorem is that point P is any point on the circumcircle, not some special location.

You have found that a circle can be constructed through all four vertices of a quadrilateral only when the opposite angles are supplementary. These quadrilaterals are called *cyclic* quadrilaterals.

In general, a circle cannot be constructed about the typical polygon with more than four sides. One set of polygons for which a circumcircle can always be found, however, is the set of all *regular polygons*. For the hexagon in Figure 6.32, all of the six triangles are congruent and equilateral. The three diagonals of the hexagon connecting opposite vertices are concurrent

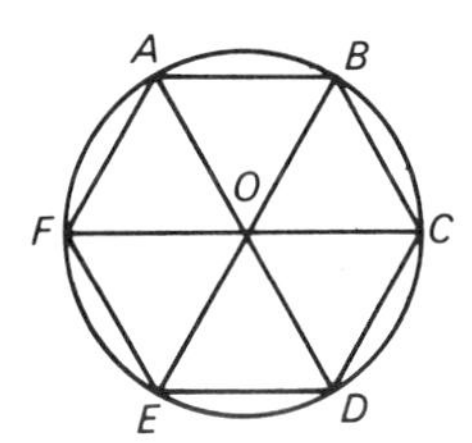

Figure 6.32. Inscribed hexagon

at a point. The point O is the center of the circumcircle. $\overline{AD}$, $\overline{BE}$, and $\overline{CF}$ are diameters of the circle. Furthermore, each side of the hexagon has a measure equal to the radius of the circle.

Exercise 6.3

1. Construct an equilateral triangle and locate the centroid, the orthocenter, and the incenter. What can you conclude about these three points in an equilateral triangle?
2. Construct a triangle with one obtuse angle. Construct the orthocenter.
3. What can you conclude about the orthocenter of an obtuse triangle?
4. Can the incenter of a triangle ever be outside the triangle? Why?
5. In Figure 6.23 (on p. 138), if the measures in degrees for angles A, B, and C are 70, 30, and 80, respectively, find the measures of $\angle AIB$ and $\angle AIC$.
6. Experiment by constructing examples to see if the theorem concerning Figure 6.25 (on p. 140) is applicable to an isosceles triangle.
7. Construct a triangle. Then construct all the bisectors of all the angles, as in Figure 6.27 (on p. 141).
8. Prove that the internal bisector and the external bisector at a vertex of a triangle are perpendicular.
9. What seems to be the location of the Lemoine point of an equilateral triangle?
10. Construct a triangle and three symmedians.
11. Draw a triangle with two sides not congruent. Compare the two angles at vertices opposite these two sides. Which of the angles seems to have a larger measure? Try other examples and state your conclusion in the form of a tentative theorem (a conjecture).
12. Draw a triangle and locate these three points: the midpoint of one side, the midpoint of the altitude to that side, and the Lemoine point of the triangle. What seems to be true about these three points?
13. In Figure 6.33, line FD, called a transversal, meets the three lines containing the sides of triangle ABC in the points D, E, and F. A

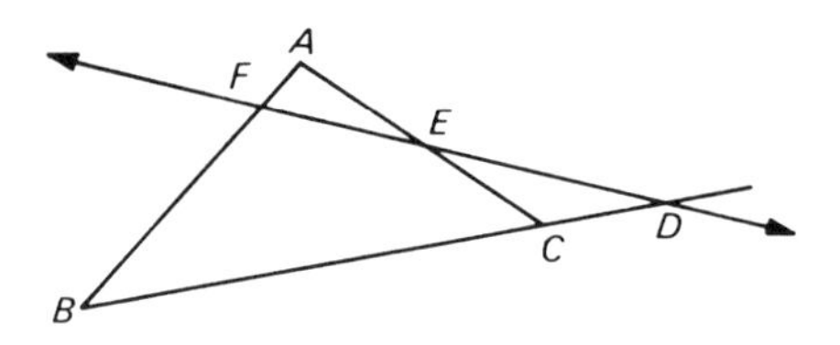

Figure 6.33

famous theorem, attributed to the Greek mathematician Menelaus of Alexandria, who lived less than a century after the time of Christ, concerns the six segments AF, FB, BD, CD, CE, and AE. Like the theorem of Ceva, it states that the product of three of the measures equals the product of the other three measures. Experiment to see if you can find the two equal products.

14. From a point outside a circle, tangents can be drawn to the given circle. Suppose that a fixed length (from an outside point to a point of tangency) is given for a tangent. Describe the set of all points from which tangents can be drawn with this same length.
15. A given triangle has how many: (a) incircles? (b) circumcenters? (c) circumcircles?
16. Construct a triangle; then construct its incircle.
17. Construct a triangle; then construct its circumcircle.
18. Given a triangle inscribed in a circumcircle, how many Simson lines can be determined?
19. If you know that two consecutive angles of a cyclic quadrilateral are 60 and 50 degrees, what are the measures of the other two vertices?

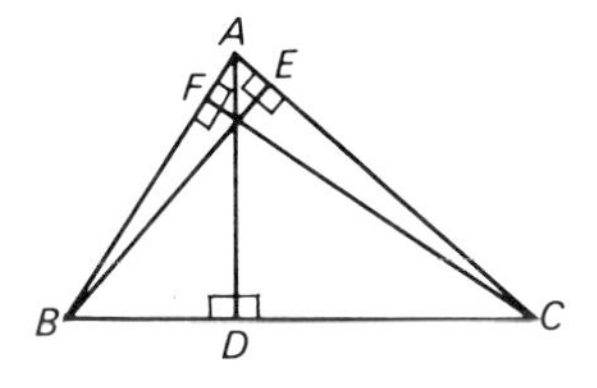

Figure 6.34

20. In Figure 6.34, how do you know that points E and D lie on a circle with $\overline{AB}$ as diameter?
21. In Figure 6.34, name two other cyclic quadrilaterals, using the labeled points.
22. Suppose that a regular octagon is given. Explain how its circumcircle could be constructed.

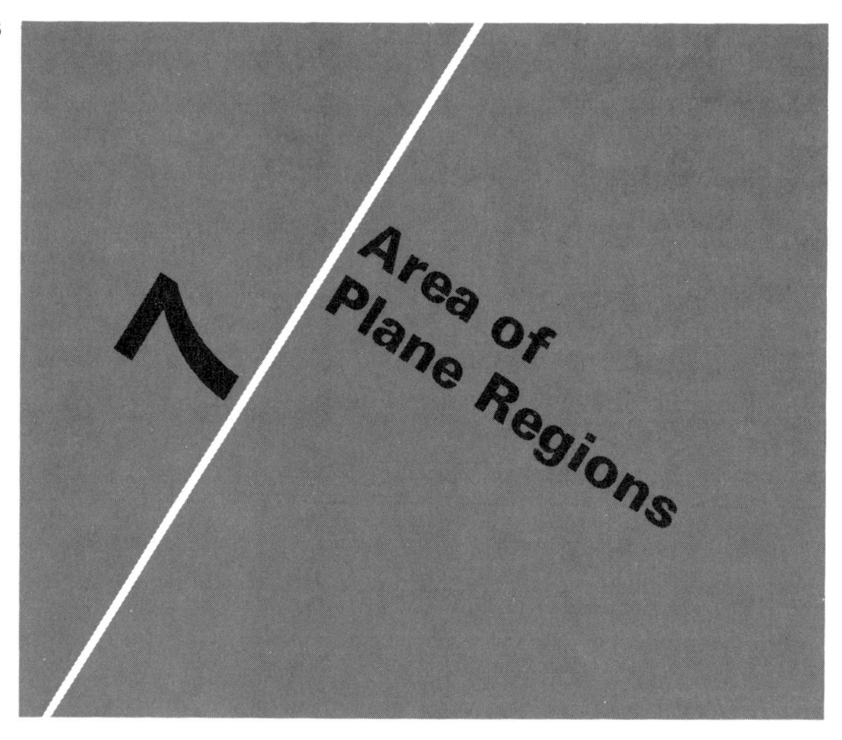

The union of a simple closed curve and its interior is a set of points called a *plane region*; area is a metric property of a plane region. In this chapter, you will first study the concept of area. This concept will help make clear formulas for finding the measure of area of the interior of a rectangle, a parallelogram, a triangle, other polygons, a circle, and other sets of points.

The Concept of Area

The question "How large is it?" may be asked about flat surfaces and objects. You may need to know the measure of area of a rug, a house, a lot, or a state. Area concerns a set of points in a region that has a boundary, such as a square, a rectangle, a circle, or some less familiar shape. A statement of the measurement of area of a plane region will give information about a metric property of the set of points whose boundary is a simple closed curve.

Many of the principles that apply to measurement of length and angles also apply to the measurement of area. One of the first requirements is a unit of measurement that is of the same nature as the thing being measured. In other words, you need to find a unit of area that is itself a region. Figure 7.1 shows a unit of area used to measure rectangular region

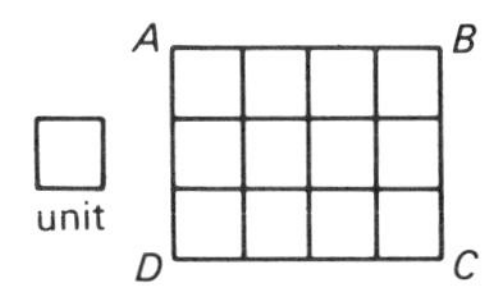

Figure 7.1. Area of rectangular region

ABCD. The unit of area in this case is a region bounded by a square. To find the measure of area of the rectangular region (often called for convenience "finding the measure of area of the rectangle") means to determine how many of the units will completely fill the interior. The mathematical basis for the measurement of area is the relation of congruence. The rectangle and its interior are partitioned into 12 square regions. Each square is congruent to the square that is the boundary of the unit. Each of these 12 square regions has a measure of area of 1, and the measure of area of rectangular region *ABCD* is 12.

Having developed the concept of a unit of arbitrary size for the measurement of area, you would expect to establish next some standard units of measurement. In most cases common standard units of measurement of area are related to common linear units. Examples are one square foot, one square yard, and one square mile. In the metric system other examples are one square millimeter, one square centimeter, one centare (one square meter), and one square kilometer.

The relationships between several standard units of area are shown below.

144 sq in. = 1 sq ft

9 sq ft = 1 sq yd

43,560 sq ft = 1 acre

640 acres = 1 sq mi

It should be pointed out that an acre is not necessarily square. The first two relationships can easily be demonstrated by a drawing, such as Figure 7.2 for the second one. This drawing is not the correct size, but it will help you understand an idea. A picture of a square yard can be partitioned into nine square regions, each representing one square foot.

The final element that should be expected in the development of the concept of area is the use of some scale for measuring area. Such a scale corresponds to the use of a ruler for length and a protractor for angles. Here, however, is where we encounter one interesting difference between the concepts of measurement. Although it is possible to construct workable scales, in practice the measure of area is normally found by using formulas that depend on linear measurements. For example, the area of a

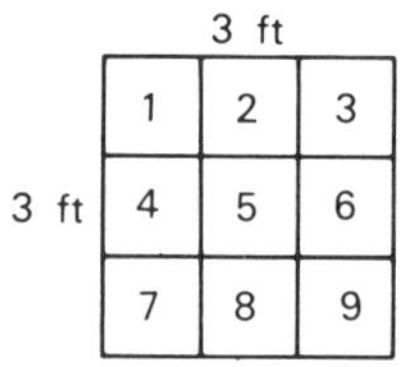

3 ft

1	2	3
4	5	6
7	8	9

3 ft

Figure 7.2. Picture of square yard

rectangular region can be found by using the measures of its length and width, and the area of a circular region can be found by using its radius.

In Figures 7.1 and 7.2, the measure of area was found by counting the number of square units in the region to be measured. It is easy to see that the measure could have been found by adding or by multiplying. For example, in Figure 7.1, you could add $4 + 4 + 4$, where 4 is the number of square units in each row. Since there are three rows, you could have found the measure by multiplying the number in each row by the number of rows. A generalization of this procedure, allowing the measure of a side to be any positive real number, results in the formula

$$A = lw,$$

where A is the measure of area, l is the measure of length, and w is the measure of width of a rectangular region.

The structure of Euclidean geometry needs to be extended in this chapter by several assumptions about plane measurements.

Axiom 19

The measure of area of a rectangular region is the product of the measures of the length and width.

Examples

l	w	A
9 in.	7 in.	63 sq in.
.4 m	.7 m	.28 sq m

Although the formula for area yields exact results as long as it is used with measurements in the mathematical model, it yields approximate results when it is used to obtain the surface area of actual objects.

Exact results come from counting elements of a set, such as eggs in a carton. This is an example of what is called a discrete variable. Exact counting is not possible when determining the length of a stick or the amount of water in a glass. This is an example of what is called a continuous variable. Not only are the concepts of discrete and continuous a part of abstract mathematics, but the real world has different kinds of entities to be measured too.

If the region to be measured is a square region, then l and w in the formula represent the same number, and the formula may be written as

$$A = s^2,$$

where s is the measure of one side of the square. The *exponent* 2 indicates that s is to be used as a factor two times.

Example

If a square has a side with a measure of 4, then the measure of area for the square region is $4^2 = 4 \times 4 = 16$.

An understanding of the concept of area can be further extended by a study of non-square units of area. In Figure 7.3, the unit of area is a rectangular region with length 1 unit and width $\frac{1}{2}$ unit. The measure of area of the larger region is 40, using this unit region. In the same figure, you could use different unit regions, such as one that is 5 units long and 1 unit wide, to find different measures of area.

For measuring some rectangular regions, units in the shape of rectangular or even triangular regions would be convenient. Circular units of measure would present difficulties, but many other polygonal shapes are possible.

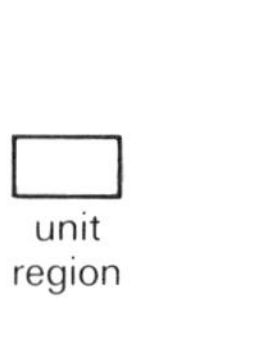

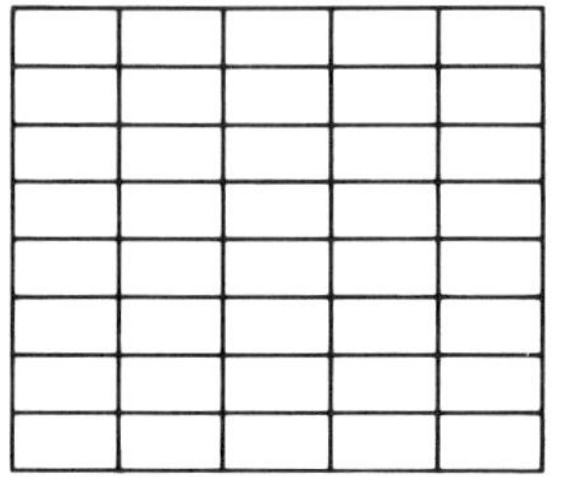

Figure 7.3. Non-square units of area

In each case the number of congruent unit regions covering the region to be measured is the measure of area.

Exercise 7.1

1. Complete the relationships of standard units of measurement of area.
   (a) 1 sq yd = ? sq in. (b) 1 acre = ? sq mi
   (c) 1 centare = ? sq cm (d) 1 sq km = ? sq cm
   (e) Since 1 acre = 160 sq rds, then 1 sq mi = ? sq rds.
   (f) Since 1 hectare (about $2\frac{1}{2}$ acres) = 10,000 sq m, then 1 hectare = ? centares.
2. Complete the table of measures for rectangular regions, using the formula $A = lw$.

l	w	A
5	7	___
2.1	3.4	___
$2\frac{1}{2}$	$3\frac{3}{4}$	___
6	___	30
3.2	___	7.8
___	$7\frac{1}{4}$	$9\frac{2}{3}$

3. Find the measure of area of a square region with the following measure of length.
   (a) 16 (b) $2\frac{1}{2}$ (c) .46 (d) $7\frac{3}{4}$
4. Suppose a rectangular region has a length of 8 in. and a width of 6 in. Find the measure of area in terms of rectangular unit areas of the following sizes.
   (a) 1 in. by 1 in. (b) 1 in. by 2 in. (c) 2 in. by 2 in.
   (d) 2 in. by 3 in. (e) 4 in. by 6 in. (f) 8 in. by 6 in.
   (g) 8 in. by 12 in. (h) $\frac{1}{2}$ in. by $\frac{1}{2}$ in. (i) $\frac{1}{8}$ in. by 2 in.

Area of Region Whose Boundary Is a Parallelogram or a Triangle

Although it is difficult to think of small square units being placed in the interior of parallelograms or triangles—because of the slanting sides—we can approach the study of the areas of these regions more easily by modifying the formula that has already been established for a rectangular region. A logical sequence of extensions of that formula will result in formulas for the area of the region whose boundary is a parallelogram, a triangle, or some other polygons. Within the mathematical model, results will be interpreted as exact. The formulas for the plane regions give exact measures of area.

In Figure 7.4, consider parallelogram $ABCD$. Any side of a parallelogram may be called the *base*. The perpendicular distance from the base to the

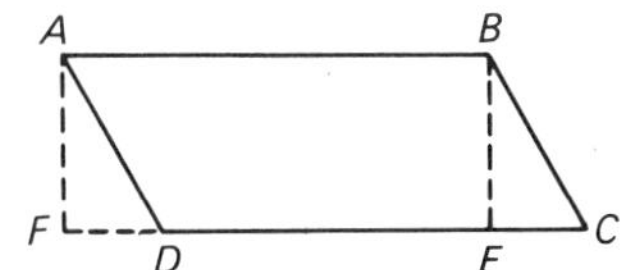

Figure 7.4. Area of region for parallelogram

opposite side is called the *altitude* of the parallelogram (any segment along which this distance is measured is likewise called an altitude). If $\overline{CD}$ is the base, then $\overline{BE}$ and $\overline{AF}$ are altitudes. Right triangle AFD is congruent to $\triangle BEC$, since $\overline{AF} \cong \overline{BE}$, $\overline{AD} \cong \overline{BC}$, and the third sides are congruent by the Pythagorean Theorem. More simply, two right triangles are congruent if the hypotenuse and side of one are congruent to the hypotenuse and corresponding side of the other.

You can imagine that the parallelogram and its interior, if cut out of paper, could be made into a rectangular region by cutting off $\triangle BEC$ and moving the piece of paper to the position of $\triangle AFD$. Furthermore, the measure of area of the region of the parallelogram is equal to the measure of area of the new rectangular region. Also, $\overline{FE} \cong \overline{DC}$, since $\overline{FD}$ and $\overline{EC}$ are corresponding sides of congruent triangles.

The outcome of the preceding argument is that the measure of area of a region whose boundary is a parallelogram is equal to the measure of area of a corresponding rectangular region. The two sets of points both have the same measures of base and altitude. You may write a formula to express the generalization that the measure of area for the parallelogram is equal to the product of the measure of the base and the measure of the altitude to that base. In symbols, this generalization may be written

$$A = bh,$$

where b is the measure of the base and h is the measure of the altitude to that base. The area will be expressed in square units, whereas the other measures will be in linear units.

Example

If the length of the base of a parallelogram is 3 in., and if the altitude to that base is 4 in., then the area of the region is 12 sq in.

The formula for the area of a region with a parallelogram was derived from the formula for the area of a rectangular region. In an analogous manner, the formula for the area of a triangular region may be derived from the formula for the region with a parallelogram as boundary. Triangle ABC is shown in Figure 7.5. Any side of a triangle may be considered as a

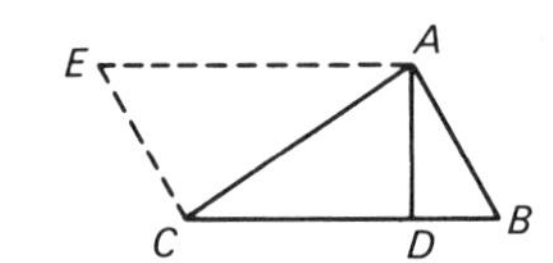

Figure 7.5. Area of triangular region

base, and the perpendicular distance from the opposite vertex to this side (extended, if necessary) is the altitude to that side. (The segment along which the distance is measured is also called the altitude.) If $\overline{CB}$ is the base, then $\overline{AD}$ is the altitude to that base.

In Figure 7.5, $\overline{AE}$ is drawn parallel to $\overline{CB}$ and $\overline{CE}$ is drawn parallel to $\overline{AB}$, so that $\angle BAC \cong \angle ECA$. Then $EABC$ is a parallelogram, and $\triangle AEC \cong \triangle CBA$.

Axiom 20

If two triangles are congruent, then the respective triangular regions have the same area.

The area of triangular region ABC is the same as the area of triangular region CEA. The measure of the entire area in the region for the parallelogram $EABC$ can be found; the measure of area of ABC will be half this measure. Also, the base and altitude of the triangle are the base and altitude of the parallelogram. The results may be summarized in a formula for the measure of area of a triangular region:

$$A = \frac{1}{2}bh,$$

where b is the measure of the base of the triangle and h is the measure of the altitude. The area will be expressed in square units, whereas the measures of the segments will be in linear units.

Examples

Find the area of a triangular region with the measurements given.
(a) base 6 in., altitude 4 in.

$$\begin{aligned} A &= \frac{1}{2}bh \\ &= \frac{1}{2}(6)(4) \\ &= 12 \end{aligned}$$

The area is 12 sq in.
(b) base .7 cm, altitude .2 cm

$$\begin{aligned} A &= \frac{1}{2}bh \\ &= \frac{1}{2}(.7)(.2) \\ &= \frac{1}{2}(.14) \\ &= .07 \end{aligned}$$

The area is .07 sq cm.

The formulas of measure of area of a square region and a triangular region are used in various proofs of the Pythagorean Theorem. The measure of area of the large square region in Figure 7.6 is $(a + b)^2$, and the measure of area of the small square region is c^2. The measure of area of each triangular region is $\frac{1}{2}ab$. Then

$$c^2 + 4\left(\frac{1}{2}ab\right) = (a + b)^2,$$

$$c^2 + 2ab = a^2 + 2ab + b^2,$$

$$c^2 = a^2 + b^2.$$

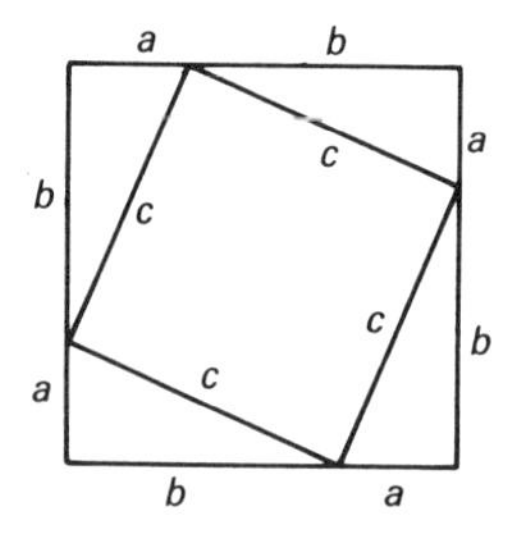

Figure 7.6. Proof of Pythagorean Theorem

The larger square was given. It should be established that the smaller figure actually is a square. This proof is required as an exercise.

Exercise 7.2

1. Duplicate, using paper or cardboard, the experiments suggested by:
   (a) Figure 7.4 (b) Figure 7.5
2. Complete the table of measurements for a parallelogram and its interior.

b	h	A
9 in.	14 in.	____
3.6 ft	4.1 ft	____
1.2 m	3.9 m	____
2 mi	____	8 sq mi
6.1 km	____	19.6 sq km
____	$3\frac{1}{3}$ yd	12 sq yd

3. Complete the table of measurements for a triangle and its interior.

b	h	A
7 in.	13 in.	____
8.4 ft	7.2 ft	____
3.1 mm	7.3 mm	____
9 mi	____	36 sq mi
4.5 cm	____	89 sq cm
____	$4\frac{1}{2}$ yd	32 sq yd

In Exercises 4–7, for a parallelogram, what would be the effect on the measure of:

4. The area, if the measure of altitude were doubled.
5. The area, if both the measure of base and the measure of altitude were halved.
6. The base, if the measure of altitude were doubled and the measure of area were left unchanged.
7. The altitude, if both the measure of the base and the measure of area were doubled.

8–11. Answer the questions of Exercises 4–7 for a triangular region.

12. Prove, in Figure 7.6, that the inner figure is actually a square.
13. The theorem stated for Figure 6.25 (p. 140) may be proved by using areas of triangular regions. Write this proof. (*Hint*: Construct $\overline{XA}$, $\overline{XC}$, and $\overline{XB}$ and use the areas of the original triangle and triangles ABX, BCX, and CXA.)

Areas of Other Polygonal Regions

A logical sequence of steps leads from the formula for the measure of area of a rectangular region through that for a parallelogram and a triangle. The next step in that sequence is the development of the formula for the region of a *trapezoid.*

Let $\overline{AB}$ and $\overline{CD}$ be the parallel sides in the trapezoid in Figure 7.7. The parallel sides of a trapezoid are its bases, and the perpendicular distance between them is the altitude. Diagonal $\overline{AC}$ partitions the trapezoid and its interior into two triangular regions. The parallel sides of the trapezoid may be considered as the bases of the two triangles, and the altitude of the trapezoid is the altitude of both triangles.

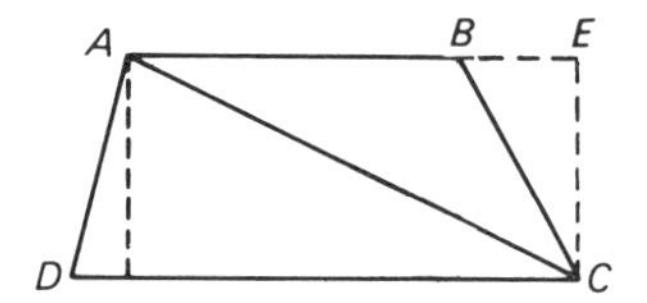

Figure 7.7. Area of trapezoidal region

The measure of area for the trapezoidal region is simply the sum of the measures of area for triangles ABC and ADC. If b_1 and b_2 are the measures of the parallel sides, and if h is the measure of the altitude, then

$$A = \frac{1}{2}b_1h + \frac{1}{2}b_2h.$$

The formula may be expressed in a somewhat simpler form by using the distributive property of multiplication over addition: $a(b + c) = ab + ac$. Then

$$A = \frac{1}{2}h(b_1 + b_2).$$

Example

If the measures of the bases are 7 and 9, and if the measure of the altitude is 6, then the measure of the area is $\frac{1}{2}(6)(7 + 9) = 3(16) = 48$.

Any convex polygonal region may be partitioned into triangular regions. One way to do this partitioning would be to draw all possible diagonals from any one vertex. You could find the measure of area of the polygonal region by adding the measure of area of several triangular regions. In some way you would need to know the base and altitude of each of the triangles, which is not always practical.

Two additional axioms clarify what must be assumed about areas of polygonal regions.

Axiom 21

There exists a correspondence that associates the number 1 with a certain polygonal region and a unique positive real number with every convex polygonal region.

Axiom 22

If a polygonal region is the union of two polygonal regions whose interiors do not intersect, then the measure of the area of the union is the sum of the measures of area of the two regions.

In the special case of a regular polygon, the situation is simplified. A regular polygon has a center—that is, a point the same distance from each of the vertices. In Figure 7.8, the regular hexagonal region is partitioned

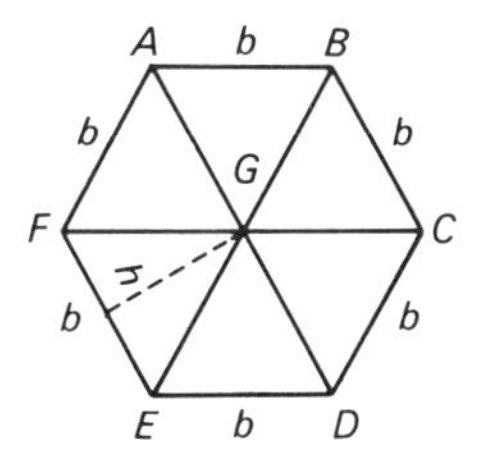

Figure 7.8. Area of regular hexagonal region

into six congruent triangular regions, by segments drawn from point G, the center, to each of the vertices. In each case the sides may be considered as the bases of the triangles. The perpendicular distance from any side to point G is the common altitude for all the triangles. Then the measure of area for the region is

$$A = 6\left(\frac{1}{2}bh\right).$$

Since the measure of the perimeter of the hexagon is $6b$, the measure of area can also be expressed as

$$A = \frac{1}{2}ph,$$

where p is the measure of the perimeter. This is the formula for the measure of area of any regular polygonal region.

Example

Find the measure of area of a regular polygonal region if the measure of the perimeter is 26 and the perpendicular distance from the center to a side is 4.

$$A = \frac{1}{2}ph$$
$$= \frac{1}{2}(26)(4)$$
$$= 52$$

Exercise 7.3

1. Complete the table of measures for a trapezoidal region.

h	b_1	b_2	A
12	9	4	___
3.1	6.4	5.7	___
$2\frac{1}{2}$	$3\frac{1}{4}$	$7\frac{3}{4}$	___
___	15	2	60
___	$\frac{3}{4}$	$\frac{1}{2}$	40

2. Complete the table of measures for a regular polygonal region.

number of sides	*measure of each side*	*distance from center to side*	*measure of area*
5	4	$2\frac{3}{4}$	___
6	3	2.6	___
6	5.7	___	85
8	5.2	7.8	___

3. Use Figure 7.9 to derive the formula for the area of a trapezoidal region from the formula for a region with a parallelogram. In the figure, $ACDF$ is a parallelogram.

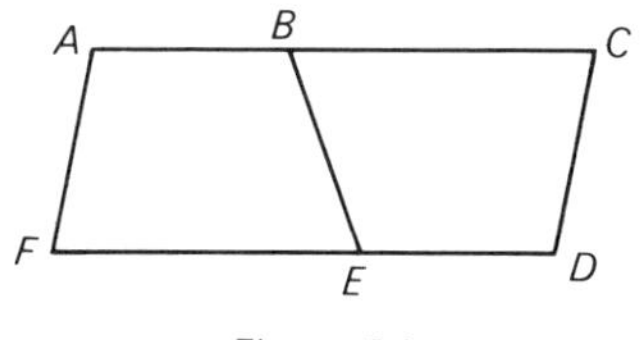

Figure 7.9

4. Find the area of the region shown in Figure 7.10.

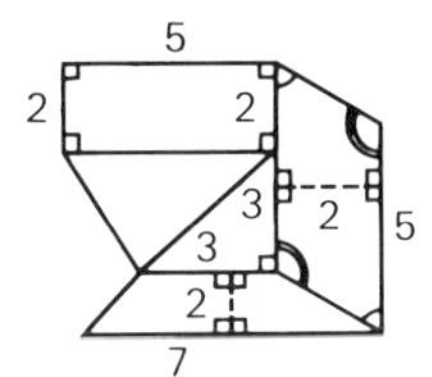

Figure 7.10

5. Use the congruent polygonal regions shown in Figure 7.11 to prove the Pythagorean Theorem in the way attributed to Leonardo da Vinci. The figure shows a right triangle and the squares on the sides.

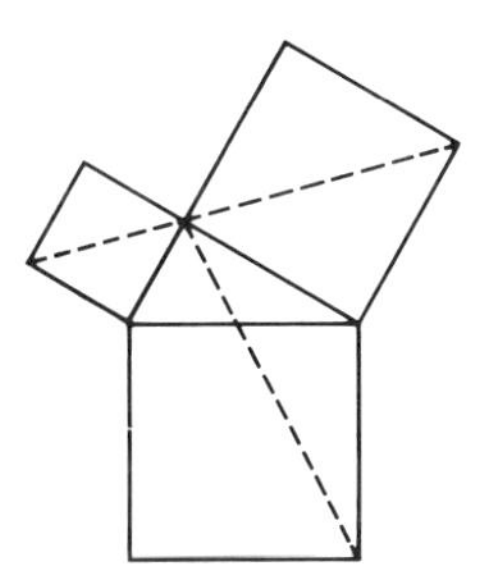

Figure 7.11

Areas of Circular Regions and Other Plane Regions

The actual development of the formula for the area of a circular region is normally found in a course in calculus. In intuitive geometry, since it is essential to use the common formula, a plausible argument is substituted for a rigorous development.

In Figure 7.12a, a circular region is shown partitioned into six *sectors* (the name given to each of the small pie-shaped pieces). In Figure 7.12b, these six sectors are arranged to resemble a parallelogram and its interior. The length of the curve from A to D is half the circumference of the circle; $\overline{BE}$ is the radius of the circle.

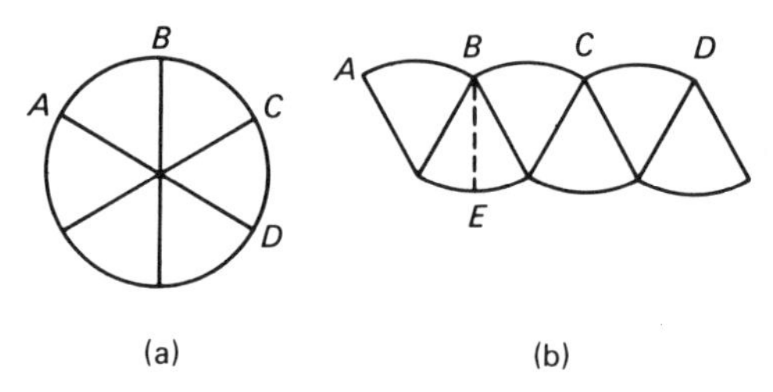

(a) (b)

Figure 7.12. Circular sectors

Suppose, instead of partitioning the circle into six sectors, you had partitioned it into 12 sectors. The arrangement, as in Figure 7.12b, would then have looked more like a parallelogram. If you had partitioned it into 24 sectors, the arrangement would have looked still more like a parallelogram. The more sectors there are, the more the arrangement resembles a parallelogram. As the number of sectors increases, the resulting arrangement approaches a parallelogram. The measure of the base of this parallelogram is half the circumference of the circle; the altitude of the parallelogram is the radius of the circle. Then, for the parallelogram, since $b = \frac{1}{2}(2\pi r)$ and $h = r$, $A = \frac{1}{2}(2\pi r)(r)$, or

$$A = \pi r^2,$$

which is the formula for the measure of area of a circular region.

Example

Find the measure of area of a circular region if the measure of length of the radius is 7 in.

$$\begin{aligned} A &= \pi r^2 \\ &= \pi(7)^2 \\ &= 49\pi \end{aligned}$$

The area is 49π sq in.

Example

Find the measure of area of a circular region if the measure of length of the radius is 2.3 in. Use 3.14 for π, and give the answer correct to one place past the decimal point.

$$
\begin{aligned}
A &= \pi r^2 \\
&\approx (3.14)(2.3)^2 \\
&\approx 16.6106 \\
&\approx 16.6
\end{aligned}
$$

The area is approximately 16.6 sq in.

The powerful and basic concept of a *limit* is used in mathematics to explain the development of the formula for the area of a circle. In Figure 7.12b, the measure of area of the sectors approaches the measure of area of a particular parallelogram as the number of sectors increases. The limit of the measure of area of the sectors, as the number of sectors continues to increase, is the measure of area of the parallelogram. You can make the difference in measures of area as small as you please by choosing enough sectors.

You have not proved that the formula $A = \pi r^2$ is correct; you have simply followed a series of steps that makes it seem reasonable. As in the formula for the circumference, however, it should be pointed out that the significance of the measure of area depends only on the measure of the radius. For example, if the radius of a circle is 5 inches, you can find that the area of the circular region will be 25π square inches. Substitution of some rational number for pi will, of course, result in an approximate answer.

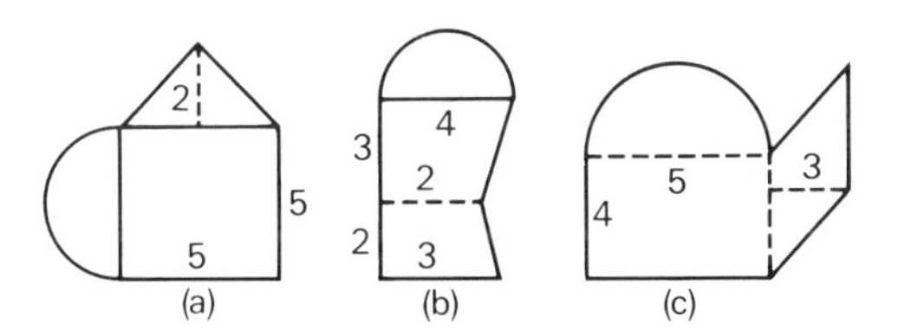

Figure 7.13. Composite figures

Application of more than one of the formulas for regions inside polygons or circles makes it possible to find the measure of area of such composite regions as those shown in Figure 7.13.

Finding the exact measure of area for a region whose boundary is a more general simple closed curve requires calculus. Archimedes, in about 225 B.C., anticipated the modern approach by investigating what is called the problem of *quadrature*. Today, an instrument called a *planimeter*, shown in Figure 7.14, is sometimes used to measure area approximately. The mathematics necessary to explain why a planimeter works is beyond the scope of this text. However, you should understand that using this instrument is really quite mechanical and that very little knowledge of mathematics is required to learn how to use it. It is used much more often by those who are interested in applied mathematics than by mathematicians. A

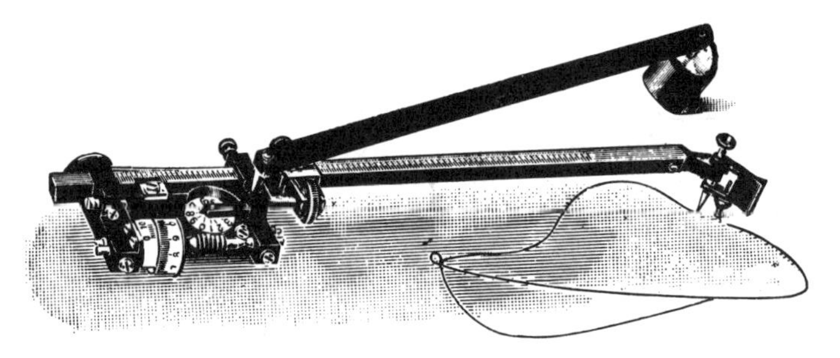

Figure 7.14. Planimeter. (Illustration by permission of Yodor Instruments, East Palestine, Ohio.)

planimeter consists of two arms and a recording mechanism. Normally, one end of an arm is placed outside the region whose area is to be measured. At the end of the second arm is a tracer point that is moved along the outline of the figure. A rolling wheel calculates the measure of area, which is then presented on a dial.

An intuitive procedure can provide approximate answers to area problems and is sometimes practical to use. Suppose the problem is to find the measure of area for the interior of a simple closed curve such as the one pictured in Figure 7.15a. Figure 7.15b shows a grid superimposed over the region. You can arrive at an estimate of the measure of area by counting the number of small regions of the grid. The need for approximation arises, however, because some of the squares of the grid are partly inside and partly outside the region. You may decide, for example, to count $\frac{1}{2}$ each time the interior includes at least a part of the square region of the grid. This procedure will give a reasonable estimate for the measure of area. One estimate for Figure 7.15b is 7.

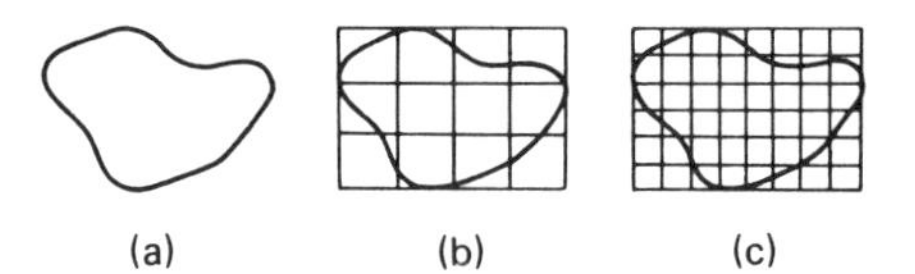

Figure 7.15. Grids used to measure area

The estimate for the measure of area within a simple closed curve usually can be improved by choosing smaller units of area. In Figure 7.15c, a different grid is superimposed. An estimate for Figure 7.15c is $29\frac{1}{2}$.

Since the unit of measure used in Figure 7.15b has a measure of four times that of Figure 7.15c, it is possible to convert the measure recorded for Figure 7.15c into the larger unit so that a direct comparison of the two measures is possible. The measure in Figure 7.15c, using the unit in Figure 7.15b, is $29\frac{1}{2}$ divided by 4, which is approximately 7.4. If you kept on drawing grids with smaller and smaller units, you could get an approximation of the actual measure of area that would be satisfactory for many practical uses.

The approximation of measure of area by grids leads to a brief consideration of how to find area using calculus. The measure of area below the curve in Figure 7.16 may be thought of as approximately the measure of

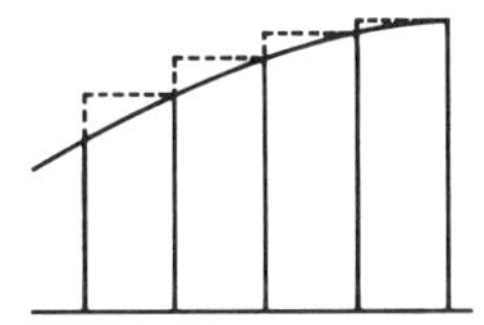

Figure 7.16. Approximation of area by rectangles

area of the rectangles shown. The approximation can be improved, as in Figure 7.17, by partitioning the regions into narrower rectangles.

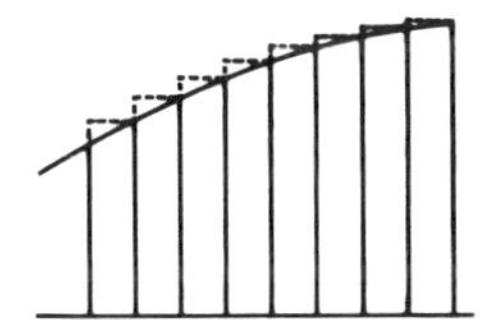

Figure 7.17. Closer approximation of area by rectangles

Suppose the process is continued indefinitely, with the number of rectangles becoming larger and larger and with the width of each becoming smaller and smaller. In calculus, the measure of area under the curve is defined to be the limit of the sum of the measure of areas of all the rectangles, as the number gets larger and larger.

Exercise 7.4

1. Carry out the experimentation suggested by Figure 7.12 (p. 160). Construct a large circle on paper to begin. Then cut it into 6 sectors, then 12.
2. Leaving the symbol for pi in the answer, find the measure of the circular area if:
   (a) $r = 5.1$ (b) $r = 3\frac{5}{6}$ (c) $r = 2.34$ (d) $r = 2\frac{7}{8}$
3. Complete this table for circular areas. Use 3.14 for pi, but express all answers to as many decimal places as are given for r or A.

r	A
7	____
12.5	____
____	6.8
____	246

4. For a circular region, what is the effect on the measure of:
   (a) Area, if the measure of radius is doubled?
   (b) Radius, if the measure of area is doubled?
5. The ancient Egyptians sometimes used an approximate formula for the area of a circular region, assuming it was equal to the area of a square region whose side was 8/9 the diameter of the circle. Use this formula to find the Egyptian approximation to π.
6. Find the measure of area of the region in
   (a) Figure 7.13a (p. 161). (b) Figure 7.13b. (c) Figure 7.13c.
7. Estimate the measure of area for each region in Figure 7.18 by using a grid in which each square is $\frac{1}{4}$ in. by $\frac{1}{4}$ in.

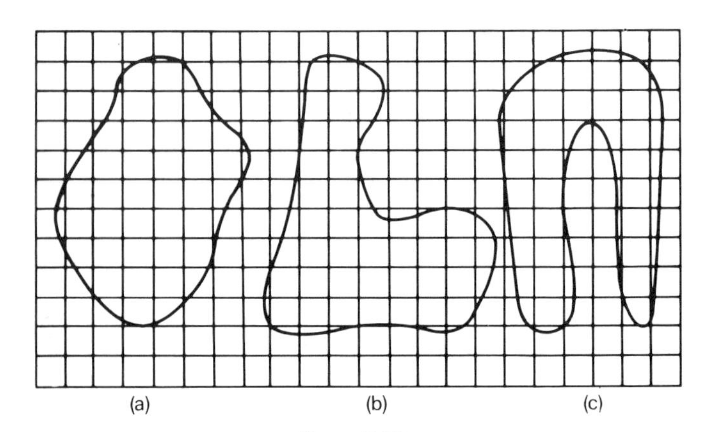

Figure 7.18

8. Estimate the measure of area for Figure 7.18 by using a grid in which each square is $\frac{1}{8}$ in. by $\frac{1}{8}$ in.
9. Convert the estimate in Exercise 8 to the unit used for Exercise 7, and compare with the answer for Figure 7.18a in Exercise 7.
10. Carry out the experimentation suggested by Figures 7.16 and 7.17 (p. 163).
    (a) Construct similar figures on sheets of paper.
    (b) Cut out the portions of the rectangles not included under the curve.
    (c) Observe that the total amount of paper cut out decreases rapidly as the number of rectangles is increased.

Two triangles are congruent if pairs of corresponding angles and pairs of corresponding sides are congruent. Congruent triangles look alike; they have the same size and the same shape. However, in observing common physical objects, you also notice pairs of objects that, although they seem to look alike, do not have the same size. For example, you see a large box and a smaller box with the same shape; you see a picture and an enlargement of the same picture; or you see a store display of socks that seem to differ only in size. In all these cases what you have observed may be explained by saying that the objects appear to be *similar*. In this chapter the intuitive idea of similar figures will be investigated through studying sets of points in the mathematical model and arriving at general statements that provide a mathematical explanation for the relationships in the physical world.

Examples of Similar Figures

Triangles ABC and DEF, in Figure 8.1, are similar. They seem to have the same shape but not the same size; this idea is the intuitive explanation of what is meant by two triangles being similar.

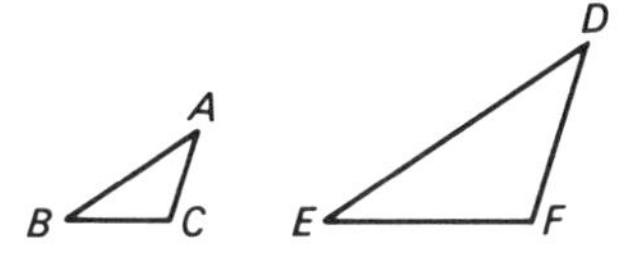

Figure 8.1. Similar triangles

How can you tell whether two triangles are similar? What must be the same about the two triangles if they are to have the same shape? One thing that is obvious is that corresponding sides are no longer congruent. But what about pairs of corresponding angles? Does it seem that angles A and D are congruent? Angles B and E? Angles C and F? Try drawing two triangles, one with a base of 3 inches and the other with a base of 2 inches, as represented in Figure 8.2. Construct congruent corresponding angles. Do the two triangles seem to be similar?

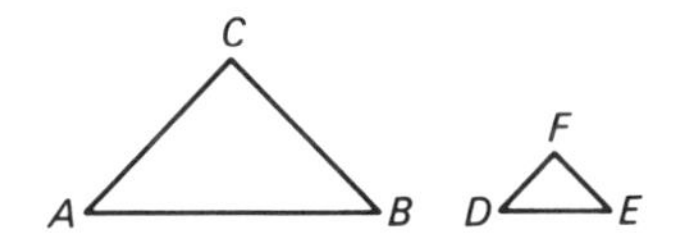

Figure 8.2. Similar triangles

Two triangles are similar if pairs of corresponding angles are congruent. Since the sum of the measures of the angles in any triangle is 180, knowing that just two pairs of corresponding angles are congruent is sufficient for you to conclude that the triangles are similar. In Figure 8.2, if angle A is congruent to angle D and angle B is congruent to angle E, then the triangles are similar, since the third pair of angles must also be congruent.

Similarity may be explained as a kind of mapping or transformation that preserves shape but not necessarily size. A *similarity* is a transformation of the plane onto itself so that the distance between each two points is multiplied by the same real number. This number is the *ratio of similarity*. In Figure 8.2, the ratio of similarity of the first triangle to the second is 3 to 1, or 3/1.

A *homothety* is a special kind of similarity such that the distance from a fixed point O to a point P is multiplied by a real number. In Figure 8.3, $OP' = r \cdot OP$, where r is the ratio of similarity. In a homothety, unlike some similarities, a segment is transformed into a parallel segment. Two homothetic figures are not only similar, but corresponding sides are parallel. The significance of a homothety lies partly in the fact that each similarity may be thought of as a homothety followed by a motion. For

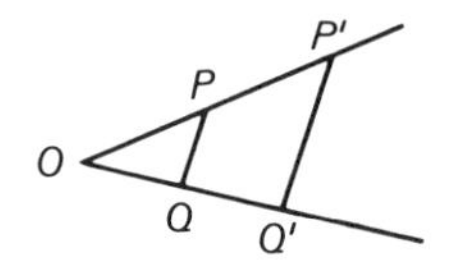

Figure 8.3. Homothetic figures

example, if you begin with a triangle and attempt to explain how it can be changed into a similar triangle, you can change it into a triangle of the right size by a homothety and then can use a motion to get it in the right position. See Figure 8.4.

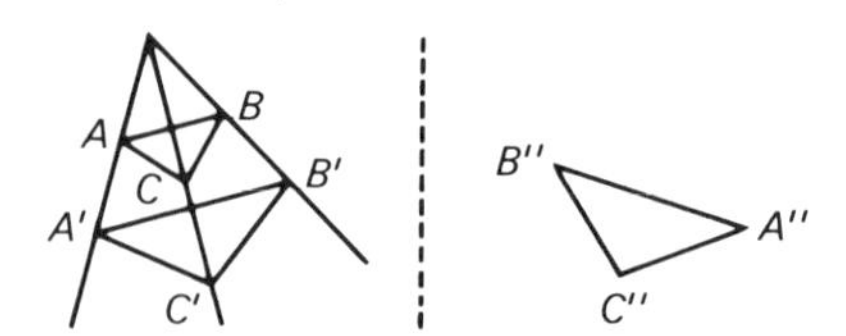

Figure 8.4. Similarity is product of homothety and motion

A notation somewhat like that used for congruence is used for similar triangles and indicates how the vertices are paired. For Figure 8.2, you could write $\triangle ABC \sim \triangle DEF$. It is significant to point out that the relationship of congruence may now be considered simply a special case of the more general relationship of similarity. Two congruent triangles are also similar, but two triangles may be similar without being congruent.

By definition, similar triangles have three pairs of corresponding congruent angles. Corresponding sides are not necessarily the same length, but something can be said about the corresponding sides. In Figure 8.1, compare the measures of sides AB and DE. How does the ratio of the measure of $\overline{AB}$ to $\overline{DE}$ seem to compare to the ratio of $\overline{BC}$ to $\overline{EF}$? Of $\overline{AC}$ to $\overline{DF}$? In Figure 8.2, compare the ratios of the measures of the pairs of corresponding sides. What seems to be the ratio in each case? In Figure 8.5, compare the ratios of the measures of the pairs of corresponding sides of similar triangles GHI and KLM. What seems to be the ratio in each case? Do you agree that in all similar triangles the ratios of the

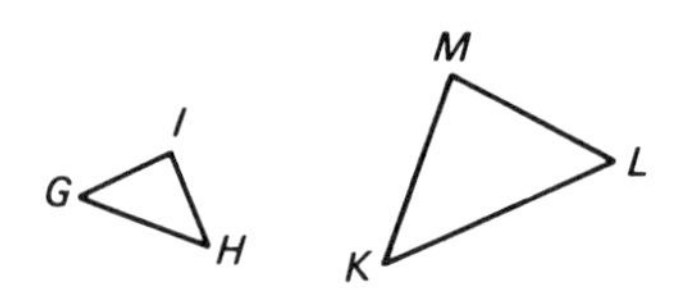

Figure 8.5. Ratios of pairs of corresponding sides

measures of pairs of corresponding sides of similar triangles are equal? This statement could serve as another way of defining similar triangles.

Suppose that in $\triangle ABC$, Figure 8.6, $\overline{DE}$ is parallel to $\overline{BC}$. Then triangles ADE and ABC are similar, since corresponding pairs of angles are congruent. The ratio of the measures of $\overline{AD}$ and $\overline{AB}$ is equal to the ratio of the measures of $\overline{AE}$ and $\overline{AC}$. Also, the ratio of the measures of $\overline{AD}$ and $\overline{AE}$ is equal to the ratio of the measures of $\overline{AB}$ and $\overline{AC}$.

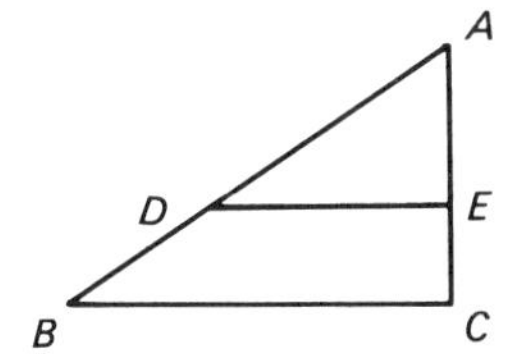

Figure 8.6. Parallel lines and similar triangles

If you know that a segment partitions two sides of a triangle so that the ratios between corresponding pairs of segments are equal, as in Figure 8.6, then you can conclude that the segment is parallel to the third side of the triangle. That is, if you know that the ratio of the measures of $\overline{AD}$ and $\overline{AE}$ equals the ratio of the measures of $\overline{AB}$ and $\overline{AC}$, you can conclude that $\overline{DE}$ is parallel to $\overline{BC}$.

The discussion in the preceding two paragraphs suggests another way of telling if two triangles are similar. Consider triangles ADE and ABC in Figure 8.6. These two triangles have a common angle at A, and the two sides of one are proportional to two corresponding sides of the other. Two triangles are similar if two sides of one are proportional to two sides of the other and the included angles are congruent.

A well-known theorem about right triangles is the following:

Theorem

> The altitude on the hypotenuse of a right triangle forms two triangles similar to each other and to the given triangle.

In Figure 8.7, right triangles ADB and ABC have $\angle A$ in common; hence $\triangle ADB \sim \triangle ABC$. Right triangles BDC and ABC have $\angle C$ in common;

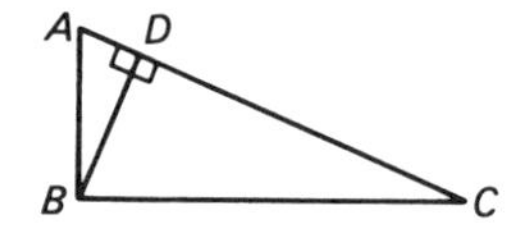

Figure 8.7. Right triangle with altitude to hypotenuse

hence $\triangle BDC \sim \triangle ABC$. Two triangles similar to the same triangle are similar to each other, so $\triangle ADB \sim \triangle BDC$. This theorem is closely related to the Pythagorean Theorem. Note that both involve relationships among the sides of a right triangle.

Sometimes proofs in geometry use similar triangles, even when the theorem itself does not seem to be related to similarity.

Theorem

If two chords intersect within a circle, the product of the measures of the segments of one chord is equal to the product of the measures of the segments of the other.

In Figure 8.8, let a, b, c, and d represent the measure of the four segments. $\triangle ADE \sim \triangle CBE$. Why? Then $a/c = d/b$, or $ab = cd$, as was to be established.

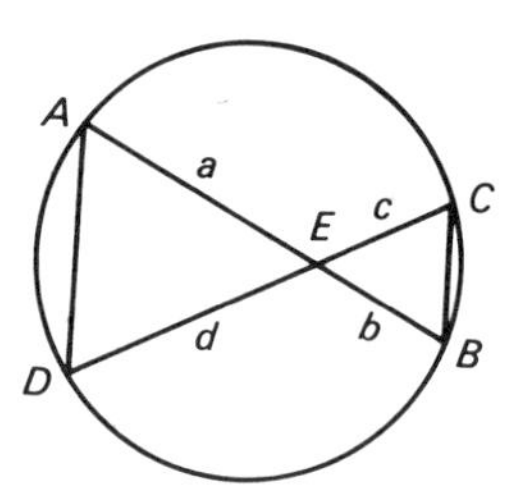

Figure 8.8. Segments of intersecting chords

Knowing that two triangles are similar and knowing some of the measurements of the triangles make it possible to find others. For example, in Figure 8.5, if you know that the measure of $\overline{KL}$ is 12 and the measure of $\overline{GH}$ is 3, you can determine the ratio of similarity, which is 1 to 4 (considering GHI as the first triangle) or 4 to 1 (considering KLM as the first triangle). A ratio of 1 to 4 may also be written 1:4 or $\frac{1}{4}$. Suppose you also know that the measure of $\overline{ML}$ is 16. Then you could find the measure of $\overline{IH}$, which is $\frac{1}{4}(16) = 4$.

The geometric study of similar figures depends on the basic ideas of ratio and proportion from elementary algebra. If two ratios are equal, then two products associated with them (traditionally called the products of the means and extremes) are equal.

$$\text{If } \frac{a}{b} = \frac{c}{d}, \quad \text{then} \quad ad = bc.$$

$$\text{If } ad = bc, \quad \text{then} \quad \frac{a}{b} = \frac{c}{d}.$$

Examples

$$\text{If } \frac{3}{4} = \frac{6}{8}, \quad \text{then} \quad (3)(8) = (4)(6).$$

$$\text{If } (6)(7) = (21)(2), \quad \text{then} \quad \frac{6}{21} = \frac{2}{7}.$$

The expression of equality of two ratios is called a *proportion*. If three of the numbers in a proportion are known, the fourth can be determined.

Examples

(a)

$$\frac{5}{x} = \frac{7}{4}$$

$$20 = 7x$$

$$\frac{20}{7} = x$$

(b)

$$\frac{14}{3} = \frac{9}{x}$$

$$14x = 27$$

$$x = \frac{27}{14}$$

Exercise 8.1

1. List other pairs of physical objects that suggest similar figures.
2. What does the ratio of similarity seem to be in: (a) Figure 8.1 (p. 167)? (b) Figure 8.2 (p. 167)?

3. What is the ratio of similarity for two triangles that are congruent?
4. If two triangles have one pair of corresponding angles congruent and one of the pairs of adjacent sides congruent, are they necessarily similar?
5. If two triangles have two pairs of corresponding sides congruent, are they necessarily similar?
6. If two triangles have two pairs of corresponding sides whose measures have the same ratio, are they necessarily similar?
7. Could a triangle and a square possibly be homothetic?
8. Could a trapezoid and a parallelogram possibly be homothetic?

In each proportion in Exercises 9–12, solve for the unknown number.

9. $\frac{x}{5} = \frac{3}{20}$

10. $\frac{2}{x} = \frac{5}{9}$

11. $\frac{2}{3} = \frac{x}{7}$

12. $\frac{3}{5} = \frac{8}{x}$

13. In Figure 8.2 (p. 167), if the measure of $\overline{EF}$ is 1.5, what is the measure of $\overline{BC}$?
14. The measures of the angles of two similar triangles are given in Figure 8.9. Write the notation showing similar triangles, pairing the vertices correctly.

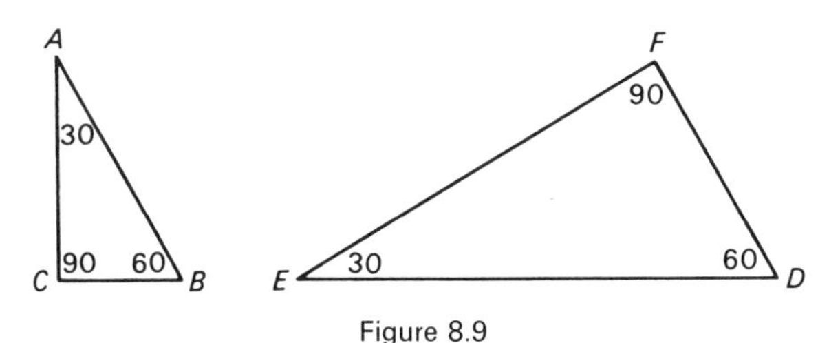

Figure 8.9

In Exercises 15–16, use Figure 8.6 (p. 169). Assume that $AD = 7$, $AE = 4$, $DE = 6$, and $AC = 7$. (The notation AD means the measure of $\overline{AD}$.)

15. Find AB.
16. Find BC.

17. In Figure 8.7 (p. 170), name all pairs of similar triangles.
18. In Figure 8.8 (p. 170), assume that $a = 7$, $b = 3$, and $c = 3.4$. Find d.

Scale Drawings

Similarity is the mathematical basis for *scale drawings*. For example, a map of a city may show the entire city on a small piece of paper yet indicate the shape of the actual city. The floor plan for a house indicates the shape of the house, but the size is different.

A mapping of one set of points onto a second set of points may result in similar figures with properties of shape preserved. For similar figures composed of segments, pairs of corresponding angles have the same measure and pairs of corresponding segments have measures with the same ratio. A constant ratio exists between the measures of linear measurements in the drawing and linear measurements of the actual objects. The ratio of similarity is the scale stated for the drawing.

Example

If 1 in. on a map represents 50 mi, then the scale could be stated as 1 in. : 50 mi.

The scale drawing in Figure 8.10 represents a room.

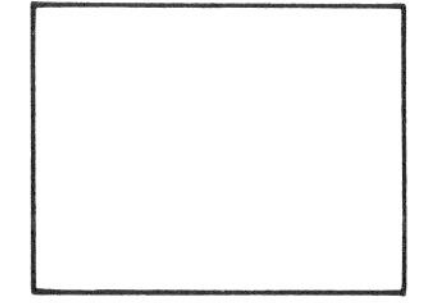

Figure 8.10. Scale 1 in.: 12 ft

You can use the scale to find the actual size of the room. Since the length of the picture is 1 inch, the length of the room is 12 feet. The width of the picture is $^3/_4$ inch, so the width of the room is 9 feet, since $^3/_4 \times 12 = 9$. Since the units of measurement are not the same, the ratio of similarity is not 1 : 12 in this case. Actually, the ratio of similarity could be given as

1:(12 × 12). The measures of the room in inches would be 144 times the measures in the picture.

Another example of similar sets of points is all regular polygons with the same number of sides, since all angles are congruent. To make a scale drawing of an object in the shape of a regular hexagon with each side 6 inches, so that each side in the drawing is $^1/_2$ inch, would require a scale of $^1/_2$ inch:6 inches, or a ratio of 1 to 12.

Problems concerning scales often involve simple mathematical sentences.

Example

Suppose that you wanted to use a scale of 1 in:4 in. for another picture of a regular hexagon with each side 6 in. Then you could find the measure of the side by solving the mathematical sentence

$$1/x = 4/6,$$
$$6 = 4x,$$
$$^3/_2 = x,$$

and each side of the drawing would be $^3/_2$ in. long.

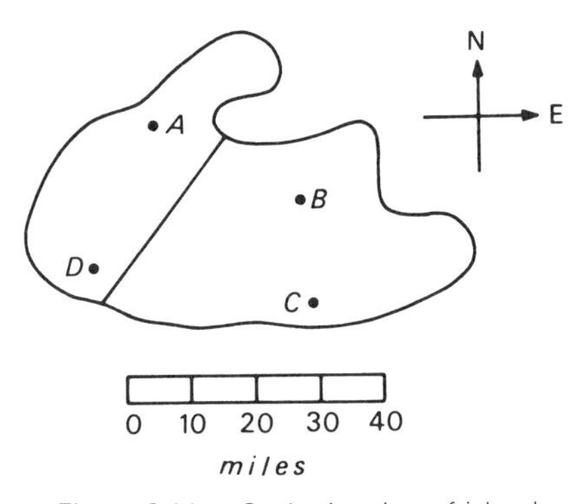

Figure 8.11. Scale drawing of island

Figure 8.11 shows an island with four towns indicated. The scale for this picture is given in a form commonly found on such maps—a ruler with markings to show how many miles the distance on the map represents. Do you agree that the island is about 70 miles long from east to west at the greatest possible length and that it is about 42 miles long from north to south? Do you agree that towns *A* and *D* are about 24 miles apart? You will recognize that these results are approximate and can be obtained by comparing the scale shown to an ordinary ruler and placing the ruler over the correct part of the picture to be measured.

Exercise 8.2

1. Write a scale:
   (a) If 1 in. in the drawing represents 10 ft.
   (b) If 1 in. in the drawing represents 20 yd.
2. Find the measurements of the actual objects, using the scale drawings and the scales shown in Figure 8.12.
   (a) Figure 8.12a (b) Figure 8.12b (c) Figure 8.12c
   (d) Figure 8.12d.

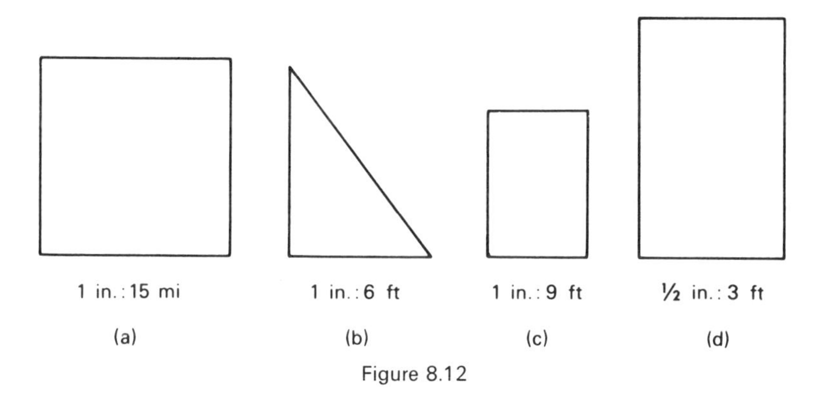

Figure 8.12

3. Find the measurement of a side of a drawing of a regular polygon if the ratio of similarity is 1 to 20 and the actual length of the side is:
   (a) 10 in. (b) 35 in. (c) 16 ft (d) 19 m
4. An airplane model was constructed with a ratio of similarity of 1/72. If the wingspan of the model is 17 in., find the wingspan, in feet, of the actual plane.
5. Use the map and the scale in Figure 8.13 to estimate the actual distances.
   (a) The distance between the mountain peaks at B and C.
   (b) The distance across the lake, from east to west.
   (c) The direct distance from town A to town D.
   (d) The direct distance from town A to mountain peak B.
6. Two maps have scales of 1:100,000 and 1:50,000, where the units are all the same. On which map would the drawing of an actual object appear larger?

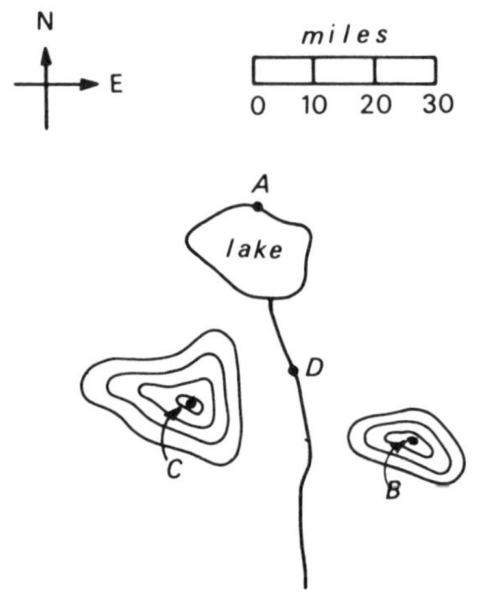

Figure 8.13

7. It is desired to draw a house floor plan on a 9″-by-11″ piece of paper. The actual house is rectangular and measures 24 ft by 60 ft. Which one of these scales would be acceptable, so that the plans would fit onto the page?
   (a) 1 in.:4.ft (b) 1 in.:5 ft (c) 1 in.:6 ft (d) 1 in.:7 ft
8. Prepare a scale drawing of the floor plan of an actual house.

Indirect Measurement

Actually making a physical measurement normally includes using some scale, such as a ruler placed alongside the object to be measured. *Indirect measurement* implies making a measurement some other way, without direct use of a scale. Examples of indirect measurement include finding the distance from the earth to the moon without actually going to the moon, measuring the height of a flagpole without actually climbing to the top, and finding the distance between two points on opposite sides of a lake without actually crossing the lake. In all these cases, the direct application of the scale is impossible or impractical. Similar figures (usually similar triangles) often can be used to help you locate sets of points with the same shape but different sizes. Solving a mathematical sentence will then result in the actual distances.

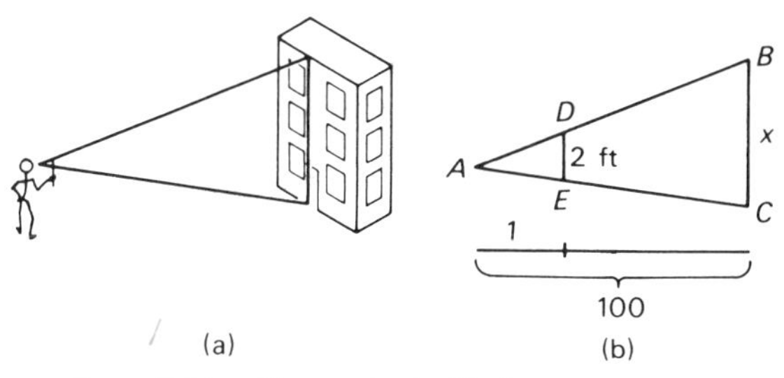

Figure 8.14. First method of indirect measurement

One of the simplest examples of the use of indirect measurement is illustrated in Figure 8.14. The problem is to measure the height of a building indirectly. The results of indirect measurement are approximate, and sometimes the relative error is rather large. On the other hand, the fact that a measurement is indirect does not imply that the error must be large. Figure 8.14b is an abstraction from the real-life event to the mathematical model. The man is 100 feet from the building. He holds a yardstick up in front of him, so that he can sight over one end of it to the building top. He holds the stick one foot from his eye, and finds that the bottom of the building can be sighted in line with the mark on the ruler indicating 2 feet.

Triangles ADE and ABC are similar, with a ratio of similarity of $1/100$, since the altitudes have the same ratios as the sides. Then, $2/x = 1/100$, $x = 200$, and the building is approximately 200 feet high.

The statement that the ratio of corresponding sides of similar triangles is equal to the ratio of the altitudes can be established, using the notation of Figure 8.15. If $\triangle ABC \sim \triangle A'B'C'$, then $\triangle ABD \sim \triangle A'B'D'$; hence $AB/A'B' = BD/B'D'$.

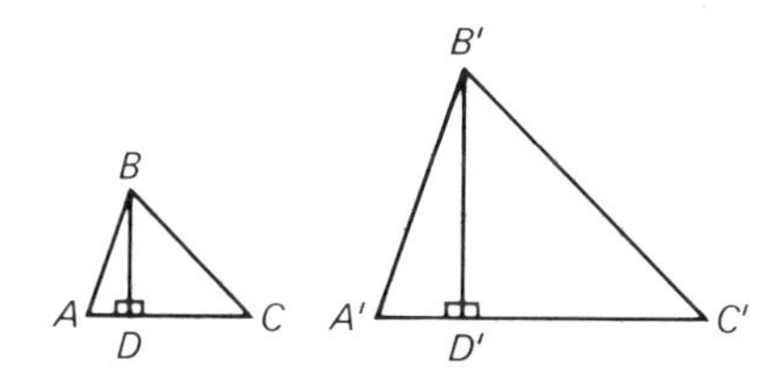

Figure 8.15. Ratios of sides and altitudes

A second method of using similar triangles for indirect measurement is illustrated in Figure 8.16. The problem is to find the distance between points D and E on opposite sides of a small lake. $\overline{DA}$, $\overline{EA}$, $\overline{AB}$, and $\overline{AC}$ are

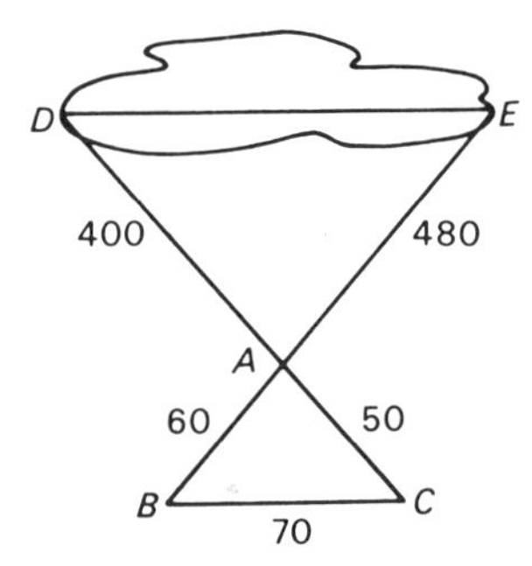

Figure 8.16. Second method of indirect measurement

laid off and measured, so that triangle *ABC* is constructed similar to triangle *AED*, with a ratio of similarity of 1 to 8. Notice that angles *BAC* and *EAD* are vertical angles, so that there is no problem of having to measure angles in this case. Then the measurement of $\overline{BC}$ is found to be approximately 70 feet, which means that the measurement of $\overline{DE}$ is approximately 560 feet.

$$50/400 = 70/x$$

$$50x = 28{,}000$$

$$x = 560$$

A third method of using similar triangles for indirect measurement requires some way of measuring angles. In the first example using Figure 8.14, one of the necessary items of information was how far you were from the object. For the present method you do not need to know how far away the object to be measured is, but you must measure two angles. The method is illustrated in Figure 8.17. The problem is to measure the height

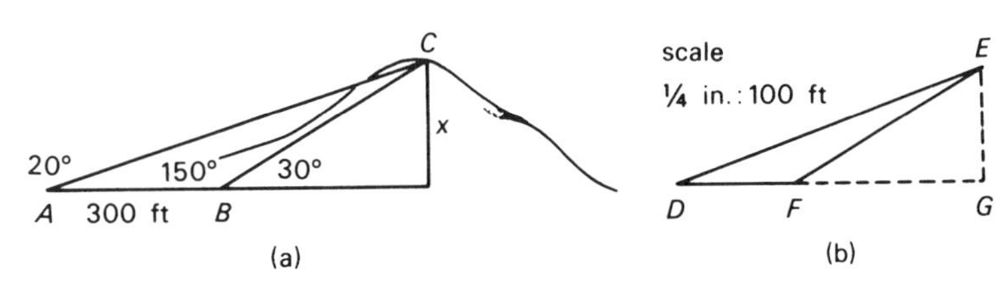

Figure 8.17. Third method of indirect measurement

of a small hill. At point *A*, representing the eye of the observer, you determine that angle *BAC*, called the angle of elevation, has a measurement of 20 degrees. After moving 300 feet closer to the hill, to point *B*, you determine that the angle of elevation of the top of the hill is now 30 degrees. For triangle ABC, you now know the measurement of one side and the two angles with vertices at either end (20 degrees and 150 degrees).

The solution to the problem consists of making a scale drawing of triangle ABC on a piece of paper and interpreting the results. In Figure 8.17b, an arbitrary scale of $\frac{1}{4}$ in. : 100 ft is chosen. In the scale drawing, the length of $\overline{EG}$, corresponding to x, is found to be approximately $\frac{3}{4}$ inch. The top of the hill is approximately 300 feet above eye level.

A simple instrument for finding approximate measures of angles of elevation, such as those needed for this example, is illustrated in Figure 8.18a. It

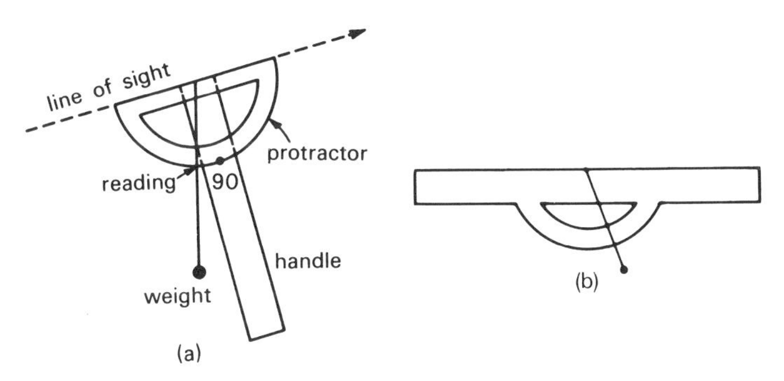

Figure 8.18. Instruments for finding angles of elevation

consists of a protractor taped to a ruler or other handle as illustrated. A small weight (such as a button) is attached with a thread at the zero point on the base of the protractor. Then, when you sight along the base of the protractor, the deviation of the string from the vertical reading of 90 degrees will indicate the measurement of the angle of elevation. For example, if the string crosses the protractor at the 70-degree mark, the angle of elevation will be 20 degrees, since $90 - 70 = 20$.

Exercise 8.3

1. Give other practical examples of the need for specific indirect measurements.
2. Use Figure 8.19 to find the height of the object represented by segment x
   (a) in Figure 8.19a. (b) in Figure 8.19b.

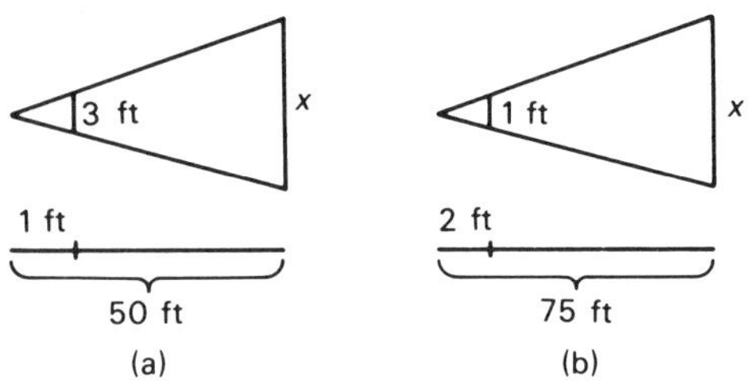

Figure 8.19

3. Use Figure 8.20 to find approximately the distance between the two points indicated.
   (a) in Figure 8.20a. (b) in Figure 8.20b.

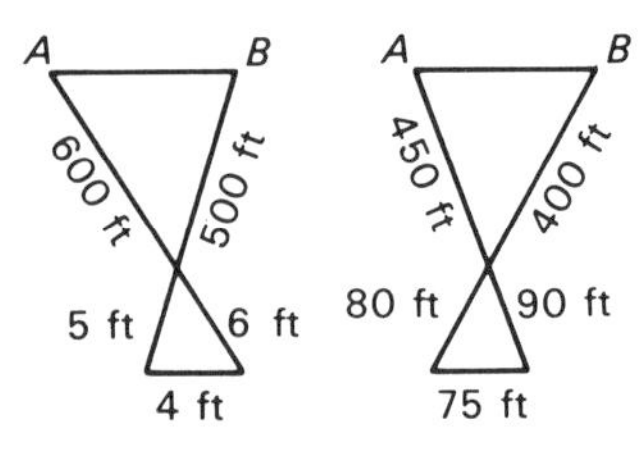

Figure 8.20

4. Use the method of Figure 8.17 (p. 178), to make a scale drawing, and estimate the height, using the length and the two angles of elevation given:
   (a) 25°, 400 ft, 40° (b) 30°, 600 ft, 50°

Exercises 5–11 suggest mathematics laboratory activities or field work utilizing the techniques of indirect measurement discussed in this section.

5. Use the procedure of Figure 8.14 (p. 177) to measure the height of a wall of a room.
6. Use the procedure of Figure 8.14 to measure the height of a building.
7. Use the procedure of Figure 8.14 to measure the height of a flagpole.
8. Use the procedure of Figure 8.16 (p. 178) to measure the length of a building.
9. Use the procedure of Figure 8.16 to measure the length of an automobile.
10. Use the procedure of Figure 8.17 (p. 178) to measure the height of a building.
11. Use the procedure of Figure 8.17 to measure the height of a hill or mountain.

Constructions and Additional Applications of Similar Figures

The study of scale drawings led to the observation that any linear distances corresponding in the drawing and the original object had measures in the constant ratio of similarity. For example, the corresponding altitudes, medians, and angle bisectors of two similar triangles have measures with the same ratio as the measures of the corresponding sides. This idea may be extended to the perimeters of two similar triangles (or other polygons). The measure of perimeter of two similar polygons has the same ratio as the measures of corresponding sides.

Do you think it is also true that the measures of area of two similar triangles have the same ratio as the measures of two corresponding sides? All three of the triangles in Figure 8.21 are similar. The ratio of similarity

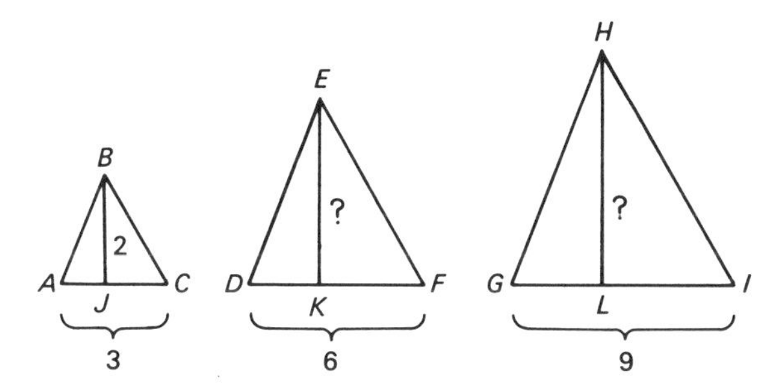

Figure 8.21. Measure of area of similar triangles

of $\triangle ABC$ to $\triangle DEF$ is 1 to 2, and the ratio of similarity of $\triangle ABC$ to $\triangle GHI$ is 1 to 3. Since the ratio of measures of the altitudes is the same as for the sides, the measure of the altitude of $\triangle DEF$ is 4. The measure of area of $\triangle ABC$ is 3; however, the measure of area of $\triangle DEF$ is 12, which tells you that the answer to the question at the beginning of this paragraph is "no."

Is there any relationship between the measures of area of similar triangles? For $\triangle ABC$ and $\triangle DEF$, the ratio for the sides was 1/2, and the ratio for area was 1/4. For $\triangle ABC$ and $\triangle GHI$, you find that the area of $\triangle GHI$ is 27, so that the ratio of sides is 1/3, whereas the ratio of areas is 1/9. Is a pattern becoming evident? It is not the ratios of the measures of the sides, but the ratios of the squares of the measures of the sides, that should be compared.

For example, the ratio of the squares of the measures of $\overline{AC}$ and $\overline{DF}$ is $9/36 = 1/4$, which is the same as the ratio of measure of area for these two triangles. The ratio of the squares of the measures of $\overline{AC}$ and $\overline{GI}$ is $9/81 = 1/9$, which is the same as the ratio of measure of area for the two triangles. Examine triangles *DEF* and *GHI* to see if they follow this generalization: the measures of area for two similar triangles have the same ratio as the ratios of the squares of the measures of two corresponding sides. Another way to state the same generalization is to say that the ratio of measures of area for two similar triangles is the square of the ratio of measures of corresponding sides.

Example

If the ratio of corresponding sides of similar triangles is $^3/_5$, what is the ratio of areas?

$$\left(\frac{3}{5}\right)^2 = \frac{9}{25}$$

Example

If the ratio of areas of similar triangles is 25/169, what is the ratio of corresponding sides?

$$\sqrt{\frac{25}{169}} = \frac{\sqrt{25}}{\sqrt{169}}$$
$$= \frac{5}{13}$$

Constructions with similar figures often depend on one basic construction not explained previously—that of dividing a segment in a given ratio. This construction is explained in Figure 8.22. Segment *AB* is to be divided by a point *E* into the given ratio *m*/*n*.

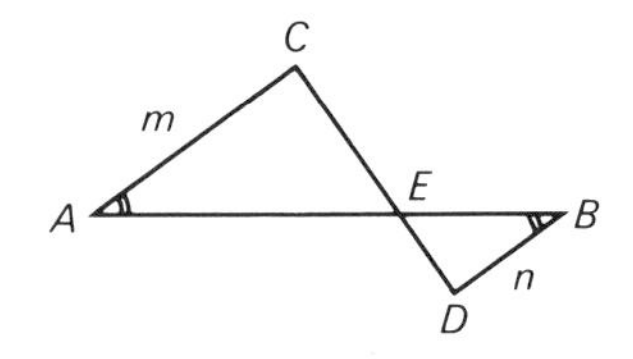

Figure 8.22. Division of a segment

Construct any line $\overleftrightarrow{AC}$ through A, with $\overleftrightarrow{BD}$ a parallel line through B. Lay off m units on $\overleftrightarrow{AC}$ and n units on $\overleftrightarrow{BD}$. Then $\overleftrightarrow{CD}$ intersects $\overline{AB}$ at the required point.

The proof depends on the fact that $\triangle ACE \sim \triangle BDE$ (why?); hence $AE/EB = AC/BD = m/n$.

Example

The construction is shown in Figure 8.23 for dividing a segment in a ratio of $^3/_5$.

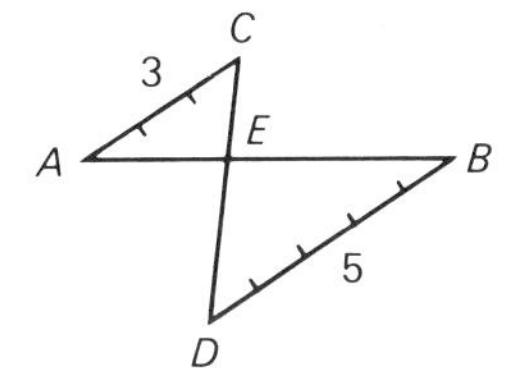

Figure 8.23. Division of segment in ratio 3/5

A modification of the previous construction makes it possible to divide a segment into any number of congruent segments. An example is shown in Figure 8.24.

Example

Construct points such that a given segment is divided into six congruent segments. Let $\overleftrightarrow{AD}$ be any line through A.

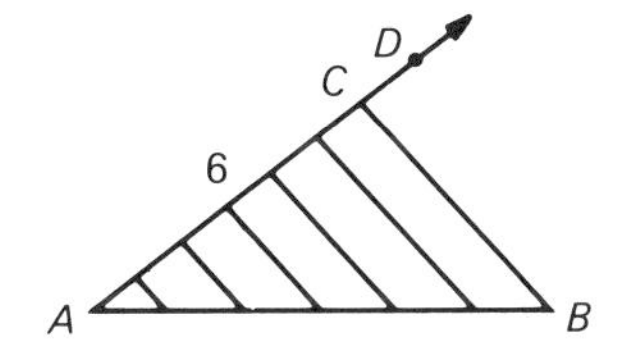

Figure 8.24. Division of segments into congruent segments

On $\overleftrightarrow{AD}$ construct six congruent segments, using any arbitrary unit. If C is the last endpoint of the segments, then lines drawn parallel to $\overline{BC}$ through the remaining five points determine the desired segments on $\overline{AB}$.

Note that this construction relates to the basic concepts of measurement introduced in Chapter 3.

The theorem stated in connection with Figure 8.25 is the basis for the construction of a segment whose length is the square root of the length of a given segment. The analysis figure is shown in Figure 8.25. If $\overline{AB}$ is the

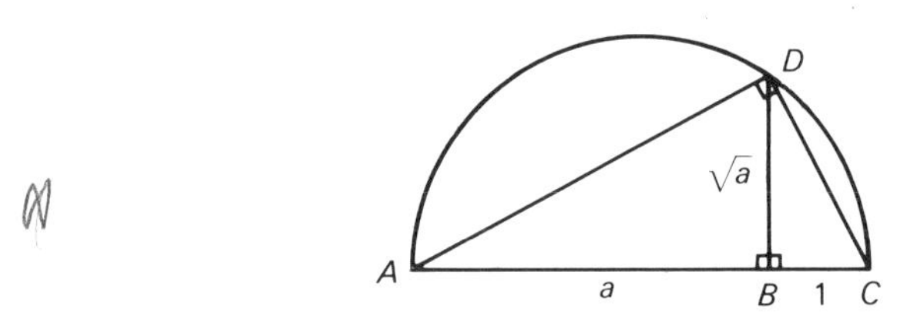

Figure 8.25. Construction of square root

given segment, extend it so that $\overline{BC}$ is the arbitrary unit length 1. Then construct a circle with $\overline{AC}$ as diameter. At B, construct a perpendicular intersecting the circle at D. Segment $\overline{BD}$ is the required segment. The proof follows from the use of similar triangles. $\triangle ADB \sim \triangle DCB$, so that $a/BD = BD/1$. Then $a = (BD)^2$ and $\sqrt{a} = BD$.

Some applications of geometry depend on the physical property illustrated in Figure 8.26. If $\overline{DE}$ represents a mirror, with A a source of light, then

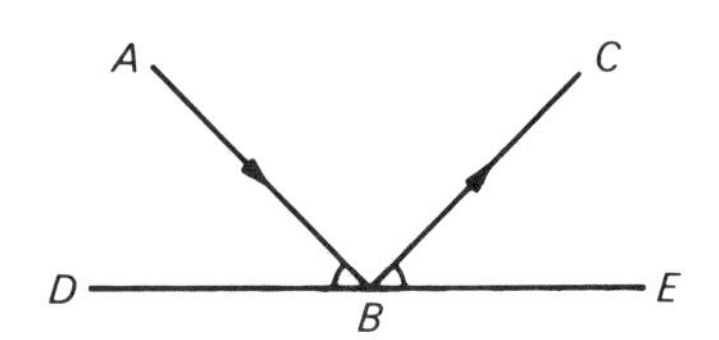

Figure 8.26. Reflection from a surface

$\angle DBA \cong \angle EBC$. This same congruence holds for a physical object bouncing off a horizontal wall.

Suppose that P and Q in Figure 8.27 represent balls on a pool table, but there are other balls between them. Where should ball P hit the edge $\overline{AB}$ of the table in order to rebound and strike ball Q? The interesting answer is that the ball should be aimed at the point Q', the image of Q in a reflection about the side of the table. This is true since the angles indicated at R are congruent, and a ball bouncing against the wall behaves just like a light ray striking a mirror. That is, the angle that $\overline{PR}$ makes with the side is the

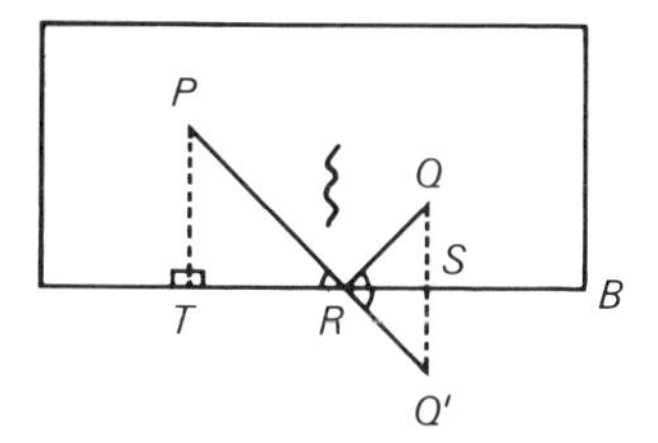

Figure 8.27. Caroms

same $\overline{RQ}$ makes with the side. Note that $\triangle PRT \sim \triangle Q'RS$ and $\triangle Q'RS \cong \triangle QRS$. The word *caroms* is often used in discussing this specific application of geometry. Figure 8.28 shows how the ball at P can be bounced off two sides to hit a ball at Q. Q is first reflected to Q', then to Q'', and the ball is aimed at Q''.

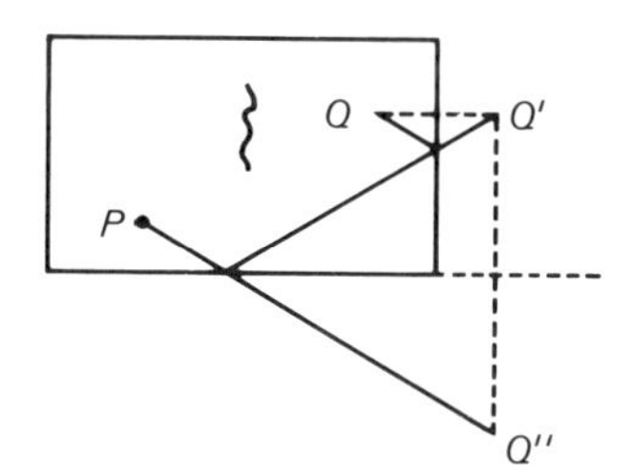

Figure 8.28. Carom for two sides

Exercise 8.4

1. Explain how to construct a triangle similar to a given triangle with one side with a measure of 5, a second side with a measure of 8, and the included angle with a measure of 50°, if the ratio of similarity of the given to the constructed triangle is 1 to 2.

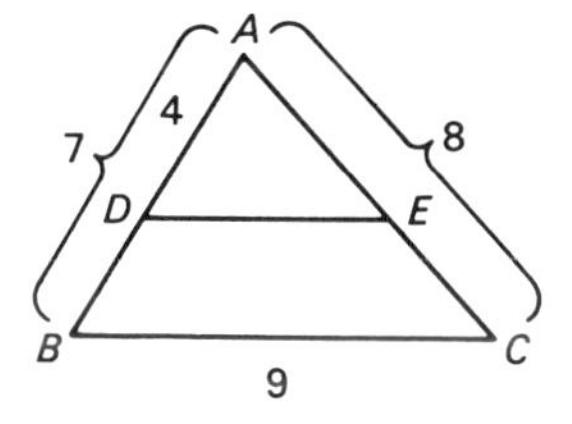

Figure 8.29

2. In Figure 8.29, if the measure of the median from vertex D of $\triangle DAE$ is 3, what is the measure of the median from vertex B of triangle BAC?
3. In Figure 8.29, what is the ratio of the measures of area of triangles DAE and BAC?
4. If the ratio of the measures of area of two similar triangles is 4 to 9, what is the measure of the side of the second triangle corresponding to a side of the first that has a measure of 100?
5. If the ratio of areas of similar triangles is 16/196, what is the ratio of corresponding sides?

For Exercises 6–9, perform the construction so that the given segment $\overline{AB}$ is divided in a ratio of:

6. 1 to 2
7. 2 to 3
8. 5 to 7
9. 7 to 5

For Exercises 10–12, perform the construction of dividing a given segment into n congruent segments, if n is equal to:

10. 3
11. 4
12. 5
13. Construct a segment whose length is the square root of the length of a given segment.

For Exercises 14–17, copy Figure 8.30, representing pool balls on a table.

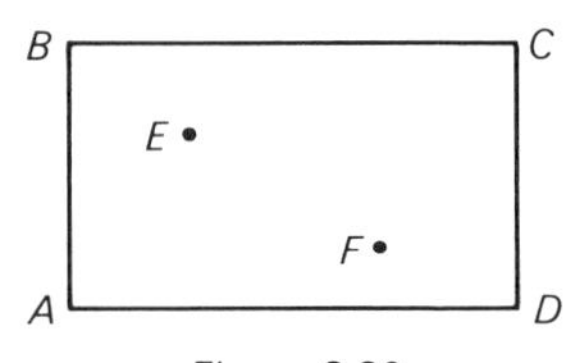

Figure 8.30

14. Construct the path if ball E is to carom off side $\overline{AD}$ and strike ball F.
15. Construct the path if ball E is to carom off side $\overline{BC}$ and strike ball F.
16. Construct the path if ball E is to carom off $\overline{AB}$, then $\overline{AD}$, to hit ball F.
17. Construct a path so that ball E caroms off three sides before it hits ball F.

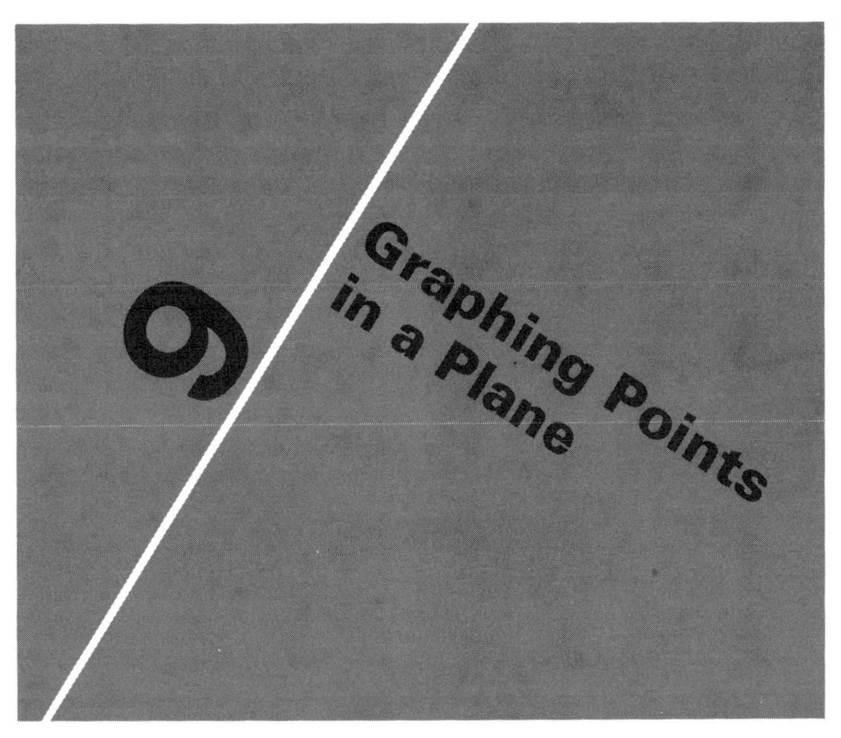

In this chapter we will introduce the idea of using an ordered pair of numbers to designate a specific point in a plane. This powerful tool makes it possible to describe sets of points in mathematical sentences, to introduce a mathematical explanation of *slope,* to discuss the meaning of trigonometric ratios, to apply trigonometric ratios to problems in indirect measurement, and to introduce vectors.

Coordinates of Points

A point in a plane may be designated by using an *ordered pair* of numbers. The method for accomplishing this is illustrated in Figure 9.1 and is probably familiar to many readers. In the plane, two perpendicular number lines, one horizontal and the other vertical, are indicated. The lines are positioned so that the point representing zero on each line is the point of intersection, called the *origin.* The horizontal line has numerals indicating numbers that increase as you read to the right, whereas the numerals on the vertical line indicate numbers that increase from bottom to top.

Axiom 23

There is a unique coordinate system on a line that assigns two distinct points two given real numbers.

This axiom implies that numbers may be assigned arbitrarily to any two points on a line but that all numbers for other points will then be determined. Numbers indicated to the left of zero on the horizontal line are negative, as are those below zero on the vertical line.

Each point in the plane may be located in reference to the two number lines (called the horizontal and vertical axes) by using an ordered pair of real numbers, called *coordinates* for the point. For example, in Figure 9.1,

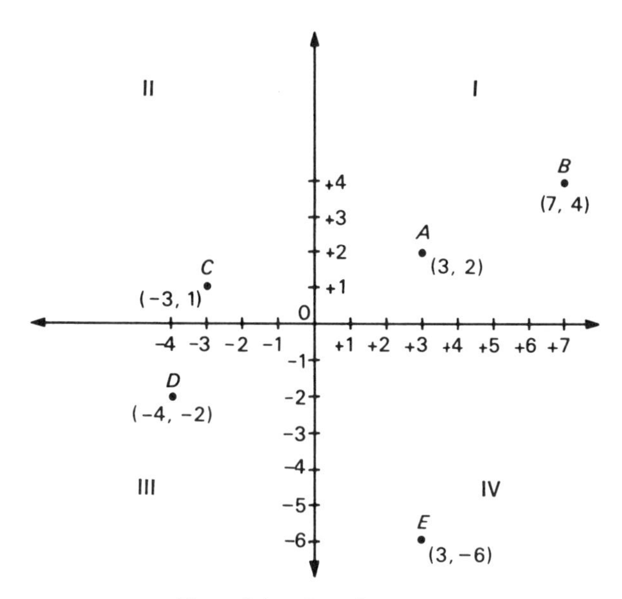

Figure 9.1. Coordinate axes

the coordinates of point A are 3 and 2, written (3, 2) or (+3, +2). The 3 indicates that the point is three units to the right, measured along the horizontal axis from the vertical axis, and the 2 indicates that the point is two units above the horizontal axis, measured along the vertical axis.

Point C has coordinates $(-3, 1)$. The negative 3 indicates that the point is three units to the left of the vertical axis, and the 1 indicates that it is one unit above the horizontal axis. The meaning of the coordinates of points B, D, and E may be stated in a similar way. The coordinate axes partition the plane into four *quadrants* numbered as illustrated in Figure 9.1.

Corresponding to any point in the plane is a unique ordered pair of coordinates that refer to a fixed pair of axes; corresponding to each ordered pair of real numbers is exactly one point in the plane. The subject of *analytic geometry* explores the details of rclating scts of points and sets of number pairs through the use of algebraic equations.

Two of the mathematicians given much of the credit for the early development of analytic geometry were René Descartes (1596–1650) and Pierre de Fermat (1601?–1665). The significance of analytic geometry is that it provides a second distinct way of studying subsets of points in a plane; for example, the condition that the points of a line or a circle satisfy may be designated algebraically. Some elements of analytic geometry are discussed in this chapter. This section concludes with the discussion of three basic topics—midpoint of a segment, distance between two points, and slope.

If C is the midpoint of $\overline{AB}$ in Figure 9.2, then its x-coordinate is halfway

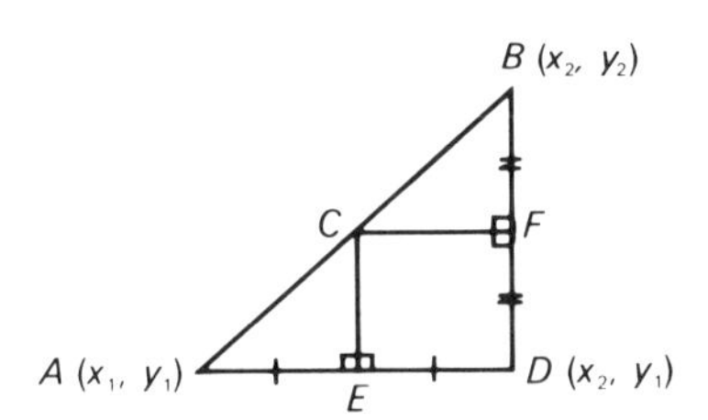

Figure 9.2. Midpoint of a segment

between the x-coordinates of A and B and its y-coordinate is halfway between the y-coordinates of A and B.

One way to show this is to use similar triangles. In Figure 9.2, $\triangle ACE \sim \triangle ADB$ with a ratio of similarity of 1 to 2, so that the ratio of AE to AD is also 1 to 2 and E is the midpoint of the horizontal segment $\overline{AD}$. Also, $\triangle BFC \sim \triangle BDA$ with a ratio of 1 to 2, so that F can be shown to be the midpoint of vertical segment $\overline{BD}$. This leads to a general expression for the coordinates

$$\left(\frac{x_1 + x_2}{2}, \frac{y_1 + y_2}{2}\right)$$

of the midpoint of a segment with endpoints (x_1, y_1) and (x_2, y_2).

Example

The midpoint of the segment with endpoints (3, 4) and (−1, 7) has coordinates $(1, {}^{11}/_{2})$.

The Pythagorean Theorem can be used to find a formula for the distance between two points.

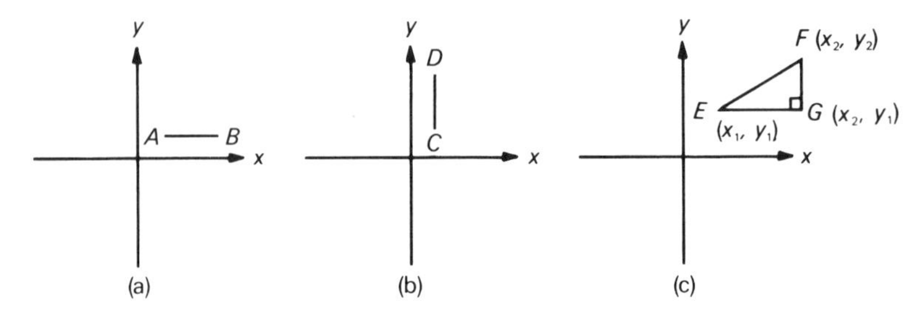

Figure 9.3. Distance between two points

The distance between two points on a horizontal line, as in Figure 9.3a, is the positive difference of their x-coordinates. For example, the distance between (4, 2) and (7, 2) is 3 units. The *absolute value* of a real number x is defined as

$$|x| = x \text{ for } x > 0$$

$$|x| = 0 \text{ for } x = 0$$

$$|x| = -x \text{ for } x < 0.$$

In other words, the absolute value of a number is always positive or zero, and shows the number of units from zero on a number line. The symbol for absolute value may be used to define the distance d between two points (x_1, y_1) and (x_2, y_1) on the same horizontal line.

$$d = |x_2 - x_1|$$

Similarly, the distance between two points (x_1, y_1) and (x_1, y_2) on the same vertical line, as in Figure 9.3b, is

$$d = |y_2 - y_1|.$$

Now in Figure 9.3c, the distance from E to F, EF, can be found by the Pythagorean Theorem.

$$\begin{aligned} EF &= \sqrt{(EG)^2 + (FG)^2} \\ &= \sqrt{(x_2 - x_1)^2 + (y_2 - y_1)^2} \end{aligned}$$

Examples

The distance d between (3, 4) and (3, 7) is

$$\begin{aligned} d &= |4 - 7| \\ &= |-3| \\ &= 3. \end{aligned}$$

The distance d between (5, 2) and (8, 7) is

$$\begin{aligned} d &= \sqrt{(8 - 5)^2 + (7 - 2)^2} \\ &= \sqrt{3^2 + 5^2} \\ &= \sqrt{34}. \end{aligned}$$

The distance d between $(-2, 1)$ and $(5, -2)$ is

$$\begin{aligned} d &= \sqrt{(5 + 2)^2 + (-2 - 1)^2} \\ &= \sqrt{7^2 + (-3)^2} \\ &= \sqrt{49 + 9} \\ &= \sqrt{58}. \end{aligned}$$

In everyday language, you say that the slope of the house roof is greater than the slope of the garage roof if the house roof seems steeper. For example, in Figure 9.4, the slope of the segment connecting points A and B seems greater than the slope of the segment connecting points A and C. For $\overline{AC}$, the line goes up two units as it goes to the right one unit. In other words, it goes vertically twice as fast as it goes horizontally. The

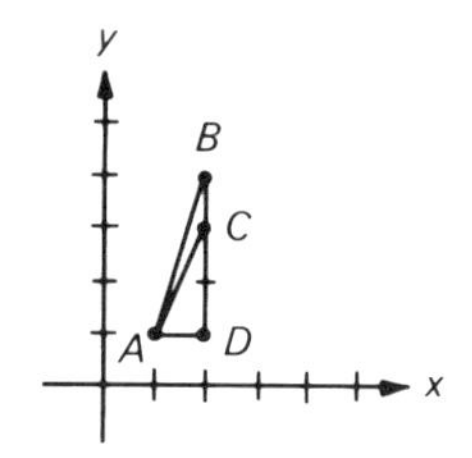

Figure 9.4. Slope of segments

ratio of the measures of the vertical change to the horizontal change is 2 to 1, so that

$$\frac{DC}{AD} = \frac{2}{1} = 2.$$

For $\overline{AB}$, the line goes up three units as it goes to the right one unit. It goes vertically three times as fast as it goes horizontally. The ratio of the measures of the vertical change to the horizontal change is 3 to 1, so that

$$\frac{DB}{AD} = \frac{3}{1} = 3.$$

These intuitive ideas about slope make it possible to give a more precise mathematical meaning to the term. For Figure 9.4, the slope of segment AC, or of line AC, is said to be 2. The slope of segment AB, or of line AB, is said to be 3. The definition of slope as the ratio of vertical change to horizontal change assigns an exact numerical value to the slope and makes comparison of slopes possible.

Segments that go upward toward the right have a positive slope, whereas segments that go downward to the right have a negative slope. The negative number comes from the fact that the vertical change must be given as a negative number.

Examples

See Figure 9.5. The slope of segment AB is $-{}^{1}/_{2}$. The slope of segment CD is 1. The slope of segment EF is $-{}^{1}/_{4}$.

By taking two general points with coordinates (x_1, y_1) and (x_2, y_2), a general formula for the slope m of a segment can be derived. See Figure 9.6.

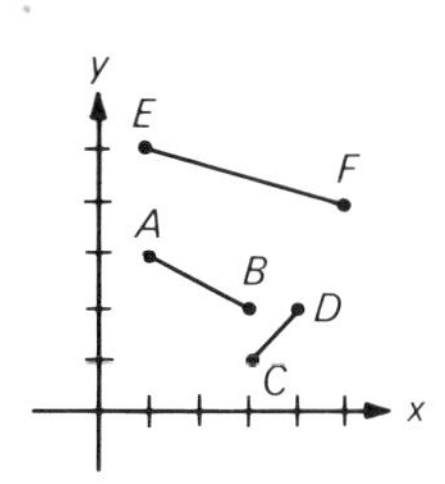

Figure 9.5. Positive and negative slopes

$$m = \frac{y_2 - y_1}{x_2 - x_1}$$

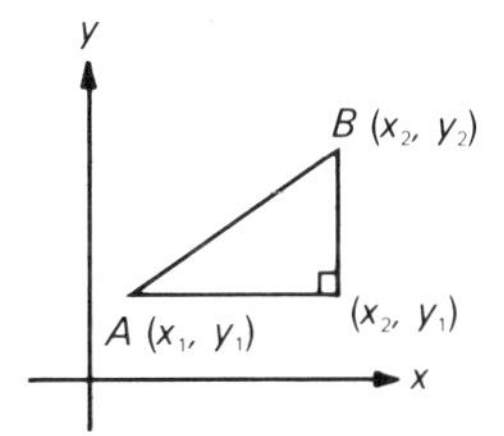

Figure 9.6. Slope of line through two points

Example

Find the slope of the line through points (4, 7) and $(-2, -3)$.

Let the first point be (x_1, y_1).

$$m = \frac{-3 - 7}{-2 - 4}$$

$$= \frac{-10}{-6}$$

$$= \frac{5}{3}$$

Exercise 9.1

1. Draw a set of axes and locate the points:
   (a) (4, 4) (b) $(-2, 4)$ (c) $(-1, -3)$ (d) $(5, -2)$

2. Draw a set of axes and locate the points:
   (a) (1, 1) (b) (2, 2) (c) (3, 3) (d) (4, 4)
3. In which quadrant is each point located?
   (a) (−2, −1) (b) (−1, 3) (c) (2, 1) (d) (1, −2)

For Exercises 4–7, find the midpoint of a segment with the endpoints given.

4. (2, 3), (5, 7)
5. (3, −1), (4, 5)
6. (6, −2), (5, −3)
7. (7, −3), (−2, −1)

For Exercises 8–11, find the length of the segment in:

8. Exercise 4
9. Exercise 5
10. Exercise 6
11. Exercise 7

For Exercises 12–15, find the slope of the segment in:

12. Exercise 4
13. Exercise 5
14. Exercise 6
15. Exercise 7

Analytic Geometry

For each point in Figure 9.7, the second coordinate is twice the first. Notice also that all the points appear to be on the same straight line. If x represents the first coordinate and y the second, then y is twice x, and a general statement of the relationship for each point pictured can be written as $y = 2x$. The set of ordered pairs in Figure 9.7 is an example of a function. For each first element, there is a unique second element. The equation $y = 2x$ represents this same function.

In Figure 9.8, the horizontal axis is labeled the x-axis and the vertical axis is labeled the y-axis. The first coordinate for a point can be thought of as the x-coordinate (the distance measured horizontally from the vertical

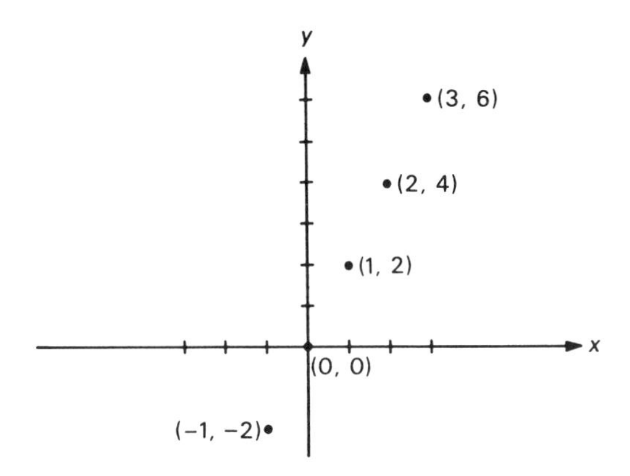

Figure 9.7. Second coordinate twice the first

axis). The second coordinate is the y-coordinate (the distance measured vertically from the horizontal axis). The line in Figure 9.8 represents all the points with coordinates satisfying the general expression $y = 2x$, where x and y represent real numbers. It is customary to say that the line *is* the line $y = 2x$ and that the graph of the equation $y = 2x$ is a line. Each of the points in Figure 9.8 lies on this line, but you can designate many other points lying on the same line simply by writing number pairs such as (5, 10) and ($^1/_2$, 1) in which the second coordinate is twice the first. For example, to emphasize that $y = 2x$ is actually a set of points, it is helpful to use the *set-builder notation*. The expression $\{(x, y)|y = 2x\}$ is read "The set of all points with coordinates x and y such that $y = 2x$."

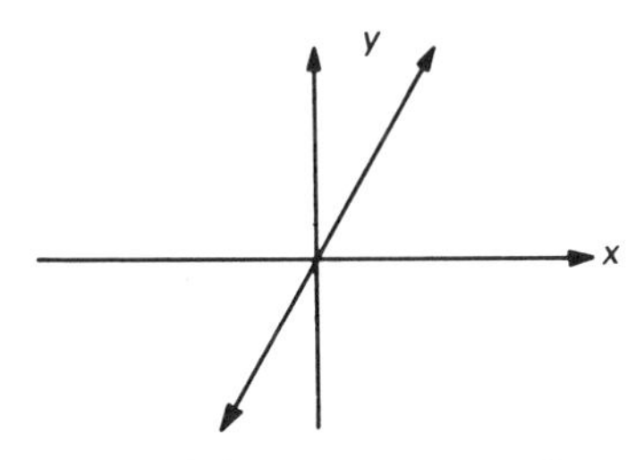

Figure 9.8. Graph of $y = 2x$

Figure 9.9 shows other lines and their equations. In each case the line passes through the origin; hence the equation is of the form $y = mx$, where m represents any real number. The equation for the y-axis is an exception, having the equation $x = 0$.

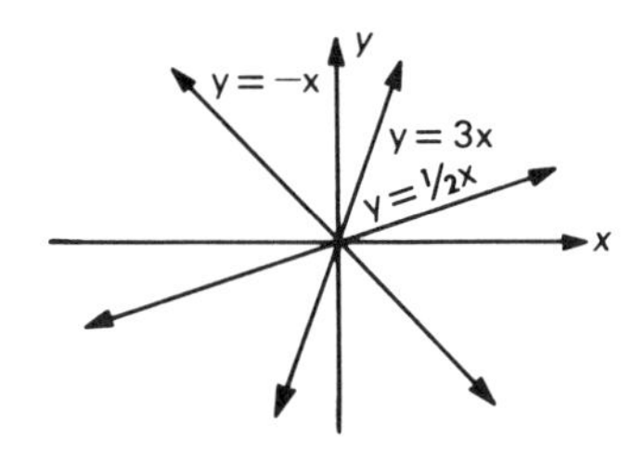

Figure 9.9. Lines and their equations

In Figure 9.10, it seems evident that the slope of $y = 3x$ is greater than the slope of $y = 2x$, which in turn is greater than the slope of $y = x$. Point A, on $y = x$, has coordinates (1, 1). The ratio of the measure of $\overline{AB}$ to the measure of $\overline{BC}$ is $^1/_1$ or 1.

Now consider the line $y = 2x$ in Figure 9.10. Point D has coordinates (1, 2), which means that the line rises twice as fast as it goes to the right. The

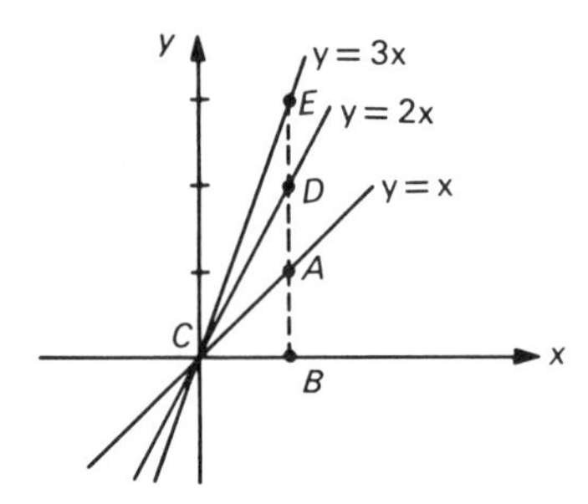

Figure 9.10. Slopes of lines

slope is $^2/_1$ or 2. It is interesting to discover that the slope can be found immediately for a line that passes through the origin, since it is the number that m represents in the equation of the form $y = mx$.

Examples

The slope of $y = 2x$ is 2, the slope of $y = 3x$ is 3, and the slope of $y = -5/2x$ is $-5/2$. This last line would slope downward and to the right.

Figure 9.11 shows $y = x$ and another line parallel to it. Two parallel lines have the same slope. For a given value of x, the y-coordinate of a point

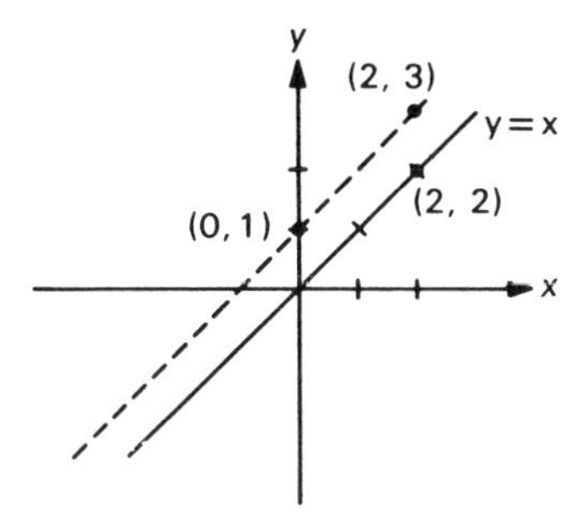

Figure 9.11. Parallel lines

on the second line is one more than the corresponding y-coordinate for $y = x$. For example, $y = x$ passes through (2, 2), but the other line passes through (2, 3). Then $y = x + 1$ or $x - y + 1 = 0$ is the equation of the second line. This is an example of the equation of a line not passing through the origin.

The equation for a straight line represents a function, except for vertical lines. Various forms of the equation are used, depending on the given information and the purpose for which the equation is to be used.

In Figure 9.12, suppose point $B(x_1, y_1)$ is a fixed point on a line with slope m, and let A be any other point on the line.

$$\frac{y - y_1}{x - x_1} = m,$$

or

$$y - y_1 = m(x - x_1)$$

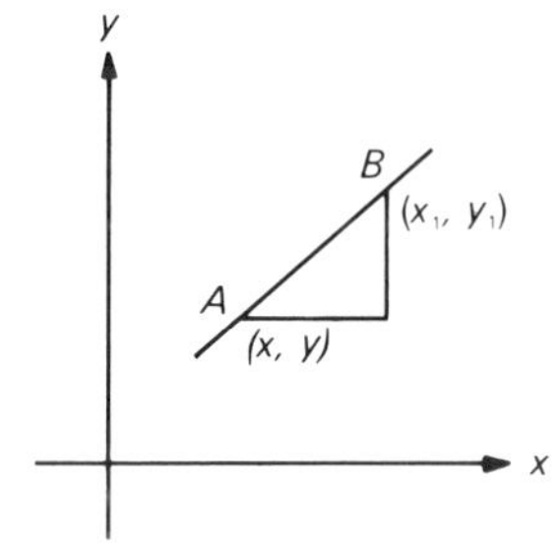

Figure 9.12. Fixed point and variable point on line

This last equation is the *point-slope* form of the equation of a straight line.

Example

Write the equation of a line with slope 3 and passing through the point $(7, -2)$.

$$y + 2 = 3(x - 7)$$

If the slope is known, and if the given point is the point (O, b) where the line intersects the y-axis, then $y - y_1 = m(x - x_1)$ can be written $y - b = mx$, or $y = mx + b$. This last equation is called the *slope-intercept* form of the equation of a straight line.

Example

Write the equation of a line with slope 2 and y-intercept $(0, -3)$.

$$y = 2x - 3$$

The *general form* of the equation for a straight line, where the line does not necessarily pass through the origin, is

$$ax + by + c = 0,$$

where a, b, and c represent real numbers, with a and b not both zero. Figure 9.13 shows two such lines and their equations.

The pictures of the lines actually show that an infinite number of points can be indicated by number pairs that satisfy any possible relationship between x and y. Also, the set-builder notation $\{(x, y)|ax + by + c = 0\}$ is used to help emphasize that a line is a set of points whose coordinates satisfy some particular linear relationship. A point is on the graph of an

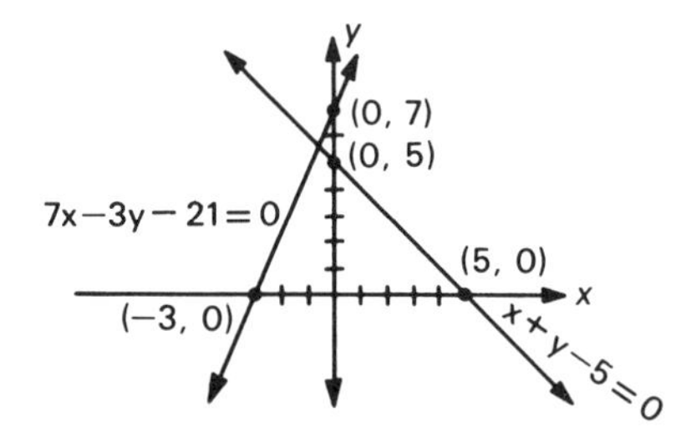

Figure 9.13. Intersecting lines

equation if and only if its coordinates satisfy the equation. Two illustrations help to clarify the idea of checking to see if a point is on a line.

1. Is the point (3, 4) on the line $7x - 3y + 21 = 0$? Substitute 3 for x and 4 for y in the mathematical sentence; then $7(3) - 3(4) + 21 \neq 0$, and the point is not on the line.
2. Is the point (3, 2) on the line $x + y - 5 = 0$? Substitute 3 for x and 2 for y in the mathematical sentence; then $3 + 2 - 5 = 0$, and the point is on the line.

The intersection of two lines can be found algebraically by considering their equations. To find the common solution of

$$7x - 3y + 21 = 0,$$
$$x + y - 5 = 0,$$

(to solve simultaneously) first multiply both sides of the second equation by 7 and subtract.

$$\begin{aligned} 7x - 3y + 21 &= 0 \\ \underline{7x + 7y - 35} &= \underline{0} \\ -10y + 56 &= 0 \\ 10y &= 56 \\ y &= \frac{28}{5} \end{aligned}$$

Substituting $y = 28/5$ in $x + y - 5 = 0$ shows that $x = -3/5$. Then the coordinates of the point of intersection of the two lines are $(-3/5, 28/5)$. The use of set notation may help to clarify the meaning of the common solution.

$$\{(x, y)|7x - 3y + 21 = 0\} \cap \{(x, y)|x + y - 5 = 0\} = \left\{\left(-\frac{3}{5}, \frac{28}{5}\right)\right\}$$

Analytic geometry also makes it possible to write mathematical sentences for sets of points that are not straight lines. For example, Figure 9.14 shows a circle with center at the origin and a radius of three units. For any point (x, y), such as the one shown, the Pythagorean Theorem can be used to find that $x^2 + y^2 = 9$. Since this equation describes the relationship between the x- and y-coordinates for each point on the circle, it is seen to be the actual equation of the circle.

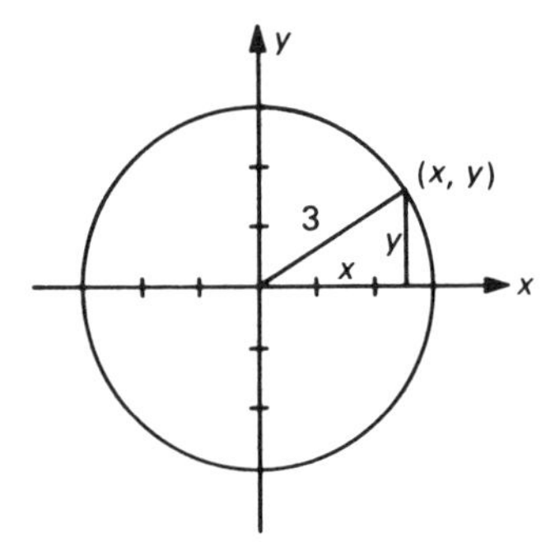

Figure 9.14. Circle of radius 3

Note that a circle is not the graph of a function. The reason for this is that two points of the graph may lie on the same vertical line, which contradicts the restriction that each first element results in a unique second element. Any set of ordered pairs, whether a function or not, is called a *relation*. Figure 9.14 is the graph of an important relation that is not a function.

Example

Use set notation to describe all points in the plane that are six units from the origin.

$$\{(x, y) | x^2 + y^2 = 36\}$$

Mathematical sentences with symbols for inequality, as well as equality, may represent sets of points you have studied. For example, $\{(x, y) | x > 2\}$ represents the half-plane shaded in Figure 9.15. The line is not a part of the graph.

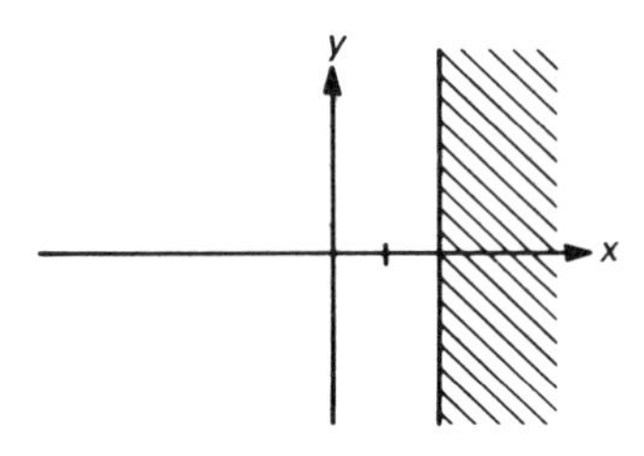

Figure 9.15. Graph of half-plane

Figure 9.15 is the graph of a relation, not a function.

Example

Use set notation to describe all the points in the plane *at least* five units from the origin.

$$\{(x, y) | x^2 + y^2 \geq 25\}$$

Practical estimates of the area of plane regions can be found by using a computer and probability. A simplified example illustrates what is known as the Monte Carlo Method.

Suppose that it is desired to find the measure of area of the shaded region under the curve in Figure 9.16. If the entire region is considered one unit, then the measure of area of the shaded region will be a fraction less than 1.

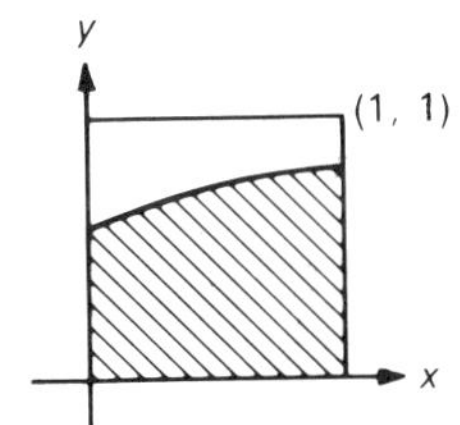

Figure 9.16. Monte Carlo Method

A computer can provide random number pairs, each representing the coordinates of a point in the square region. Furthermore, the computer can check to see whether each point is above or below the given curve. For example, (.216, .913) may be above the curve, and (.317, .248) may be below the curve.

Suppose that 7215 of the first 10,000 number pairs correspond to points in the shaded region. Then an estimate of the measure of area is 7215/10,000. If 13,816 of the first 20,000 number pairs corresponded to points in the shaded region, then 13,816/20,000 would be a better estimate of the measure of area.

Exercise 9.2

In Exercises 1–3, use set-builder notation to describe the set of points such that:

1. Each second coordinate is $^1/_3$ of the first coordinate.

2. Each second coordinate is 4 times the first.
3. Each second coordinate is 2 less than 3 times the first.
4. Which of these points are on the line $y = 2x$?
   (a) (3, 5) (b) (6, 12) (c) (12, 6) (d) (−3, −6)
5. Which of these points are on the line $x + y - 7 = 0$?
   (a) (3, 4) (b) (9, −2) (c) (0, 4) (d) (−3, −4)
6. Which of these points are on the line $2x + 3y - 1 = 0$?
   (a) (3, 5) (b) (4, −1) (c) (5, 1) (d) (1, 1)

For each line in Exercises 7–8, (a) name the slope and (b) sketch the graph.

7. $y = {}^2/_3 x$
8. $y = -4x$

For the point and slope given in Exercises 9–10, (a) write the equation of the line in point-slope form and (b) sketch the graph.

9. $(2, 3),\ m = \dfrac{1}{2}$

10. $(4, -1),\ m = \dfrac{3}{4}$

For the slope and intercept given in Exercises 11–12, (a) write the equation of the line in slope-intercept form and (b) sketch the graph.

11. $m = 3$, (0, 4)
12. $m = -{}^1/_2$, (0, −1)
13. Find the coordinates of the point of intersection of $x + y - 5 = 0$ and $x = y$.
14. Find the coordinates of the point of intersection of $x + y - 5 = 0$ and $x = 2y$.
15. Which of these points are on the circle $x^2 + y^2 = 9$?
   (a) (1, 8) (b) (3, 3) (c) (0, 3) (d) (−3, 0)
16. Sketch the graph of $x^2 + y^2 = 49$.
17. Sketch the half-plane represented by the mathematical sentence $x > 4$.
18. Sketch the half-plane represented by the mathematical sentence $y > 3$.

Trigonometric Ratios

The branch of mathematics called *trigonometry,* which literally means "the measurement of triangles," depends on an understanding of coordinates of points for its modern presentation. The Greek astronomer Hipparchus, who lived about 140 B.C., is often called the father of trigonometry. Johann Müller (1436–1476) helped to develop the subject in a more modern form. In Figure 9.17, the circle shown is a *unit circle,* with a radius

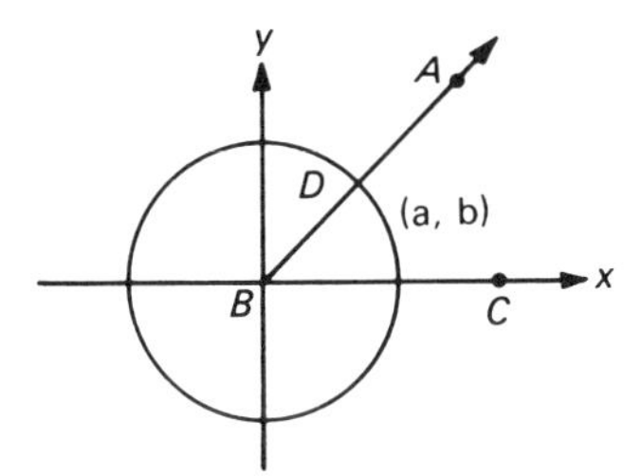

Figure 9.17. Unit circle

whose measure is 1. Point D, any point on this circle, is labeled (a, b). These coordinates are used to define two particular *trigonometric ratios.* Coordinate a is the *cosine* of angle ABC, and coordinate b is the *sine* of angle ABC. The sine and cosine of angle ABC, which is an angle whose sides are the positive end of the x-axis and the ray from the origin through point D, are defined to be the coordinates of point D. As various points are chosen on the unit circle, different angles are determined. For each of these angles, the two numbers used as coordinates for points on the unit circle are the cosine and the sine of the angle.

In the formal study of trigonometry, the implications in the diagram in Figure 9.17 are investigated more thoroughly. For some simple applications, however, it is possible to limit the study of the trigonometric ratios to those related to angles with a measure between 0 and 90 in degrees. In Figure 9.18, it is apparent that if this is done there is always a triangle such as $\triangle ABC$ determined by a point on the unit circle. The hypotenuse of this triangle has a measure of 1, since it is the radius. The measure of $\overline{AC} = b =$ sine of $\angle ABC$, and the measure of $\overline{BC} = a =$ cosine of $\angle ABC$.

For angles with measures between 0 and 90 in degrees, the cosine and sine may be interpreted as the ratio of two sides of a right triangle. From the

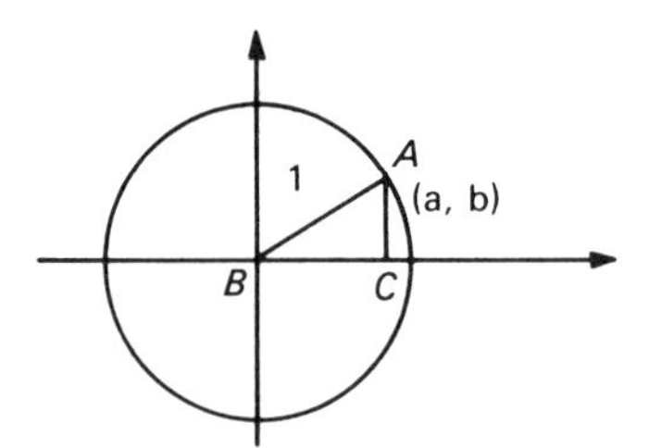

Figure 9.18. Point on unit circle

previous paragraph, you can determine that the sine of an acute angle of a right triangle is the ratio of the measure of the opposite side to the measure of the hypotenuse:

$$\text{sine } \angle ABC = \left(\frac{\text{measure of opposite side}}{\text{measure of hypotenuse}}\right) = \frac{AC}{AB},$$

and the cosine of an acute angle of a right triangle is the ratio of the measure of the adjacent side to the measure of the hypotenuse:

$$\text{cosine } \angle ABC = \left(\frac{\text{measure of adjacent side}}{\text{measure of hypotenuse}}\right) = \frac{BC}{BA}.$$

The expression *adjacent side* of an acute angle in a right triangle always refers to a side of the triangle other than the hypotenuse. Thus the adjacent side of the angle is that side of the triangle, other than the hypotenuse, lying on the side of the angle. The study of trigonometric ratios could also have been approached by first considering right triangles and the ratios of their sides.

In Figure 9.19, all of the right triangles with common vertex A are similar; the ratios a/e, b/f, c/g, and d/h are all equal. Since each of these is the sine of angle A, you can see that the sine depends only on the measurement of the angle, and not on the size of the triangle, since all the triangles are similar. In the same figure, all the ratios i/e, j/f, k/g, and l/h are equal also. Each of these is the cosine of angle A; hence the cosine depends only on the measurement of the angle.

Notice that, for the similar triangles shown, still another set of equal ratios can be found: a/i, b/j, c/k, and d/l. These are the ratios of the measure of the opposite side to the measure of the adjacent side. Each of these ratios is called the *tangent* of angle A.

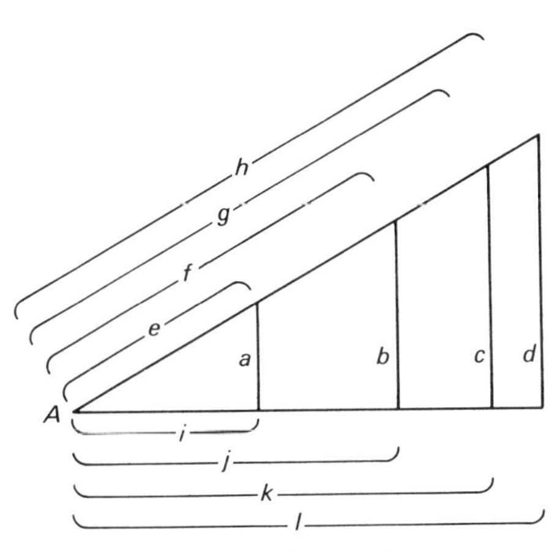

Figure 9.19. Similar triangles

The tangent of the angle a line makes with the positive x-axis is the numerical value of the slope of the line. In Figure 9.18,

$$\text{tangent } \angle ABC = \left(\frac{\text{measure of opposite side}}{\text{measure of adjacent side}}\right) = \frac{AC}{BC}.$$

The tangent, sine, and cosine of an angle are the three basic trigonometric ratios.

A trigonometric ratio of an angle may be determined approximately by measuring a drawing of a right triangle that has an acute angle of the correct measure. Tables of values for the trigonometric ratios are very common. Normally, they give approximations correct to perhaps four places past the decimal point, since, in general, the trigonometric ratios are irrational numbers. An abbreviated table of trigonometric ratios, correct to four places, is shown here.

trigonometric ratios

measurement of angle	*sine*	*cosine*	*tangent*
10°	.1736	.9848	.1763
20°	.3420	.9397	.3640
30°	.5000	.8660	.5774
40°	.6428	.7660	.8391
50°	.7660	.6428	1.1918
60°	.8660	.5000	1.7321
70°	.9397	.3420	2.7475
80°	.9848	.1736	5.6713

One observation that can be made from a study of the table is that the sine of an angle increases as the measure of the angle increases from 0 to 90 in degrees, whereas the cosine of the angle decreases. Other observations are noted in the exercises.

Trigonometric ratios have an important application in indirect measurement. In the previous examples of indirect measurement, three facts had to be known to arrive at the estimated measurement. By using right triangles and the trigonometric ratios, the answers can be determined with only two facts known. Two examples illustrate the method.

Example

If the angle of elevation of the top of a flagpole, measured at a distance of 100 ft, is 40°, find the height of the flagpole. Figure 9.20 illustrates the information given. The problem is to find the measure of x in feet.

$$x/100 = \text{tangent } 40^\circ$$
$$\approx .8391$$
$$x \approx 83.91$$

The flagpole is about 84 feet high. Detailed discussions of rules for rounding and of the possible errors involved in these problems are not presented

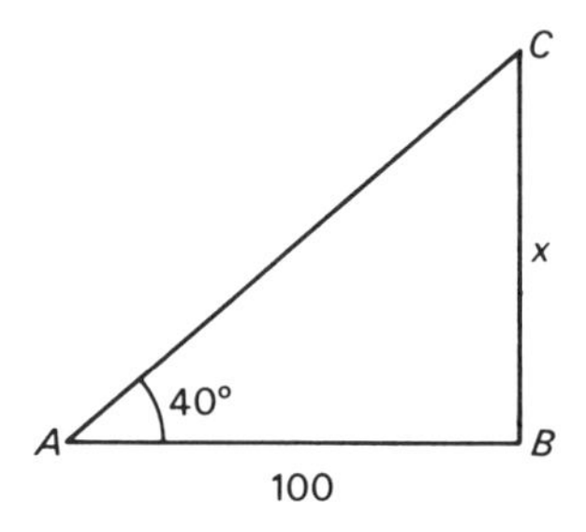

Figure 9.20. Measuring height by trigonometry

here. You should realize that the results are approximate, since they are based on physical measurements; furthermore, the tangent was given correct only to four decimal places and the angle correct only to the nearest 10 degrees.

Example

An airplane is at a position 20,000 ft high and the angle of elevation is 30°.

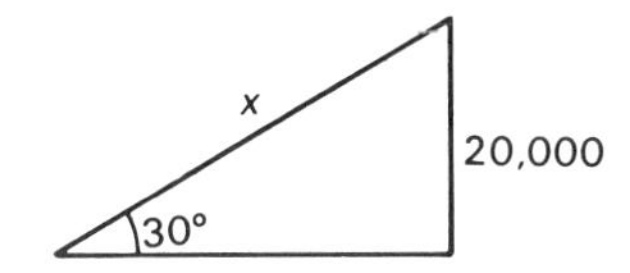

Figure 9.21. Measuring distance by trigonometry

How far away from the observer is the airplane? Figure 9.21 illustrates the problem.

$$20{,}000/x = \text{sine } 30°$$

$$20{,}000/\text{sine } 30° = x$$

$$20{,}000/.5000 = x$$

$$40{,}000 = x$$

The airplane is 40,000 feet from the observer.

The rest of this section may be omitted without loss of continuity. It gives an introduction to an application of geometry in calculus, so that the interested and prepared reader will further appreciate the value of some of the concepts presented earlier in the chapter.

The concepts of slope and tangent lead to consideration of one of the fundamental ideas of calculus—that of a *derivative*. In Figure 9.22, let $\overline{PQ}$

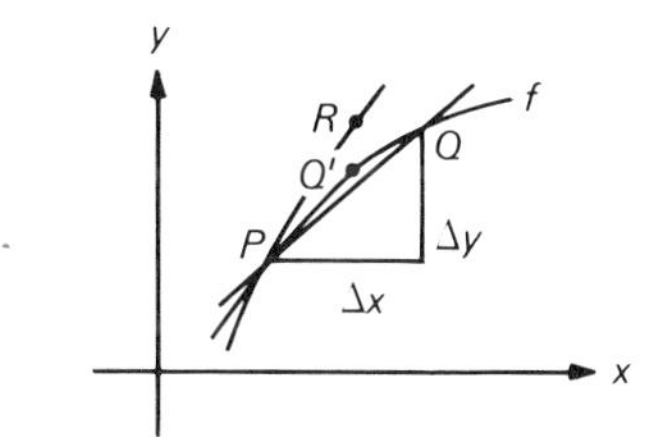

Figure 9.22. Concept of derivative

be a line intersecting the graph f of a function at two distinct points P and Q. Let Δx (delta x) and Δy be the changes in x- and y-coordinates from P to Q. Then the ratio $\Delta y/\Delta x$ represents the average rate of change and is the slope of $\overline{PQ}$.

Now think about considering points Q, Q', and so on along the curve, coming closer and closer to P. As the second point Q approaches P, the position of the line approaches the position of the tangent line $\overleftrightarrow{PR}$. The ratio $\Delta y/\Delta x$ approaches the slope of the tangent. This is expressed by saying that the limit of $\Delta y/\Delta x$, as Δx approaches zero, is the slope of the tangent. In calculus, the derivative dy/dx is defined as

$$\frac{dy}{dx} = \lim_{\Delta x \to 0} \frac{\Delta y}{\Delta x}.$$

The derivative indicates the instantaneous rate of change at a point on a function. That is, it indicates the slope of the tangent at that point.

An example will show how the concepts above can be used, but it should be pointed out that a course in calculus provides many shortcuts.

Example

Find the slope of the tangent to the curve $y = x^2$ at the point (1, 1). In Figure 9.23,

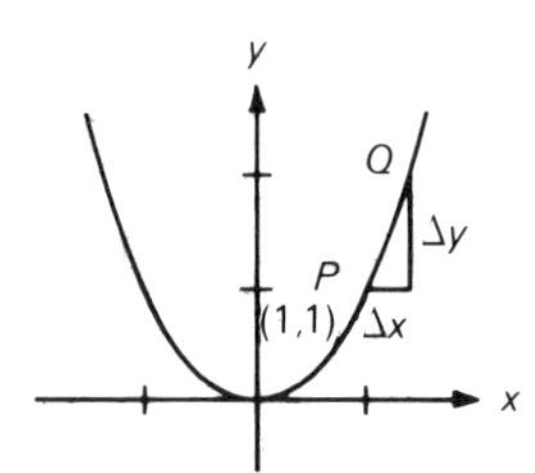

Figure 9.23. Slope of tangent at a point

$$\begin{aligned}\frac{\Delta y}{\Delta x} &= \frac{(x + \Delta x)^2 - x^2}{\Delta x} \\ &= \frac{x^2 + 2x\Delta x + (\Delta x)^2 - x^2}{\Delta x} \\ &= \frac{2x\Delta x + (\Delta x)^2}{\Delta x} \\ &= 2x + \Delta x. \qquad \text{(Divide numerator and denominator by } \Delta x.\text{)}\end{aligned}$$

The limit of $2x + \Delta x$ as Δx approaches zero is

$$\frac{dy}{dx} = 2x.$$

For $x = 1$, $dy/dx = 2$; hence the slope of the tangent at (1, 1) is 2.

Since $dy/dx = 2x$ is a general formula giving the slope of the tangent at any point on $y = x^2$, it is easy now to find that the slope at the origin is 0, the slope at (2, 4) is 4, and the slope at (3, 9) is 6. The interpretation is that the curve increases more and more rapidly toward the right. Note that for x negative, the slope of the tangent is also negative.

Exercise 9.3

1. As the measure of the angle increases from 10 to 80 degrees, does the tangent of the angle seem to increase or decrease?
2. The sine of a 40° angle is equal to the cosine of an angle with what measurement?
3. The cosine of a 70° angle is equal to the sine of an angle with what measurement?
4. What seems to be true about the sine of an angle and the cosine of its complementary angle?
5. For an angle with what measurement is the tangent equal to 1.0000?
6. What seems to be true about the ratio of the sine of an angle to the cosine of that same angle?
7. Define the tangent of an angle by using coordinates of points on a unit circle.
8. Give the tangent of the angle from the positive x-axis to the line:
   (a) $y = \frac{3}{2}x$ (b) $y = -\frac{1}{3}x$ (c) $y = 9x$
9. If the angle of elevation of the top of a hill, measured at a distance of 100 ft, is 50°, find the approximate height of the hill.
10. The top of a mountain 12,000 ft high is sighted at an angle of elevation of 20°. What is the approximate distance from the observer to the top of the mountain?
11. Seen from a point on one side of a river, a tree 30 ft high has an angle of elevation of 10°. Approximately how wide is the river?
12. A wire 50 ft long is tied to a tower and is anchored 30 ft from the base of the tower. What angle of elevation, to the nearest 10°, does the wire make with the ground?
13. The top of a mountain is sighted at an angle of elevation of 10° from a horizontal distance of 5 mi. How high is the mountain?
14. Find the slope of the tangent to the curve $y = x^2$ at the point:
    (a) (4, 16) (b) (5, 25)

Vectors

Many of the applications of geometry in physics and engineering involve the concept of a *vector*. It is often desirable to think of a force operating in a particular direction. This force can be symbolized by using a directed line segment. For example, Figure 9.24 can indicate a velocity thought of as a

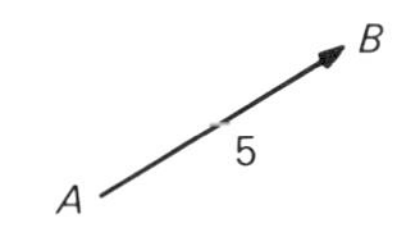

Figure 9.24. Vector for speed

speed of 5 miles per hour in the direction from A to B. A directed line segment used as an element in a mathematical system is called a vector.

The topic of vectors is a common one in European schools at a much lower level than is true in the United States at this time. Although this section can be omitted without loss of continuity, it does provide an introduction to what is becoming a topic of increasing importance in pre-high school grades.

The notation for a vector from A to B is $\overrightarrow{AB}$ or **AB**. The boldface notation will be used in this text. Point A is called the initial point, and B is the terminal point. The measure of the distance from A to B is the magnitude of the vector.

Two vectors are equal if they have both the same direction and the same magnitude. Thus all the vectors shown in Figure 9.25 are equal. Since each vector has an infinite number of other vectors equal to it, representing any member of this set of vectors by the same notation would be very desirable.

Figure 9.25. Equal vectors

A simple way of doing so is to use the coordinates of the terminal point if the initial point is at the origin. In Figure 9.26, the vector (2, 3) is shown. Note that both the direction and the magnitude can be found if the number

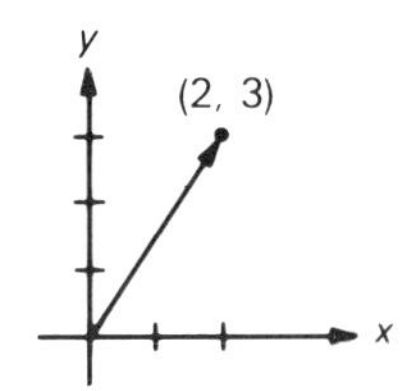

Figure 9.26. Vector (2, 3)

pair is known. In this case, the slope is $3/2$ and the magnitude is $\sqrt{2^2 + 3^2} = \sqrt{13}$.

The definitions of addition and subtraction of vectors come from the physical applications. In Figure 9.27, the combined effect of the forces

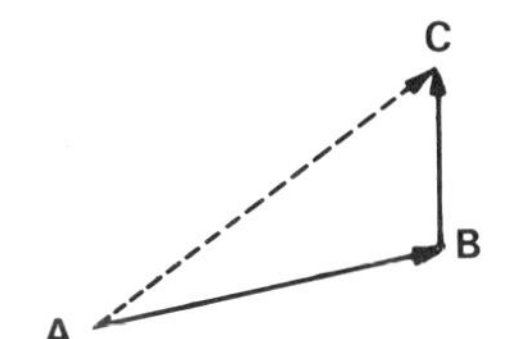

Figure 9.27. Sum of vectors

from A to B and then from B to C is a single force operating from A to C. Thus,

$$\mathbf{AB} + \mathbf{BC} = \mathbf{AC}.$$

Since any vector equal to a given vector may be substituted, vector addition can be carried out even when the original vectors are not positioned as in Figure 9.27. This idea is illustrated in Figure 9.28

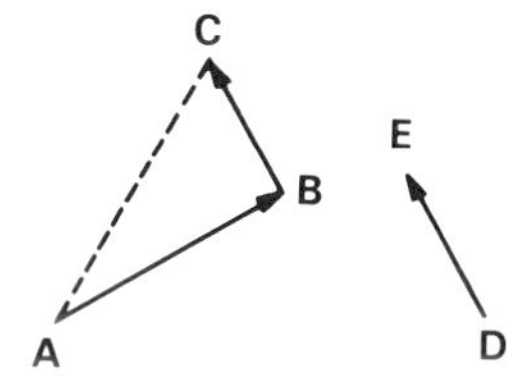

Figure 9.28. Sum of vectors without common point

$$\mathbf{AB} + \mathbf{DE} = \mathbf{AB} + \mathbf{BC}$$
$$= \mathbf{AC}$$

More than two vectors may be combined by repeating the definition of addition. Also, a small letter is often used to name a vector when no confusion is possible. Figure 9.29 shows the addition $\mathbf{a} + \mathbf{b} + \mathbf{c} = \mathbf{d}$.

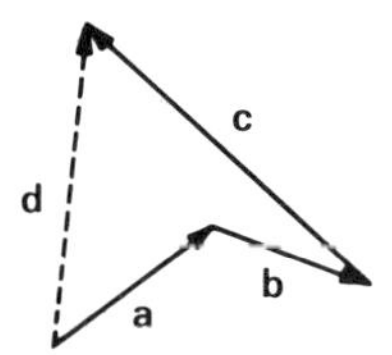

Figure 9.29. Sums of three vectors

Before reading ahead, draw various additions of vectors using a coordinate system to attempt to discover a rule for adding vectors named by ordered number pairs.

Figure 9.30 shows the addition of vectors (3, 1) and (−3, 4). The sum is (0, 5). Note that this sum can be found by adding the separate coordinates

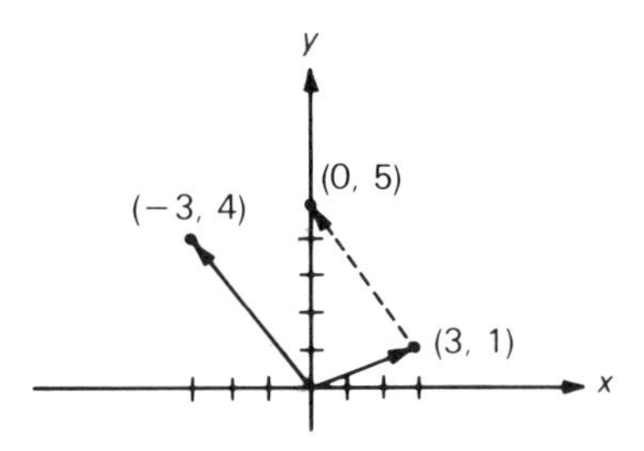

Figure 9.30. Addition of coordinates of vectors

of the given vectors, since $3 + (-3) = 0$ and $1 + 4 = 5$. Figure 9.31 shows $(3, 3) + (2, 4) = (5, 7)$. These examples lead to a general formula for the addition of vectors named as number pairs:

$$(\mathbf{a}, \mathbf{b}) + (\mathbf{c}, \mathbf{d}) = (\mathbf{a} + \mathbf{c}, \mathbf{b} + \mathbf{d}).$$

Subtraction of vectors is defined in terms of addition. In Figure 9.32, since $\mathbf{a} + \mathbf{b} = \mathbf{c}$, $\mathbf{c} - \mathbf{a} = \mathbf{b}$. Two vectors to be subtracted are chosen to have the same initial point. The vector representing the difference connects

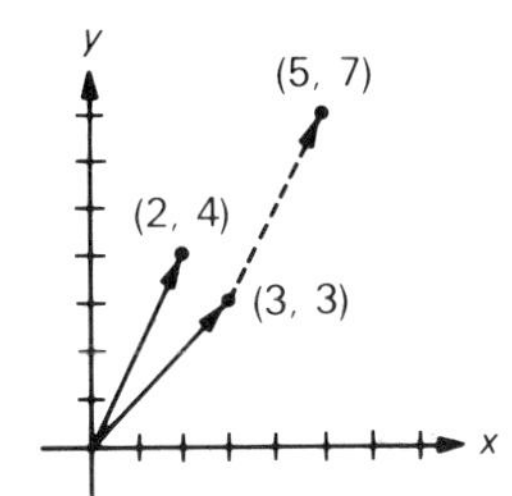

Figure 9.31. Addition of coordinates of vectors

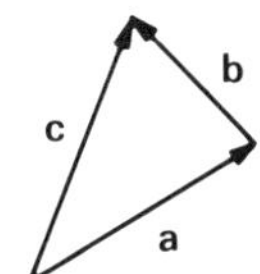

Figure 9.32. Subtraction of vectors

the terminal points, with its initial point at the terminal point of the vector to be subtracted.

When named as number pairs, vectors can be subtracted by subtracting the separate coordinates. Thus, $(5, 7) - (2, 1) = (3, 6)$.

Examples of elementary uses of vectors follow:

Example

Figure 9.33 shows two forces **a** and **b** operating on the same object.

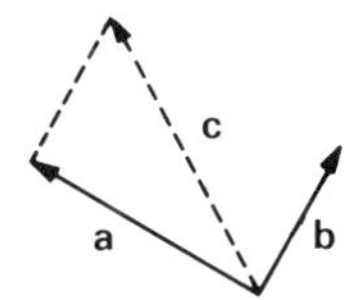

Figure 9.33. Force with combined effect

The magnitude of a single force with this combined effect is shown by drawing vector **c**, the sum of **a** and **b**.

Example

An airplane flies for one hour at a speed of 120 mph in a direction 30° east of north. It then flies for one hour at a speed of 100 mph in a direction 70° west of north. Approximately how far and in what direction from its starting point is the airplane at the end of the two hours?

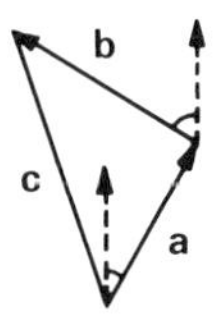

Figure 9.34. Path of airplane

The solution is shown in Figure 9.34. A convenient scale is chosen so that the lengths of the vectors represent the magnitude 120 and 100. For example, $\frac{1}{2}$ cm could be chosen to represent 20 mph. Vector **c** represents the approximate magnitude 123 mph and points approximately 13° west of north. The airplane is approximately 123 miles from the starting point and in a direction 13° west of north. Trigonometry could be used if a closer approximation were necessary.

Example

A boat moves at a rate of 4 mph (if it were in still water) across a river. The current flows at the rate of 2 mph. Find the actual direction and rate of the boat.

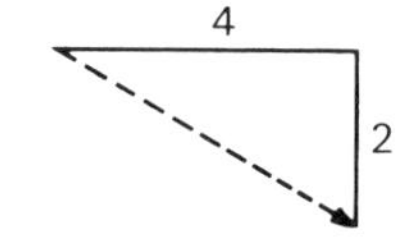

Figure 9.35. Path of boat

The solution is shown in Figure 9.35. The boat is carried downstream at an angle of approximately 27° and is actually moving at a rate of about 4.6 mph.

Exercise 9.4

1. Draw each vector indicated by its terminal point, if the initial point is the origin.
   (a) $(3, 7)$ (b) $(-1, 4)$ (c) $(-2, -5)$ (d) $(4, -2)$
2. Find the slope for each vector in Exercise 1.
3. Find the magnitude for each vector in Exercise 1.

For Exercises 4–7, copy the figure and find the sum of the vectors shown.

4. Figure 9.36a
5. Figure 9.36b
6. Figure 9.36c
7. Figure 9.36d

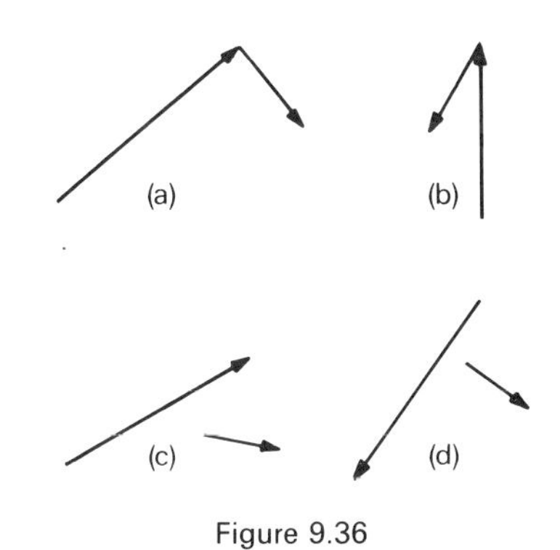

Figure 9.36

For Exercises 8–9, copy the figure and find the sum of the vectors shown.

8. Figure 9.37a
9. Figure 9.37b

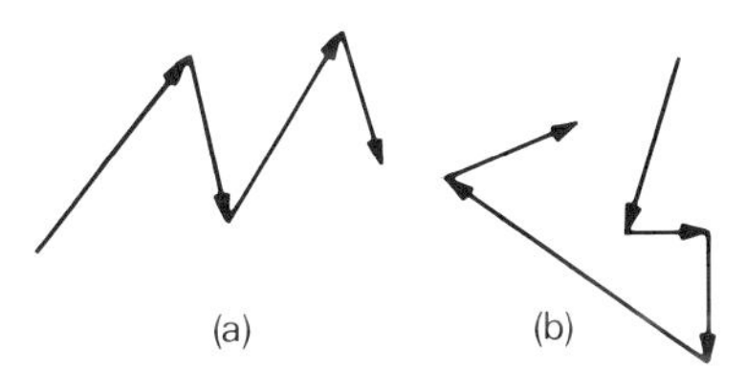

Figure 9.37

For Exercises 10–12, find the vector sum.

10. $(2, 1) + (7, -2)$
11. $(3, 5) + (-2, 1)$
12. $(6, -4) + (-2, -3)$

For Exercises 13–14, copy the figure and subtract vector **c** from vector **d**.

13. Figure 9.38a
14. Figure 9.38b

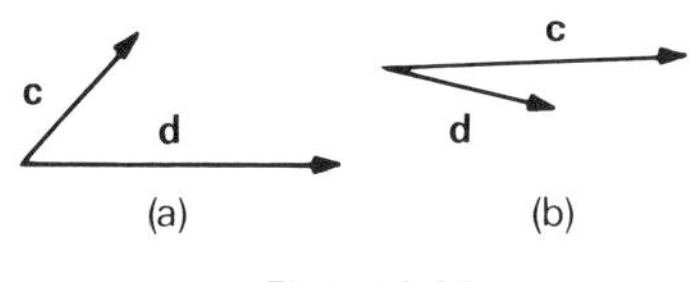

Figure 9.38

For Exercises 15–16, subtract the vectors.

15. $(9, 4) - (2, 1)$
16. $(-3, 2) - (3, -4)$

For Exercises 17–18, Figure 9.39 shows two forces operating on the same object. Copy the figure and draw a vector to represent a single force with the combined effect.

17. Figure 9.39a
18. Figure 9.39b

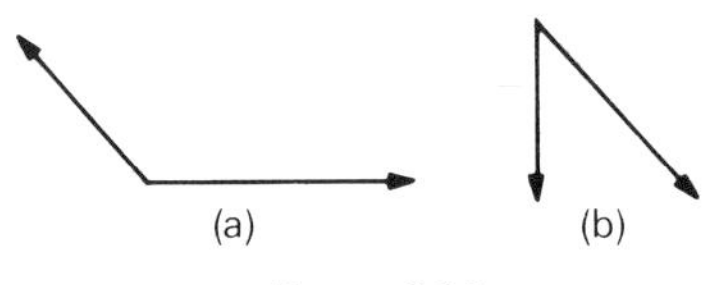

Figure 9.39

For Exercises 19–21, prepare a vector drawing to scale and then approximate the answer.

19. A boat sails for two hours at a speed of 10 mph in a direction of 20° west of north. Then it sails for 3 hours at a speed of 15 mph in a

direction of 40° west of south. Approximately how far and in what direction from its starting point is the boat at the end of the 5 hours?

20. A boat moves at a rate of 5 mph across a river. The current flows at the rate of 4 mph. Find the actual direction and rate of the boat.

21. An airplane heads north at a rate of 100 mph. A wind of 50 mph blows from the west. Find the actual speed and direction of the path of the plane.

Sets of points in a plane are an important special case of points in space. In this chapter, the non-metric geometry of space is introduced. The concepts of *skew lines, half-space, dihedral angle,* and *perpendicularity* are explained prior to the study of intersection and union of sets of points in space. A more detailed analysis of the non-metric properties of special sets of points in space, such as rectangular prisms, rectangular solids, and polyhedrons, makes extensive use of ideas developed in previous chapters.

Sets of Points in Space

To the mathematician, *space* is an undefined term. Space is described as a set of points, each of which indicates a specific location. Since no physical object lies entirely in one plane, every physical object should remind you of sets of points in space. The mathematical model of sets of points in space is an idealization from the actual world of experience. So far the set of axioms given as a part of the structure of Euclidean geometry has characterized the geometry of the plane. Additional axioms are necessary for a geometry of three-dimensional space. The first new axiom is an existence axiom.

Axiom 24

All the points of space do not lie on the same plane.

Several new concepts are introduced here before we begin to study the intersection and union of sets of points in space.

1. In a plane, two lines either are parallel or meet at a point; in space, there is one other possibility. Figure 10.1 shows two *skew lines*, $\overleftrightarrow{AB}$ and $\overleftrightarrow{CG}$. Skew lines are lines that are not parallel and do not meet. They cannot lie

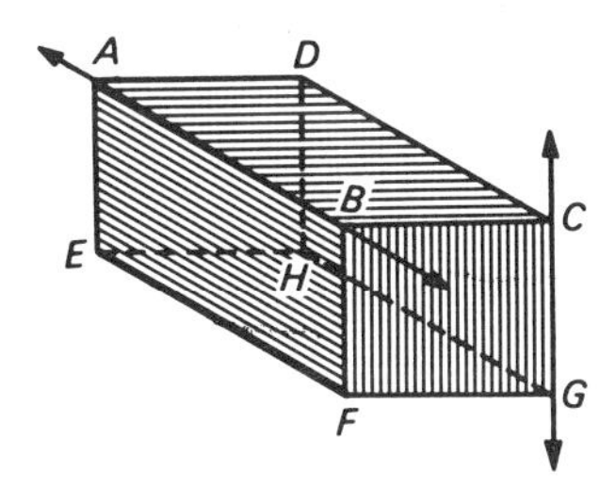

Figure 10.1. Skew lines

in the same plane, according to this definition. If Figure 10.1 represents a room with *CGFB* the front, then the skew lines could be described as the edge *CG* between the right wall and the front wall and the edge *AB* between the left wall and the ceiling. Some other pairs of skew lines in Figure 10.1 are $\overleftrightarrow{CG}$ and $\overleftrightarrow{HE}$ and $\overleftrightarrow{EF}$ and $\overleftrightarrow{DH}$.

2. The concepts of half-line and half-plane are analogous to the idea of a *half-space*. A plane partitions all the points of space into three disjoint subsets: the set of points in the plane and a set of points on each side of the plane. The set of points in space on one side of a given plane is called a half-space.

Axiom 25

A plane separates the points of space not on the plane into two convex sets such that every segment that joins a point of one of the sets to a point of the other intersects the plane.

3. A *dihedral angle* is defined by means of the half-plane concept. Figure 10.2 shows two intersecting planes with a common line AB.

Axiom 26

If two different planes intersect, the set of points in the intersection is a line.

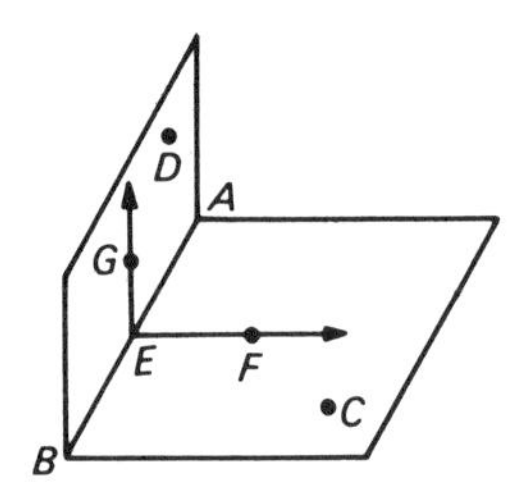

Figure 10.2. Dihedral angle

A dihedral angle is the union of the set of points in the two half-planes on intersecting planes and the set of points on the common line of the two planes. The common line of intersection determines the two half-planes. The dihedral angle shown in Figure 10.2 consists of the points on line AB, those on the half-plane containing point C, and those on the half-plane containing point D.

4. The concept of *perpendicular* must be extended if it is to be meaningful for lines and planes in space. In Figure 10.2, what does it mean to say that plane ABD is perpendicular to plane ABC? Choose any point E on $\overleftrightarrow{AB}$, and draw $\overrightarrow{EF}$ perpendicular to $\overleftrightarrow{AB}$. Then, if $\overrightarrow{EG}$ is also drawn perpendicular to $\overleftrightarrow{AB}$, and if $\angle GEF$ is a right angle, the two planes are perpendicular.

Figure 10.3 shows line AD perpendicular to plane ABC. In this case, $\overrightarrow{AD}$ and $\overrightarrow{AB}$ form a right angle, as do $\overrightarrow{AD}$ and $\overrightarrow{AC}$. Line DA is perpendicular to

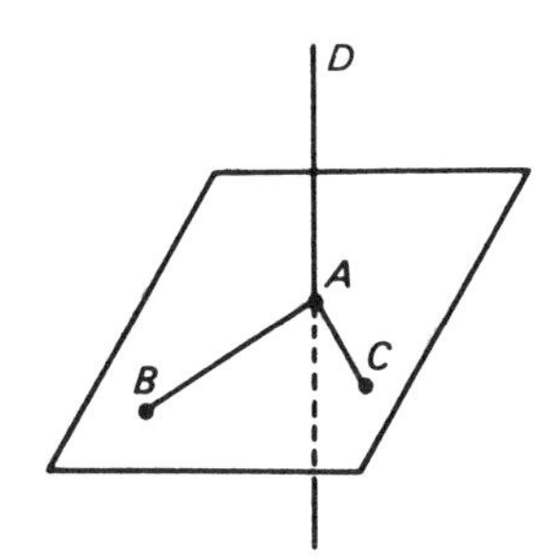

Figure 10.3. Concept of perpendicular

any line in plane ABC that passes through point A. A minimum condition to determine that $\overleftrightarrow{DA}$ is perpendicular to plane ABC is that $\overleftrightarrow{DA}$ must be perpendicular to two distinct lines through point A in plane ABC. Notice that in this case the shortest distance from point D to plane ABC is measured along $\overline{DA}$. The distance between parallel planes (planes whose intersection is the empty set) is the distance along a segment perpendicular to both planes.

Some extensions of the previous study of intersection and union of sets of points may now be made. In this chapter, sets of points in space implies that the points are not all on one plane. You have already found that two lines in space may be skew lines. Figure 10.4 shows two possibilities for the intersection of two planes in space. The intersection of two distinct planes in space is either a line or the empty set.

A line that has two points in common with a plane lies entirely in the plane. The intersection of a plane and a line not in the plane may be empty or a single point. In the first case, the line is parallel to the plane. In general, a line not in the plane will meet any plane figure, such as a triangle or circle, in one or no points.

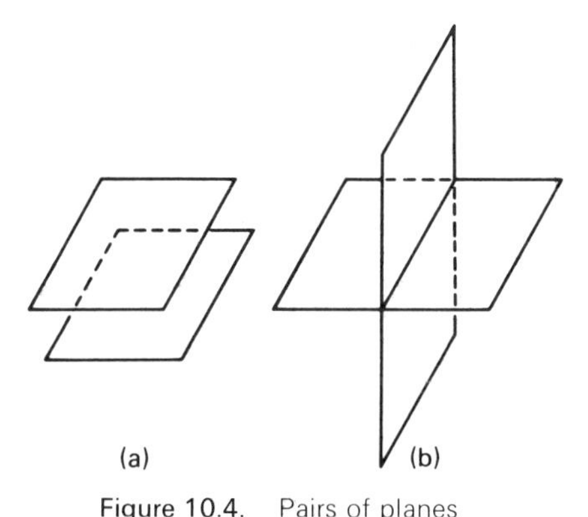

Figure 10.4. Pairs of planes

Figure 10.5 shows just three of the many interesting possibilities for intersection of a dihedral angle and other sets of points. The intersection of a dihedral angle and a line may be two distinct points, as in Figure 10.5a. The intersection of a dihedral angle and a circle not on one of the sides of the angle may be four distinct points, as in Figure 10.5b. The intersection of a dihedral angle and a rectangle not on one of the sides of the angle may be two separate line segments, like $\overline{AB}$ and $\overline{CD}$ in Figure 10.5c. In studying

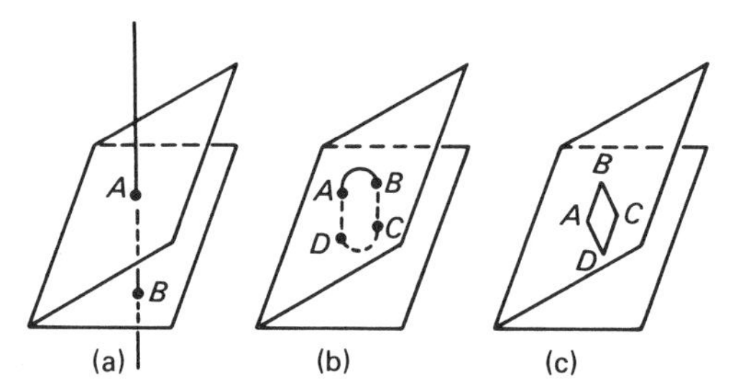

Figure 10.5. Dihedral angle and intersections

Figure 10.5, you will probably find it helpful to think of a partially open book to help you visualize a dihedral angle; a small rectangular-shaped piece of cardboard placed to hold the book open will help you visualize Figure 10.5c.

Exercise 10.1

1. For an ordinary room, which of these examples suggest skew lines?
   (a) The edge where the front wall meets the ceiling and the edge where the right wall meets the ceiling.
   (b) The edge where the floor meets the left wall and the edge where the front wall meets the ceiling.
   (c) The edge where the floor meets the front wall and the edge where the back wall meets the ceiling.
   (d) The edge where the right wall meets the ceiling and the edge where the front wall meets the floor.
2. For an ordinary room,
   (a) do all three edges meeting at a corner suggest three lines perpendicular to one another?
   (b) does each of the four vertical edges where the walls meet seem to be perpendicular to the floor?
3. In Figure 10.6 let α and β indicate the set of points on each of the two intersecting planes.
   (a) $\alpha \cap \beta =$ ____ (b) $\overleftrightarrow{AB} \cap \alpha =$ ____ (c) $\overleftrightarrow{AB} \cap \beta =$ ____
   (d) $\overleftrightarrow{AB} \cap \overleftrightarrow{CD} =$ ____ (e) $\overleftrightarrow{AC} \cap \beta =$ ____ (f) $\overleftrightarrow{AC} \cap \alpha =$ ____
4. List all possibilities for sets of points in the intersection of the sets of points given below.
   (a) Dihedral angle and a plane. (b) Dihedral angle and a ray.
   (c) Dihedral angle and a triangle. (d) Half-space and a plane.
   (e) Half-space and a ray. (f) Half-space and a line.
   (g) Three distinct planes.

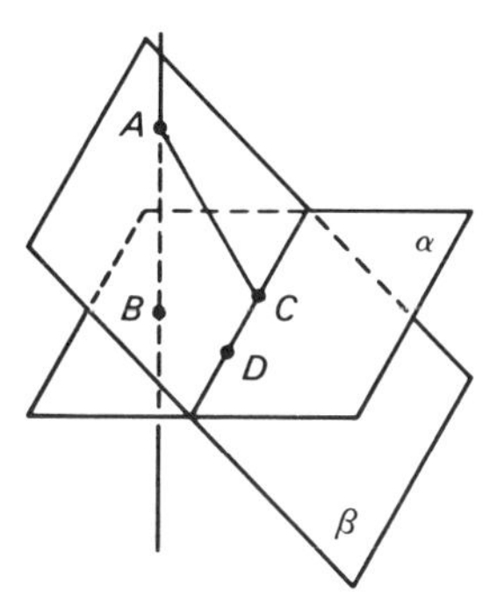

Figure 10.6

For Exercises 5–8, make a sketch of the set of points described.

5. A plane and a line parallel to it.
6. Three planes all having the same line in common.
7. A dihedral angle and a triangle with one vertex lying on one half-plane of the angle and the opposite side of the triangle lying in the second half-plane of the angle.
8. Three planes all having the same point in common.
9. In Figure 10.1 (p. 219), name four lines that are skew to $\overleftrightarrow{AD}$.

Polyhedrons

One of the important types of three-dimensional figures is the set of polyhedrons. Polyhedrons are *surfaces,* which means that the interior is not included. Polyhedrons are composed of polygonal regions. The plural of polyhedron is either polyhedrons or polyhedra.

Two particular subsets of the set of polyhedrons will be studied first, followed by a consideration of properties of the general set of points. One particular subset of polyhedrons is an idealization of many common objects: the set of all *right prisms.* Figure 10.7 shows five different right prisms. The set of points called a right prism is a surface with the interior empty.

The bases of right prisms are two congruent polygonal regions in parallel planes. The sides of right prisms are rectangular regions and are perpendicular to the bases of the prisms. The dihedral angles containing a side and a base are all right angles.

The sides and bases of a right prism are called *faces*. The union of the set of points in the bases and the sides is a right prism. Sometimes the sides are

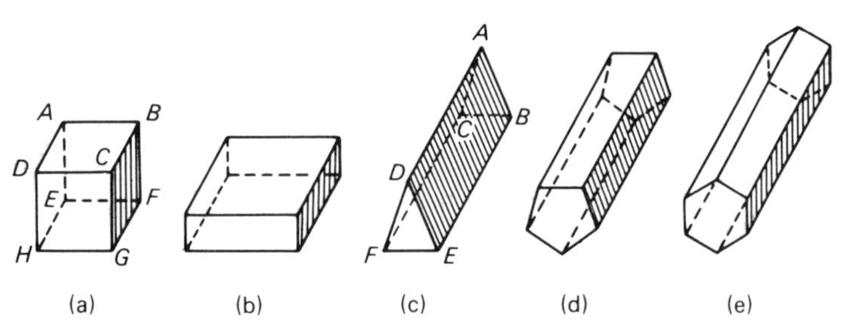

Figure 10.7. Right prisms

called *lateral faces* to distinguish them from the bases. The segments forming the boundary of the faces of right prisms are called *edges*, and the vertices are called the *vertices* of the prism.

The set of points in Figure 10.7a is called a *cube*; each face consists of a square region. Examples of common objects shaped like a cube include the surfaces of play blocks and of sugar cubes. A cube is a special case of a general *rectangular prism*, such as the one in Figure 10.7b. Right prisms are named according to the type of polygon that encloses the base. For the rectangular prism, the bases are rectangle $ABCD$ and its interior and rectangle $EFGH$ and its interior. A rectangular right prism has four sides, or lateral faces, each of which consists of a rectangle and its interior. Some common objects that suggest a rectangular prism are a shoe box and a room.

A *triangular right prism* is shown in Figure 10.7c. The bases are triangular regions ABC and DEF. Common objects that resemble triangular right prisms are wedges and troughs. Figure 10.7d shows a pentagonal right prism and Figure 10.7e a hexagonal right prism. In the first case, the bases are pentagons and their interiors; in the second case, the bases are hexagons and their interiors. Examples of objects suggesting hexagonal prisms are some pencils and some bolt heads. Although many common examples of right prisms have regular polygons as the boundary for their bases, this is not a requirement.

Right prisms are subsets of the more general sets of points called *prisms*. Prisms that are not right are *oblique*. Figure 10.8 shows two oblique prisms. Since the sides of oblique prisms are not at right angles to the bases, the sides are not rectangles; instead, they are parallelograms.

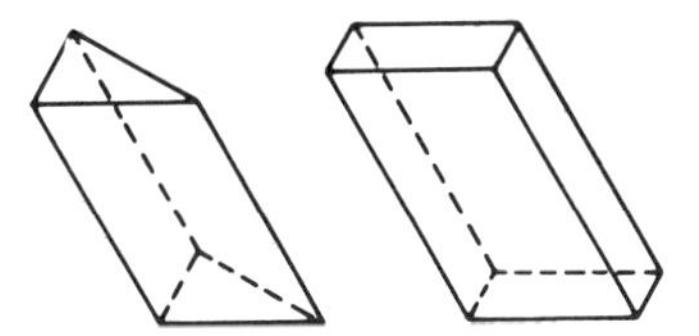

Figure 10.8. Oblique prisms

Prisms are one major subset of the set of polyhedrons. Another subset is the set of *pyramids*. Figure 10.9 shows two examples. Pyramids, like prisms, are the unions of sets of points in polygonal regions. Unlike prisms, pyramids have only one base. In Figure 10.9a, the base is rectangle $BCDE$ and its interior. In Figure 10.9b, the base is triangle EFG and its interior. Besides the base, a pyramid consists of one point (called the *apex*) not in the plane of the base and all the triangles and their interiors determined by this point and each edge of the base. In Figure 10.9a, the apex is point A.

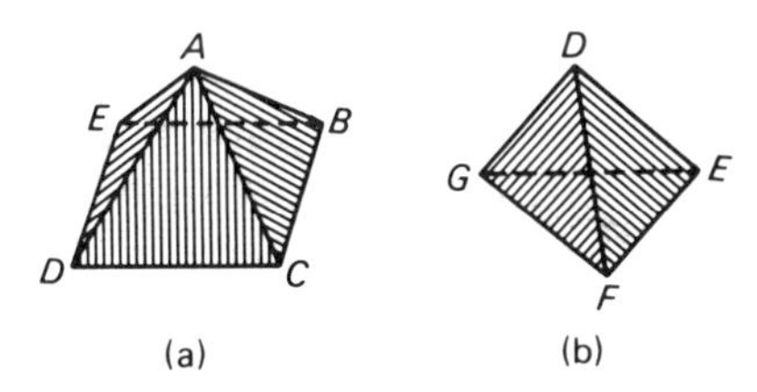

Figure 10.9. Pyramids

Four triangular faces are determined by point A and each of the four edges of the base. In Figure 10.9b, point D is the apex. Three triangular sides are determined by point D and the three edges of the triangular base. Although common examples of pyramids, such as the pyramids of Egypt, have rectangular bases, the base of a pyramid may be any polygon and its interior. Unlike the pyramids of Egypt, mathematical pyramids are not solids; they do not include the points on the inside.

Each set of points in Figures 10.7 through 10.9 is a polyhedron. The set of *polyhedrons* includes the special cases studied earlier, both prisms and pyramids, as subsets. In general, polyhedrons are surfaces whose faces are polygons and their interiors. Polyhedrons are the union of sets of points in portions of several planes. Another name for the triangular pyramid in Figure 10.9b is *tetrahedron,* which names it as a polyhedron with four faces. Figure 10.10 shows two polyhedrons that are neither pyramids nor prisms; the first one has ten faces and the second one has eight.

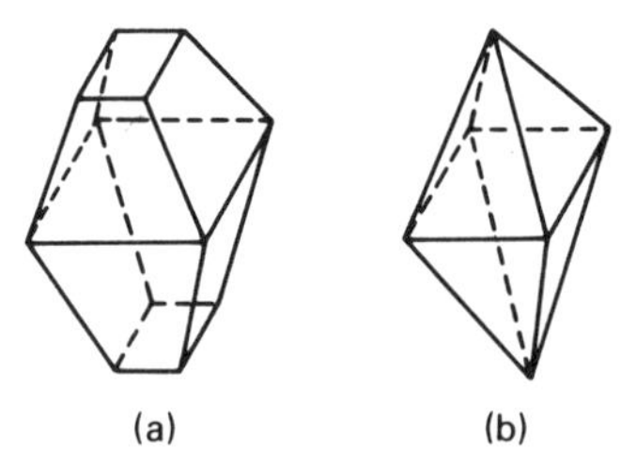

Figure 10.10. Polyhedrons

All of the polyhedrons discussed so far are convex, but Figure 10.11 shows an example of a non-convex polyhedron.

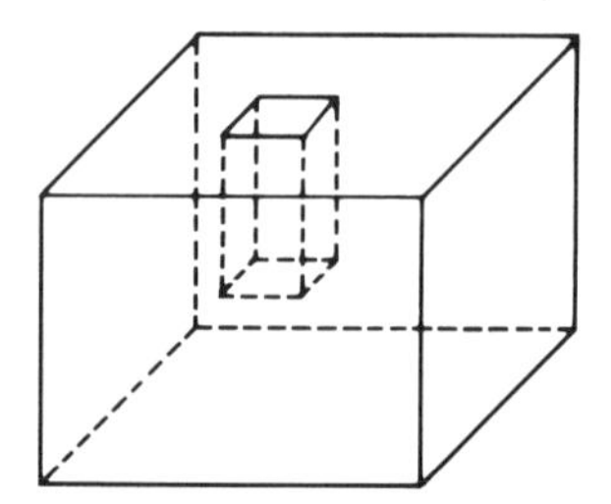

Figure 10.11. Non-convex polyhedron

From the time of the ancient Greeks, geometers have studied a particular set of polyhedrons, the *convex regular polyhedrons.* The faces are congruent regular polygons, with the angles between the faces all congruent. Oddly enough, there are only five different types of convex regular polyhedrons. These are shown in Figure 10.12.

The following table shows the name, the number of faces, and the shape of the faces for the regular polyhedrons in the figure.

	number of faces	*shape of each face*
tetrahedron (a)	4	triangular
cube (b)	6	square
octahedron (c)	8	triangular
dodecahedron (d)	12	pentagonal
icosahedron (e)	20	triangular

The mathematician Euler, who lived about 1750, is given credit for the generalization about the numbers of vertices, edges, and faces for polyhedrons. Study the following table for three of the sets of points previously

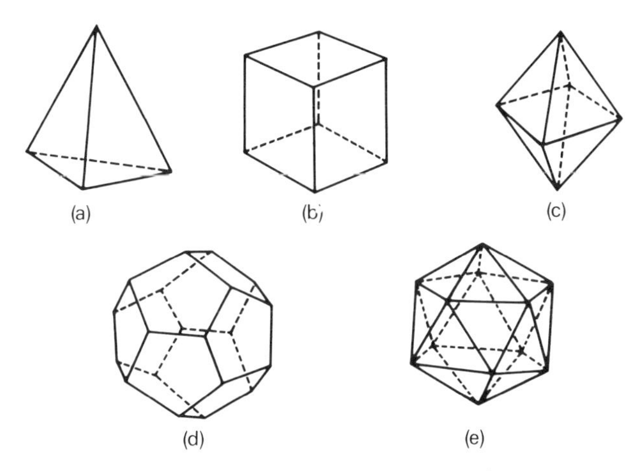

Figure 10.12. Convex regular polyhedrons

number of figure	*V*	*E*	*F*
10.7a	8	12	6
10.9a	5	8	5
10.10a	12	20	10

shown to see if you can find a mathematical sentence describing the relationship among V, E, and F, which represent the numbers of vertices, edges, and faces, respectively. Now verify that the same relationship holds for other polyhedrons. "Euler's formula" for polyhedrons may be written

$$V + F - E = 2$$

where V represents the number of vertices, F the number of faces, and E the number of edges. Did you discover it?

Much can be learned about polyhedrons by studying models. It is even more instructive to prepare models by drawing polygonal regions on a sheet of paper and then folding them to represent a polyhedron. For example, Figure 10.13 can be used to make a model for a cube. If it is desired to use glue, then additional tabs should be included.

Exercise 10.2

For Exercises 1–7, answer yes or no.

1. Every pyramid is a prism.
2. Every pyramid is a polyhedron.

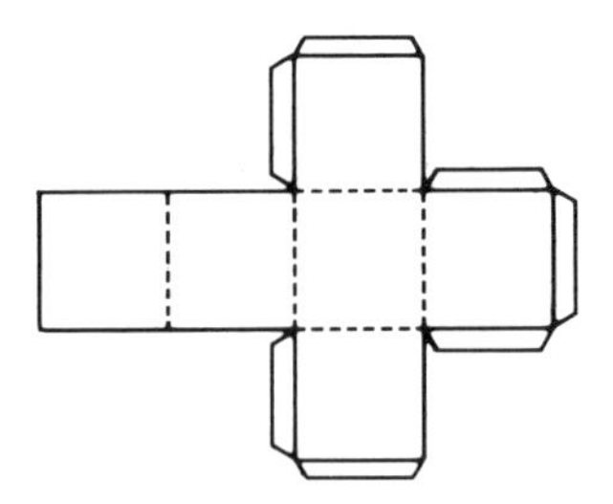

Figure 10.13. Pattern for model of cube

3. Some polyhedrons are prisms.
4. The bases of prisms may lie on planes perpendicular to each other.
5. A pyramid with seven faces has a hexagon as the border for its base.
6. Euler's formula holds for prisms and pyramids only.
7. If the sides of a prism are rectangles, then it is a right prism.
8. Consider an object in the shape of a rectangular right prism:
   (a) What is the largest number of faces you can see at one time?
   (b) What is the largest number of vertices you can see at one time (assume that you can see the vertex if you can see at least one of the edges with the vertex as an endpoint)?
   (c) What is the largest number of edges you can see at one time?
9. Complete the table, verifying Euler's formula for each set of points.

	V	E	F	$V + F - E$
Figure 10.7b	____	____	____	____
Figure 10.7c	____	____	____	____
Figure 10.7d	____	____	____	____
Figure 10.7e	____	____	____	____
Figure 10.10b	____	____	____	____

10. Find the number of vertices of each of the convex regular polyhedrons.
11. Find the number of edges for each of the convex regular polyhedrons.
12. Verify Euler's formula for each of the convex regular polyhedrons.
13. List all the possibilities for the intersection of a plane and a cube.
14. List all the possibilities for the intersection of a plane and a tetrahedron.

In Exercises 15–20, draw a figure similar to the one in Figure 10.13 that can be used to make a model for a:

15. Rectangular right prism
16. Triangular right prism

17. Hexagonal right prism
18. Pyramid with a square base
19. Tetrahedron
20. Octahedron

Simple Closed Surfaces and Solid Figures

Simple closed surfaces are sets of points in space that are analogous to simple closed curves in a plane. They are continuous and smooth, and they partition all the points of space into three disjoint sets: the set of points on the surface, the set of points in the interior of the figure, and the set of points in the exterior. Simple closed surfaces have exactly one interior.

Polyhedrons are a subset of the set of all simple surfaces. A polyhedron may be defined as a simple closed surface with faces that are polygonal regions. In common language, a polyhedron is a simple closed surface that has flat sides. Prisms and pyramids, since they are subsets of the set of polyhedrons, are all simple closed surfaces.

Three examples of simple closed surfaces that are not polyhedrons are *cylinders, cones,* and *spheres.* Each of these surfaces is a curved surface, at least in part. That is, the surface is not composed entirely of portions of a finite number of planes. Figure 10.14a shows a right circular cylinder. Some common objects that suggest right circular cylinders are tin cans, water pipes, and paper straws. A right circular cylinder has two congruent parallel bases, each of which is a circle and its interior. Segments lying on the sides of a cylinder and in a vertical position are perpendicular to the plane of the horizontal bases.

Not all cylinders are circular or right. Figure 10.14b shows a cylinder whose base is an *ellipse* and its interior. Figure 10.14c shows an oblique

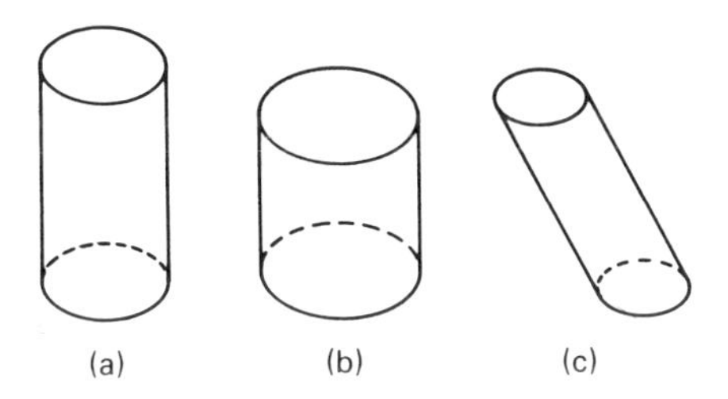

Figure 10.14. Cylinders

circular cylinder (a cylinder whose base is not perpendicular to the elements in the lateral surface). Technically, the bases of a cylinder may be any pair of simple closed curves and their interiors.

A second example of a simple closed surface not a polyhedron is a cone. A right circular cone is shown in Figure 10.15a. This cone has one base, which is a circle and its interior. The cone consists of this base and the surface determined by one point not on the base and the segments from this point to each point of the circle. In a right circular cone, the vertex is directly over the center of the base. Figure 10.15b shows an oblique

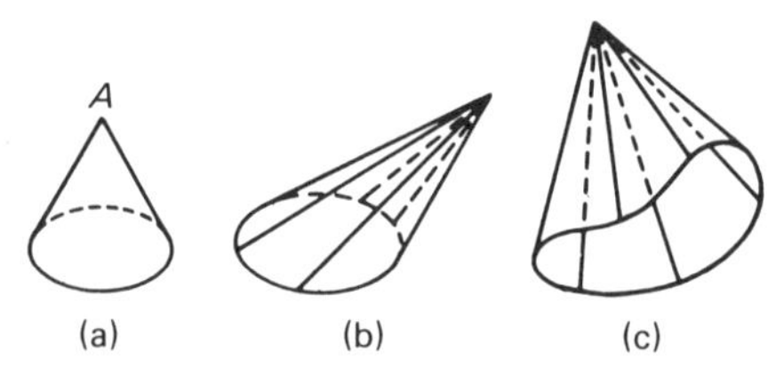

Figure 10.15. Cones

circular cone. Figure 10.15c shows a cone with an irregular simple closed curve for its base. Technically, a cone may have any simple closed curve and its interior for the base.

You should have noticed some analogies between the sets of points discussed in this section and those in the last section. A cylinder seems to be related to a prism in much the same way that a cone is related to a pyramid. Think of a right prism with a rectangular base. If you continue to increase the number of sides for the polygon enclosing the base, the prism seems to be approaching the shape of a cylinder. Similarly, if you think of a pyramid with a rectangular base and then think of increasing the number of sides for the polygon of the base, the pyramid seems to be approaching the shape of a cone. It should be emphasized that, by definition, a cylinder is not a prism; neither is a cone a pyramid, since the process of increasing the number of sides must be carried on an infinite number of times.

The third example of a simple closed surface that is not a polyhedron is a sphere, shown in Figure 10.16. A sphere consists of all the points in space at an equal distance from a point called the center. It is analogous to a circle on a plane in this respect. Some common spherical objects are a rubber ball and the surface of the earth.

Figure 10.16. Sphere

The intersection of a plane and a simple closed surface is called a *section.* The expression *cross section* will be used to indicate the set of points in the intersection if the plane is passed through the figure parallel to the bases.

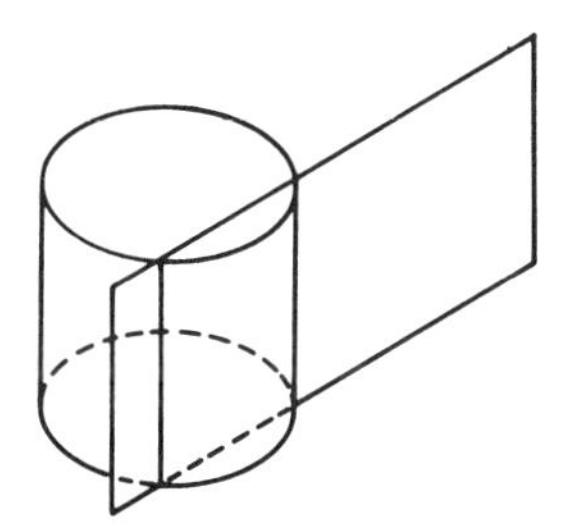

Figure 10.17. Cylinder and plane

The cross section of a circular cylinder is a circle. The section of a right circular cylinder made by a plane at right angles to the bases, as illustrated in Figure 10.17, is a rectangle.

The cross section of a right circular cone is also a circle. Figure 10.18 shows three other sections of a cone. Here, the cone is considered to have no base and the lateral sides extend indefinitely. Figure 10.18a shows a section that is an *ellipse.* The section in Figure 10.18b is a *parabola,* and the section in

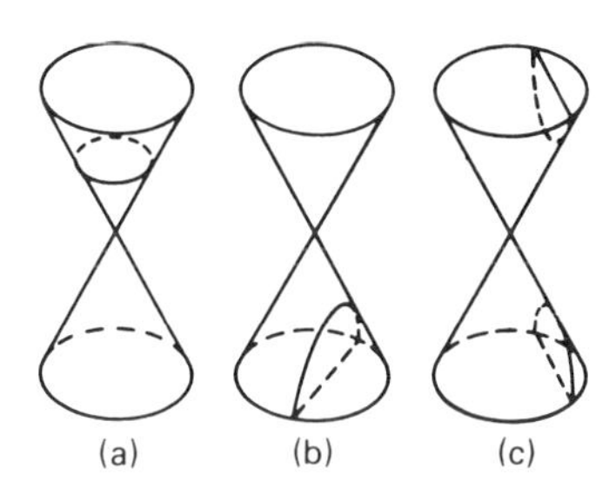

Figure 10.18. Conic sections

Figure 10.18c is a *hyperbola.* A hyperbola is an example of a plane curve with two branches. Circles, ellipses, parabolas, and hyperbolas are all conic sections. Their properties are investigated in detail in a course in analytic geometry.

Visualize various sections of a sphere, using a ball and a piece of cardboard if necessary. What seems to be true in each case? Any section of a sphere is a circle. If the plane passes through the center of the sphere, the section is called a *great circle*; otherwise, the section is a *small circle.* An example of both kinds of sections is shown in Figure 10.19. Great and small circles will be used in Chapter 11 to help locate points on the surface of a sphere.

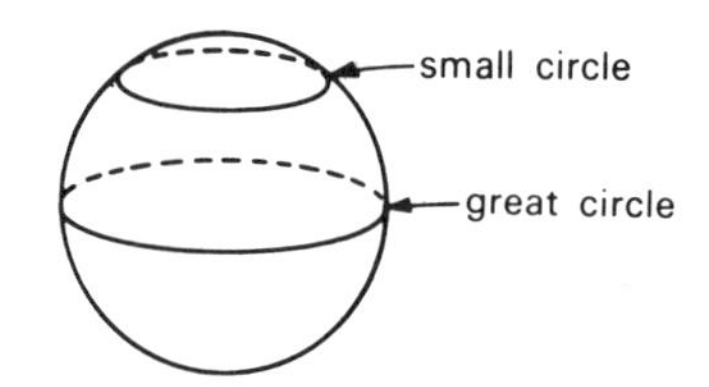

Figure 10.19. Great and small circles

Solids are sets of points in space that are the union of a surface and its interior. Each simple closed surface studied in this chapter is associated with a solid figure containing the points on the surface and the points in the interior. In general, solids are named from the surfaces with which they are associated, as shown in the following list. There is some lack of standardization, however, in the naming of solid figures.

surface	*solid made up of surface and interior*
Polyhedron	Polyhedral solid
Prism	Prismatic solid
Pyramid	Pyramidal solid
Cylinder	Cylindrical solid
Cone	Conical solid
Sphere	Spherical solid or spheroid

The section of each solid listed is, in general, a simple closed curve and the set of points in its interior. For example, the section of a spheroid is a circle and its interior.

Exercise 10.3

1. What could you conclude about a line segment with one endpoint in the interior of a simple closed surface and the other endpoint in the exterior of the same surface?
2. Name other common objects that suggest right circular cylinders.
3. Name other common objects that suggest right circular cones.
4. Name other common objects that suggest spheres.
5. Is it possible for all the points of one cone to lie in the interior of another cone?
6. Is it possible for all the points of one sphere to lie in the exterior of another sphere?
7. Is it possible for a cone to have an ellipse and its interior for a base?
8. Is it possible for a cylinder to have only one base?
9. In general, what is the name for the set of points in the intersection of a solid and a line passing through the solid?
10. Describe the set of points in the intersection of each of the surfaces pictured in Figure 10.20.

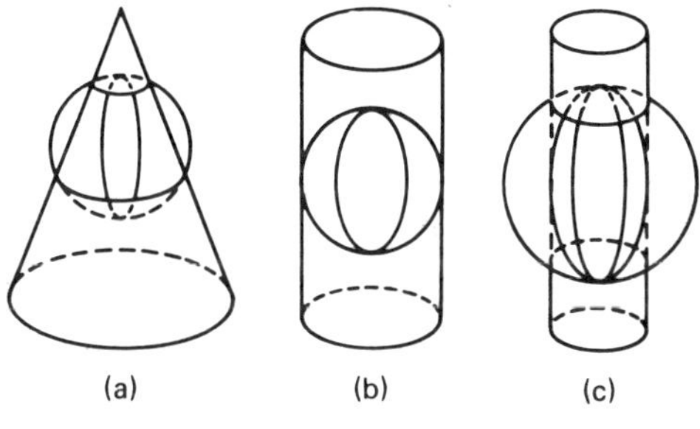

Figure 10.20

    (a) Figure 10.20a (sphere and cone).
    (b) Figure 10.20b (sphere and cylinder).
    (c) Figure 10.20c (sphere and cylinder).

11. What is the shape of the cross section of a solid whose surface is a right prism?
12. What is the shape of the section of a conical solid parallel to the base?
13. What is the shape of the section of a pyramidal solid parallel to the base?

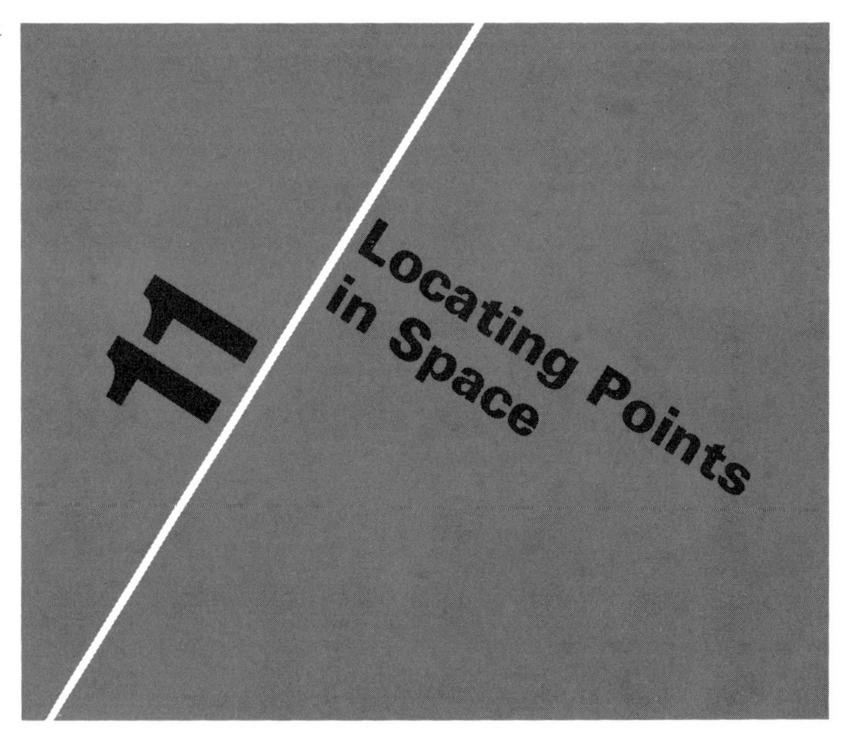

You have previously studied how to locate points on a plane by using two coordinates in reference to two fixed number lines. In this chapter, the method is extended to points in space. This extension involves understanding the mathematical concept of dimension. The chapter concludes with the special problem of locating points on the surface of the earth, which is considered as a sphere.

Coordinates for Points in Space

A line is one-dimensional; it has length only. You could think about moving up and down the line in one way only. You could not turn a 90° angle and still be on the line. A plane is two-dimensional; it has length and width. You could think of moving along on a plane, turning a 90° angle, and still being on the plane. However, if you turned another 90° angle, you would either be moving parallel to part of your previous path or be off the plane. Mathematical space, based on the space of experience, is considered here to be three-dimensional; it has length, width, and depth.

In space, you could visualize moving along a line, turning a 90° angle, then turning a second 90° angle, and still not be moving parallel to part of your previous path. Loosely speaking, you have a degree of freedom of movement for each dimension. Thus, in three dimensions, you have freedom to move in three ways.

An understanding of the mathematical significance of dimension must go far beyond this intuitive view, however, and depends on the use of coordinates. The idea of locating a small object in a room will help you

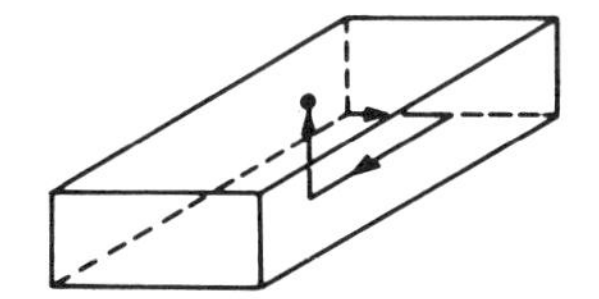

Figure 11.1. Locating object in space

understand the basis for the usual way of locating points in the mathematical model for space. The object in the room is located by starting at a particular corner. Study Figure 11.1. First you must move 3 feet along the front wall. Then you must move 4 feet from this point, parallel to the side of the room. Finally, the object is located 5 feet above the position on the floor you have located. Do you agree that you have selected exactly one location and that you have used a set of three instructions to locate it?

In the mathematical model, three reference planes, as shown in Figure 11.2, take the place of the floor and two walls of the room. These three planes meet at right angles at a single point called the *origin*. Every point in space is located by reference to the three fixed planes.

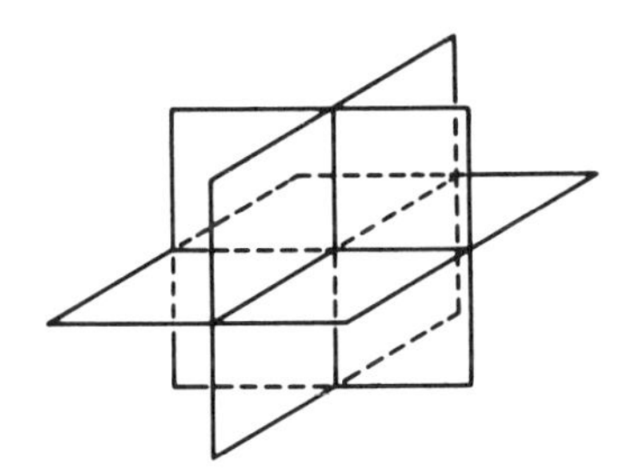

Figure 11.2. Three reference planes

Points in space are named by assigning to each point an ordered triple of coordinates (x, y, z), which serve as a set of instructions for reaching the

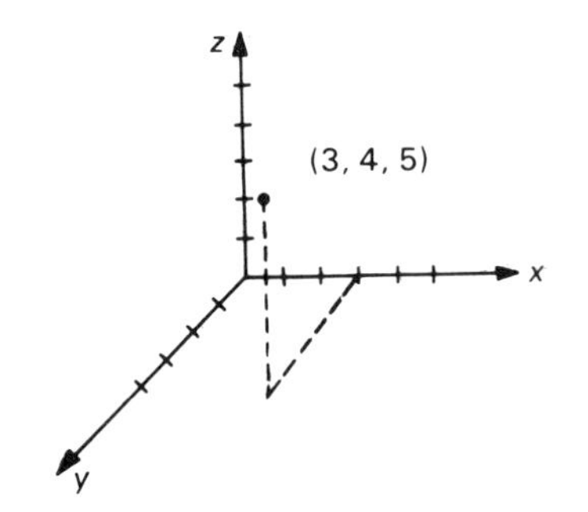

Figure 11.3. Locating a point in space

point from the origin. First, it is necessary to establish three axes (number lines) meeting at right angles. These are often drawn as they are in Figure 11.3. You should compare Figure 11.3 with Figures 11.1 and 11.2.

In Figure 11.3, the point (3, 4, 5) is located three units along the x-axis from the origin, then four units from that position parallel to the y-axis, then five units from that position parallel to the z-axis. Notice that each pair of axes determines one of the three reference planes, which are called the x-y-plane, the x-z-plane, and the y-z-plane to indicate the axes they contain.

Examples

Locate the points $(3, -4, 5)$, $(5, 3, -4)$, and $(-3, 4, 5)$.

Figure 11.4 shows the solution. The dotted lines are not necessary but are helpful as a means of visualization.

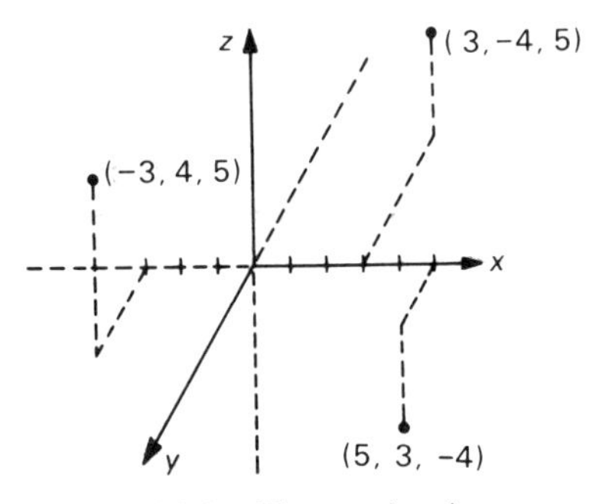

Figure 11.4. Three points in space

The use of three axes makes it possible to write mathematical sentences for sets of points in space, although the procedure is somewhat more complicated than it is for a plane. The first problem is determining a mathematical sentence that will designate the sets of points in each of the three fixed coordinate planes, using x, y, and z. Figure 11.5 shows four points, all of

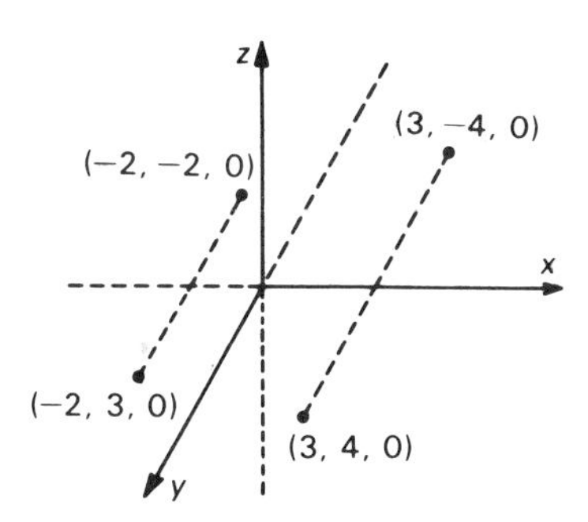

Figure 11.5. Four points in a reference plane

which lie in the x-y-plane. What do you notice about all these sets of coordinates? Do you think that any point in the x-y-plane has zero for its third coordinate? Do you think there are any other points in space that have zero for their third coordinate? It seems reasonable that the equation of the x-y-plane is $z = 0$. A set of number triples (a, b, c) represents a point in the plane $z = 0$ if and only if its third coordinate is zero. The statement may be expressed using set-builder notation as $\{(x, y, z)|z = 0\}$. In a similar way, you can arrive at the equations for the other two given planes; the equation for the x-z-plane is $y = 0$ and the equation for the y-z-plane is $x = 0$.

A second problem is finding the equation of a plane in space parallel to one of the three fixed planes. For example, Figure 11.6 shows several points

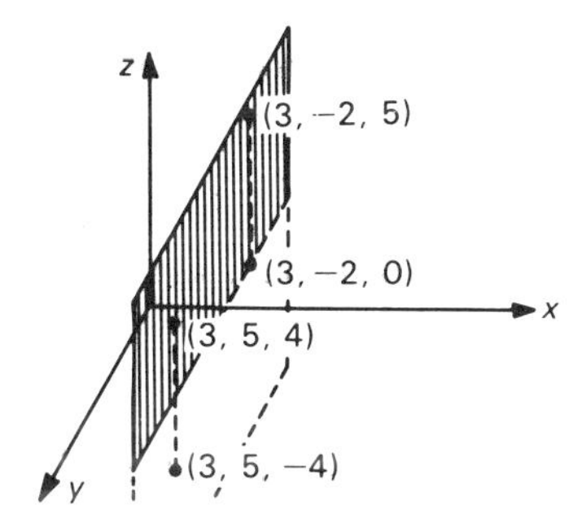

Figure 11.6. Four points in a plane

lying in a plane parallel to the plane $x = 0$. What do the sets of coordinates have in common? Do you think that every point with a first coordinate of 3 lies in this same plane? Do you think that any points with a first coordinate of 3 lie elsewhere than in this plane? Then the equation $x = 3$ represents the plane shown. The plane contains the points $\{(x, y, z)|x = 3\}$.

Other planes parallel to one of the reference planes have equations in one of the forms $x = a$, $y = b$, or $z = c$. For example, the plane $y = 5$ is parallel to the plane $y = 0$ and lies five units in front of it in a drawing. The plane $z = 2$ is parallel to the plane $z = 0$ and lies two units above it. The plane $z = -3$ is parallel to the plane $z = 0$ and is three units below it.

Examples of more general equations for planes are $2x + 7z - 5 = 0$ and $6x - y + 2z - 4 = 0$. In three dimensions, equations of planes are linear, just as equations of lines are linear in two dimensions. The general form for the equation of a plane is $ax + by + cz + d = 0$. A point lies on a plane if and only if its coordinates satisfy the equation of the plane.

Example

The point $(4, 1, -3)$ lies on the plane

$$2x + y + 2z - 3 = 0, \quad \text{since}$$

$$2(4) + 1(1) + 2(-3) - 3 = 0.$$

The formula for the distance between two points, (x_1, y_1, z_1) and (x_2, y_2, z_2) in space can be found by using the Pythagorean Theorem twice. This procedure is illustrated in Figure 11.7.

$$P_1C = \sqrt{(x_2 - x_1)^2 + (y_2 - y_1)^2}$$

$$P_2C = z_2 - z_1$$

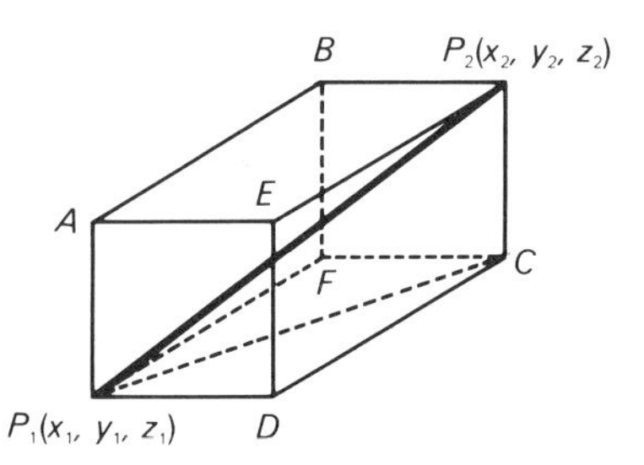

Figure 11.7. Distance between points in three dimensions

For triangle P_1CP_2,

$$P_1P_2 = \sqrt{(P_1C)^2 + (P_2C)^2}$$
$$= \sqrt{(x_2 - x_1)^2 + (y_2 - y_1)^2 + (z_2 - z_1)^2}$$

Example

Find the distance d between the points (3, 4, 2) and (−1, 2, 5).

$$d = \sqrt{(3 + 1)^2 + (4 - 2)^2 + (2 - 5)^2}$$
$$= \sqrt{4^2 + 2^2 + (-3)^2}$$
$$= \sqrt{16 + 4 + 9}$$
$$= \sqrt{29}$$

The distance formula can be used to help find the equations for other sets of points in three dimensions.

Example

Find the set of points, in three dimensions, such that the distance from the origin is three units.

In Figure 11.8, $OP = 3$, so that

$$\sqrt{x^2 + y^2 + z^2} = 3, \quad \text{or}$$
$$x^2 + y^2 + z^2 = 9.$$

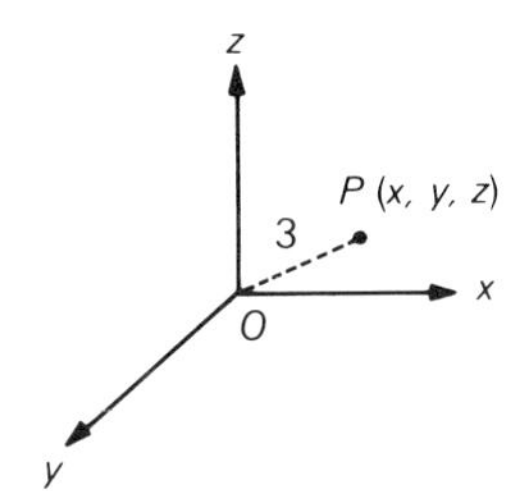

Figure 11.8. Points three units from origin

This last equation describes the set $\{(x, y, z) | x^2 + y^2 + z^2 = 9\}$ of points on a sphere.

In addition to the system of axes used in this section, other systems such as those pictured in Figure 11.9 are often used.

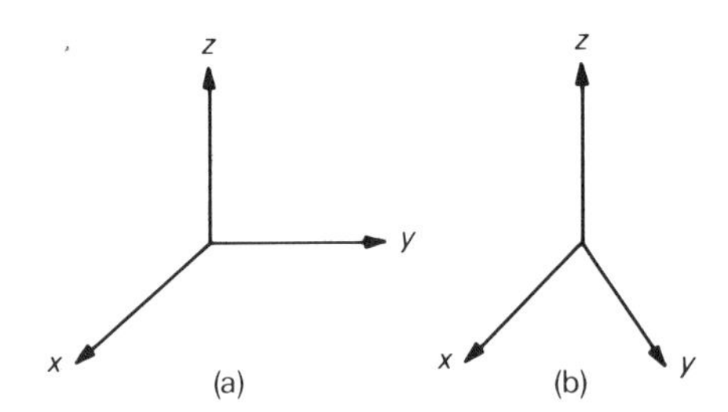

Figure 11.9. Alternate coordinate axes systems

The idea of more than three dimensions is greatly misunderstood by many people. It makes sense to talk about four dimensions as allowing another degree of freedom of movement as compared to three dimensions. In four dimensions, it is possible to make one more 90° turn without going parallel to your previous path than is possible in three dimensions. Space of more than three dimensions is not the space of naïve experience and is difficult to picture. However, by using the notation of coordinates, it is a simple matter to represent points in more than three dimensions, to write equations for sets of points in more than three dimensions, and to find practical examples of the need for equations with many more than three letters representing numbers. For example, the point (3, 5, 7, 2) is a particular point in four-dimensional space, and the point (3, 5, 7, 2, 1) is a point in space of five dimensions. The mathematician finds it necessary to use *n-tuples*, which are sets of coordinates $(a_1, a_2, \ldots, a_n)$ that can be thought of as representing points in n-dimensional space.

From an algebraic point of view, the number of dimensions may be thought of as the number of variables, rather than a number of distinct physical measurements or locations. Thus an equation of the form $ax + by + cz + dw = C$ represents a set of points in four dimensions and a relationship involving the four variables x, y, z, w. The fact that the physical world we live in seems to be three-dimensional does not keep the mathematicians from creating a meaningful mathematics of n dimensions for n greater than 3.

Exercise 11.1

1. Illustrate, as in Figure 11.1 (p. 235), the location of an object found by measuring 5 feet along the front wall, then 3 feet from this point parallel to the side of the room, then 6 feet up from the floor.

2. What are the coordinates of the origin in space?
3. Locate all these points on the same set of axes:
   (a) (1, 4, 5) (b) (2, 7, 1) (c) (3, 2, 2)
4. Locate all these points on the same set of axes:
   (a) (2, 4, 6) (b) (−2, 4, 6) (c) (2, −4, 6) (d) (2, 4, −6)
5. Which of the given points lie on the plane:
   (a) $x = 0$? (b) $y = 0$?
   (3, 4, 0) (0, −2, 3) (0, 5, 18) (1, 0, −4) (0, 5, 2) (1, 1, 1)
6. Describe the location of each plane:
   (a) $y = 3$ (b) $x = -4$ (c) $z = 9$
7. Write the equation of the plane parallel to a reference plane that contains each set of points:
   (a) (4, 2, −5), (7, 2, 6), (−9, 2, $^1/_2$)
   (b) (1, 5, 7), (1, 3, −2), (1, −4, 2)
   (c) (4, 3, $^3/_4$), ($^3/_4$, 7, $^3/_4$), (−6, −2, $^3/_4$)
8. Sketch each plane in Exercise 6.
9. Which of the given points lies on the plane $2x - y + z - 3 = 0$?
   (a) (1, 2, 3) (b) (3, 2, 1) (c) (−1, −1, 4) (d) (0, 1, 4)
10. Find the distance between the pairs of points.
    (a) (2, 4, 1) and (7, 5, 3) (b) (2, 4, 1) and (−7, 5, 3)
    (c) (2, 4, 1) and (−7, −5, 3) (d) (2, 4, 1) and (−7, −5, −3)
11. Write the equation for the sphere with radius 4 and center at the origin.
12. Describe the sets of points.
    (a) $\{(x, y, z) | x > 5\}$ (b) $\{(x, y, z) | z < 2\}$

Locating Points on the Surface of the Earth

The assumption that the surface of the earth is a sphere is the basis for the common method of indicating the position of the special set of points in space on the surface of the earth. You have learned that a *great circle* on a sphere is the set of points in the intersection of the sphere and a plane through the center of the sphere and that another plane section of the sphere will always be a *small circle.* The special sets of great circles and small circles used to locate points on the surface of the earth are called lines of *longitude* and *latitude,* respectively.

Figure 11.10 shows some of the lines of longitude, which are the set of great circles passing through the North and the South Poles. These lines of longitude are called *meridians.* Originally, the line of longitude that passed

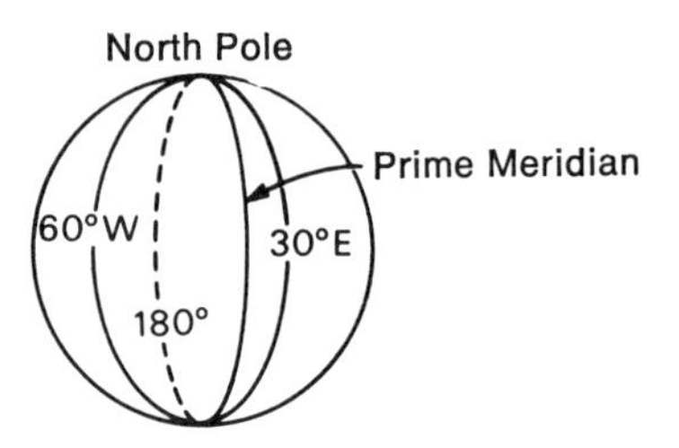

Figure 11.10. Lines of longitude

through the Greenwich Observatory, in England, was called the *Prime Meridian* and was labeled 0°. Although the observatory itself has since been moved, the original numbering system has been retained. The extension of the Prime Meridian passing through the Pacific Ocean on the other side of the earth from England is labeled 180°.

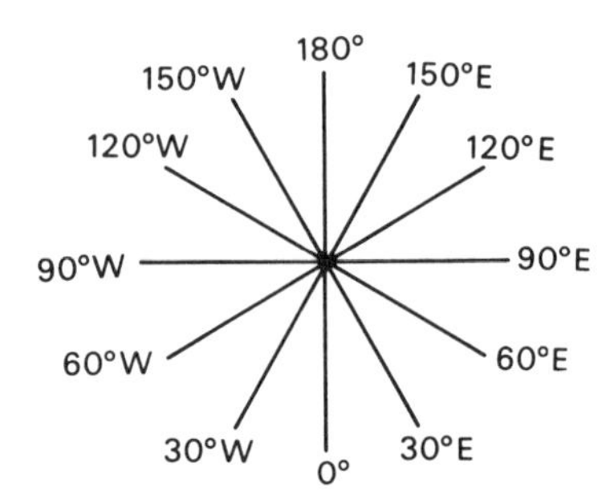

Figure 11.11. Lines of longitude at pole

The set of lines of longitude makes it possible to locate a point on the surface of the earth as so many degrees east or west of the Prime Meridian. Figure 11.11 shows how the lines of longitude intersect at each pole and how they are labeled in degrees. Check by using a globe to see that the longitude of each of the following cities is approximately that given: Naples, 15°E; Philadelphia, 75°W; Baghdad, 45°E; Memphis, 90°W.

The lines of longitude alone do not locate a specific point on the earth's surface; to complete the descripion of the location requires the use of lines of latitude. Some lines of latitude are shown in Figure 11.12. The lines of latitude run east and west around the earth. The line of latitude halfway between the North and South Poles is called the *Equator* and is, unlike the other lines of latitude, a great circle. Lines of latitude are numbered north and south from the equator, which is called 0° latitude. The North Pole is located at 90° north latitude and the South Pole is located at 90°

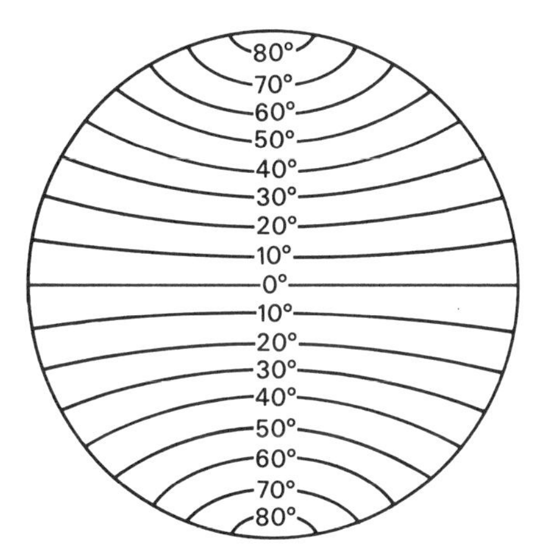

Figure 11.12. Lines of latitude

south latitude. Check by using a globe to see that the latitude of each of the following cities is approximately that given: Cairo, 30°N; Denver, 40°N; Buenos Aires, 35°S; Leningrad, 60°N.

In addition to the Equator, several other lines of latitude are given special names, as pictured in Figure 11.13. These are, from north to south, the *Arctic Circle,* about $23^1/_2$° from the North Pole; the *Tropic of Cancer,* about $23^1/_2$° north of the Equator; the *Tropic of Capricorn,* about $23^1/_2$° south of the Equator; and the *Antarctic Circle,* about $23\frac{1}{2}$° from the South

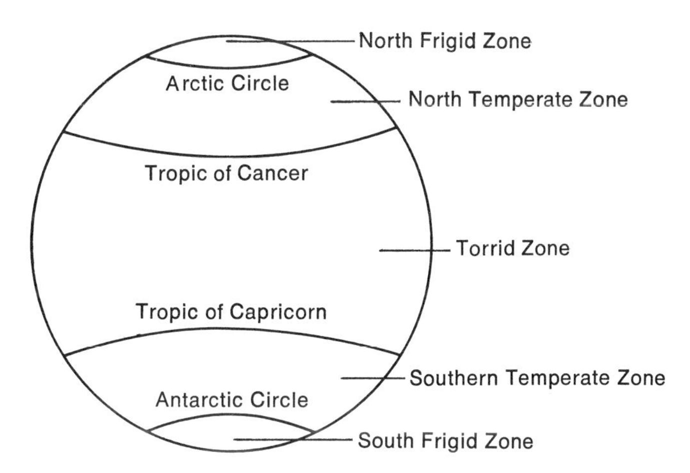

Figure 11.13. Zones on earth's surface

Pole. These four special lines of latitude partition the sphere into five zones. From the north, these zones are the *North Frigid zone,* the *North Temperate zone,* the *Torrid zone,* the *South Temperate zone,* and the *South Frigid zone.*

The location of a point on the surface of the earth is usually given by naming first its latitude and then its longitude. Actually, the degrees of measure are subdivided into minutes and seconds, but that subdivision will not be considered here. All the coordinates for the locations will be considered as approximate. For example, here are the approximate locations of several cities: Naples, 41°N, 14°E; Denver, 40°N, 105°W; Buenos Aires, 35°S, 58°W; and Melbourne, 38°S, 145°E. It is interesting to note that lines of latitude and longitude, including the designations of N, E, S, or W, make it possible to locate a point on the surface of a sphere by using only two coordinates—unlike the system of space coordinates studied previously, which requires three numbers.

Exercise 11.2

1. Find the approximate longitude of each city: Cairo, Leningrad, Madrid, New York, and New Delhi.
2. Find the approximate latitude of each city: Philadelphia, Baghdad, Memphis, Madrid, and New York.
3. Which zone contains each city listed in Exercise 1?
4. Which zone contains each city listed in Exercise 2?
5. Locate each city approximately by giving its latitude and longitude, using correct notation:
   (a) San Francisco, California (b) Caracas, Venezuela
   (c) Capetown, South Africa (d) Oslo, Norway
6. In what country would you be if you were at each location listed?
   (a) 10°S, 45°W (b) 50°N, 100°W
   (c) 45°N, 3°E (d) 50°N, 50°E
7. Would you be on land or on water at each of these locations?
   (a) 30°N, 30°W (b) 10°S, 75°E
   (c) 65°N, 150°W (d) 40°N, 100°W

The study of surface area is a direct extension of the concept of area of a plane region. The study of *volume* of three-dimensional solids requires a new type of unit of measurement, new standard units, and the development of formulas that make it possible to find volume, given linear and square measurements. Many of the ideas of measurement for curved surfaces and solids cannot be established rigorously without the use of limits and hence must be presented very informally.

Measurements of Prisms and Prismatic Solids

The *surface area* of a prism is the combined area of the various regions composing the surface. For the rectangular prism shown in Figure 12.1, the measure of the surface area in square units is the sum of the measures of the surface area in square units of each of the six faces. If the measures of the edges in linear units are given, then the measures of the faces in square units can be found.

measure of region $ABCD$	$3 \times 4 = 12$
measure of region $EFGH$	$3 \times 4 = 12$
measure of region $DAEH$	$3 \times 2 = 6$
measure of region $CBFG$	$3 \times 2 = 6$
measure of region $ABFE$	$4 \times 2 = 8$
measure of region $DCGH$	$4 \times 2 = 8$
measure of surface of prism	52

Can you explain how to take a shortcut in finding the surface area, since there are three pairs of faces with the same area?

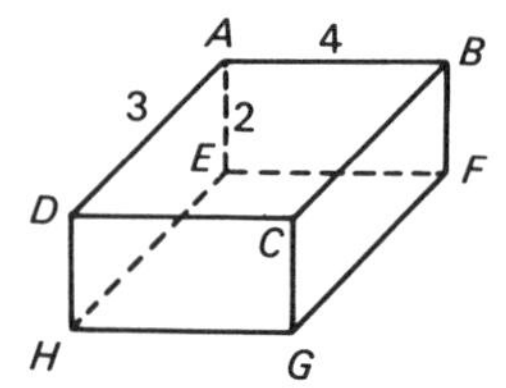

Figure 12.1. Surface area of prism

For prisms other than rectangular ones, you must still be able to find the measure of area for each base and each side and then add all the measures. For example, finding the surface area of a triangular prism would require finding the areas of the triangular bases and the three rectangular sides.

Sometimes, it is practical to find all of the surface area of the prism except for that of the two bases. This area is called the *lateral surface area.* Of course, it is still convenient to add the measures of area of each of the sides, but one interesting observation can be made. In Figure 12.1, the lateral area is

$$(4 \times 2) + (3 \times 2) + (4 \times 2) + (3 \times 2) = (4 + 3 + 4 + 3) \times 2.$$

The expression in parentheses is the measure of perimeter of the base, whereas the second factor, 2, is the measure of the altitude. A formula for the measure of the lateral surface area of a right prism can be written

$$A = ph,$$

where p is the measure of the perimeter of the base.

Example

Find the lateral surface area of a right prism if the perimeter of the base is 17 cm and the height is 4 cm.

$$\begin{aligned} A &= ph \\ &= 17 \cdot 4 \\ &= 68 \end{aligned}$$

The lateral surface area is 68 sq cm.

The problem of finding the surface area of a prism might occur in such practical situations as painting a room. But you might also wonder about how much space is in the room, or about how much space a block of wood occupies. There is, then, a second type of measurement connected with a prism and its interior, or a prismatic solid, to indicate how much space is occupied or filled by the object, or how much space is in the interior of the prism.

The measurement of space a solid occupies is called its measurement of *volume.* Since volume is a new concept of measurement, it is necessary to have a new unit of measurement of volume that will have the same nature as the thing being measured. Figure 12.2 shows a cubic unit of volume and indicates how you may think of placing these units inside a prism. Is the number of cubic units that can be placed along the long side the same as the measure of length? Is the number of cubic units on the entire bottom the same as the product lw? Is the number of layers of cubic units the same as the measure of height?

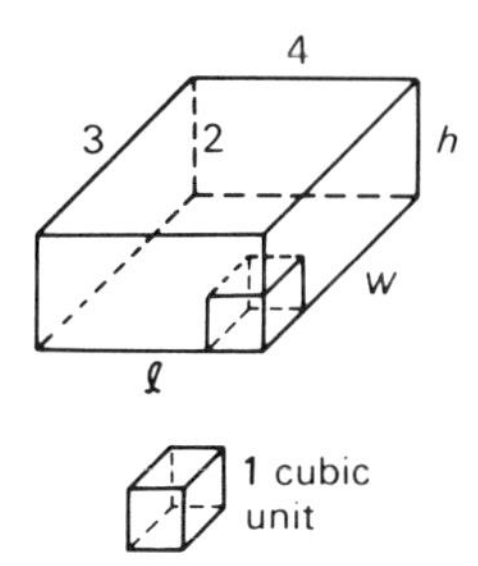

Figure 12.2. Volume of prismatic solid

In Figure 12.2, the number of small units that may be placed in the interior of the prism is $4 \times 3 \times 2 = 24$. The measure of volume is 24, using the cubic unit given. You can see that the formula for the measure of volume of a rectangular prismatic solid, in cubic units, is

$$V = lwh,$$

where l is the measure of length, w is the measure of width, and h is the measure of height.

Example

Find the volume of a prismatic solid if the length is 9 cm, the width is 4 cm, and the height is 5 cm.

$$\begin{aligned} V &= lwh \\ &= (9)(4)(5) \\ &= 180 \end{aligned}$$

The volume is 180 cu cm.

For the special case of a cube and its interior, the formula $V = lwh$ becomes

$$V = eee = e^3,$$

where e is the measure of length of one edge of the cube. The expression e^3 is read "e to the third power" or "e cubed." The exponent 3 indicates that the base, e, is to be used as a factor three times.

Example

A cubical solid 4 units along each edge has a measure of volume of 4^3 or 64.

Since, in the formula $V = lwh$, the product lw is the measure of area of the base of the rectangular prism, the formula may be written

$$V = Ah,$$

where A is the measure of area of the base. In this form the formula is more readily seen to be applicable to right prisms that are not rectangular. For example, the formula for the measure of volume of the solid whose surface is a triangular right prism is $V = Ah$, with A the measure of area of the triangular base.

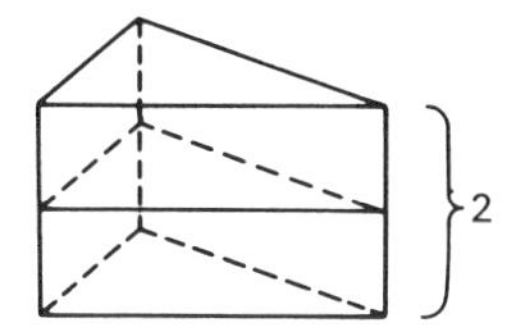

Figure 12.3. Volume of solid with triangular base

Example

If the solid in Figure 12.3 has a measure of height of 2 and a measure of area for the base of $^3/_2$, then the measure of volume is $^3/_2 \times 2 = 3$.

A further extension of the use of the formula $V = Ah$ for measure of volume can be imagined by considering Figure 12.4. Figure 12.4a shows small pieces of wood stacked at an angle; Figure 12.4b shows them stacked

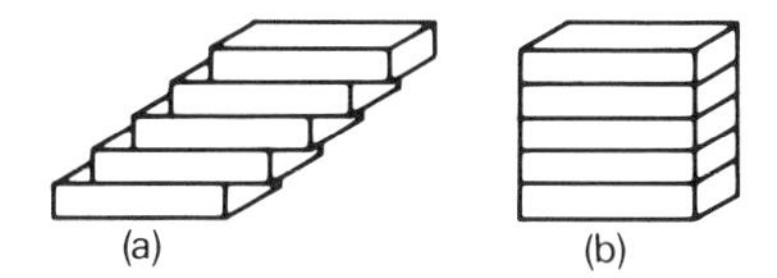

Figure 12.4. Volume of oblique prismatic solid

in a straight pile. In both cases the amount of wood is the same, and the volume of the two piles is identical. The measure of volume for an oblique prismatic solid is the same as the measure of volume for a corresponding right prismatic solid with the same height if their bases are congruent. This final observation makes it possible to generalize that, for any prismatic solid, right or oblique, the measure of volume is Ah, where A is the measure of area of the base and h is the measure of height.

The final extension of the axiomatic system for Euclidean geometry is necessary to support the statements about volume considered informally in this section.

Axiom 27

There exists a correspondence that associates the number 1 with a certain geometric solid and a unique positive real number with every geometric solid.

Axiom 28

If a polyhedral solid is the union of two polyhedral solids whose interiors do not intersect, the measure of volume is the sum of the measures of volume of the two solids.

Axiom 29

For two polyhedral solids each having a face in a given plane, if every plane parallel to the given plane has intersections with equal areas with the two regions, the volumes of the two solids are equal.

Axiom 30

If two solids are congruent, they have equal volumes.

Axiom 27 provides for unit solids to use in measuring volume. Axiom 29 is called *Cavalieri's Principle* and is a generalization of the idea shown in Figure 12.4. A more general approach is illustrated in Figure 12.5. For

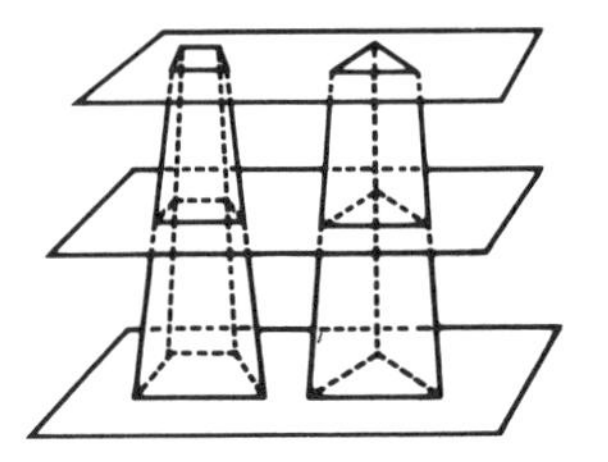

Figure 12.5 Cavalieri's Principle

any plane parallel to the two planes shown, the cross sections have the same area, although the shape is quite different. The two volumes are therefore equal.

The concept of similar figures can be extended to three-dimensional surfaces and solids. Similar figures have the same shape. The corresponding sides of similar prisms are similar plane regions. The corresponding edges of similar figures have a constant ratio.

Figure 12.6 shows two similar right prisms with a ratio of similarity of $^{2}/_{1}$.

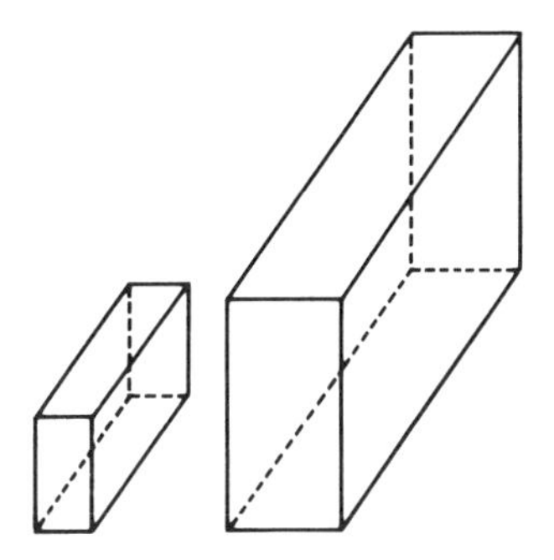

Figure 12.6. Similar right prisms

Since each of the dimensions has this same ratio, the volume of the larger prism is

$$\begin{aligned} V &= (2l')(2w')(2h') \\ &= 8V', \end{aligned}$$

where the primes refer to the smaller prism. From this formula the generalization can be stated that the ratio of volumes is the cube of the ratio of the sides.

Example

What is the volume of a cubical box if the edges are twice the length of a similar box with edges 4 in. long?

The volume of the original box is 64 cu in. The larger box has a volume eight times as great, so its volume is 512 cu in.

Volume is measured in cubic units. Three common standard cubic units are the cubic inch, the cubic foot, and the cubic yard. A cubic inch may be visualized as the amount of space occupied by a small object in the shape of a cubical solid 1 inch on each edge. This cubical solid has a measurement of volume of 1 cubic inch. The following two statements of equivalence can be explained by thinking of placing cubic-inch blocks inside the larger unit, as in Figure 12.2 (p. 247): 1 cubic foot = 12^3 cubic inches = 1728 cubic inches and 1 cubic yard = 3^3 cubic feet = 27 cubic feet.

In the metric system, statements of equivalent measurements for cubic units are easily derived, since they always involve powers of 10. For example, 1 cubic centimeter = 1000 cubic millimeters and 1 cubic meter = 1,000,000 cubic centimeters.

Exercise 12.1

1. Find the measure of surface area of each rectangular right prism.
   (a) $l = 5, w = 2, h = 6$ (b) $l = 9, w = 4, h = 8$
   (c) $l = {}^4/_3, w = {}^3/_4, h = {}^7/_4$ (d) $l = {}^2/_3, w = {}^9/_5, h = {}^2/_5$
   (e) $l = 2^1/_2, w = 4^1/_2, h = 5$ (f) $l = 5^3/_4, w = 3, h = 4^1/_4$
2. Find the measure of lateral surface area of each prism in Exercise 1.
3. Find the measure of volume of each prismatic solid, using the measures in Exercise 1.
4. Find the surface area of a cube with the measure of edge given.
   (a) $e = 8$ (b) $e = {}^3/_4$
5. Find the measure of volume for each cubical solid with an edge as indicated in Exercise 4.
6. Find the measure of volume for the solid, using the dimensions given.

	measure of area of base	*measure of height*
rectangular prismatic solid	15	4
triangular prismatic solid	17	9
hexagonal prismatic solid	$14^1/_2$	7
octagonal prismatic solid	${}^9/_4$	${}^7/_4$

7. 1 cu yd = ? cu in.
8. 1 cu m = ? cu dec.
9. Use the formula $V = lwh$ to find the unknown measurement for a rectangular right prismatic solid.

length	*width*	*height*	*volume*
6 in.	____	4 in.	12 cu in.
____	4 m	7 m	84 cu m
6 ft	9 ft	____	4 cu ft

10. What will be the effect on the volume of a prismatic solid if the measure of each edge is doubled?
11. What would be the effect on the volume of a prismatic solid if the measure of each edge is tripled?
12. Find the approximate number of cu ft in the rooms of a house with the shape shown in Figure 12.7 if the height is approximately 8 ft.
13. Find the number of sq ft in the walls of the house in Exercise 12.

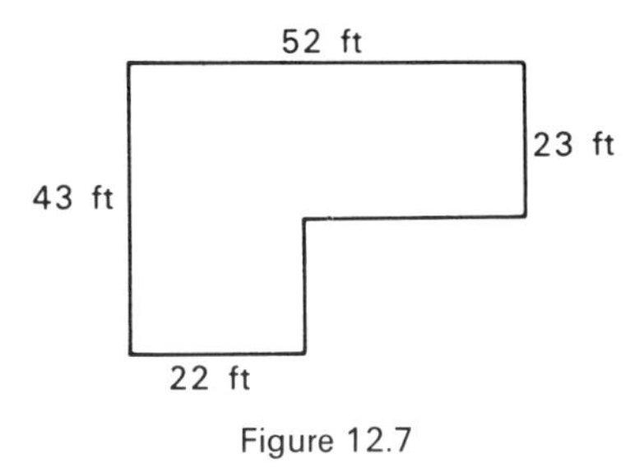

Figure 12.7

14. Find the distance between opposite corners of a cube with side 3 in.
15. Find the distance between opposite corners of a prismatic solid with length 2 in., height 3 in., and width 1 in.

Measurements of Cylinders and Cylindrical Solids

The surface of a cylinder consists of the two bases and the curved side. To find the total measure of surface area, find twice the measure of area of the base and add it to the measure of lateral surface area.

Finding the area of the circular base is familiar, but finding the measure of the lateral surface area may be new. Imagine the lateral surface of the cylinder cut along segment AB, as in Figure 12.8a, and then laid flat, as in

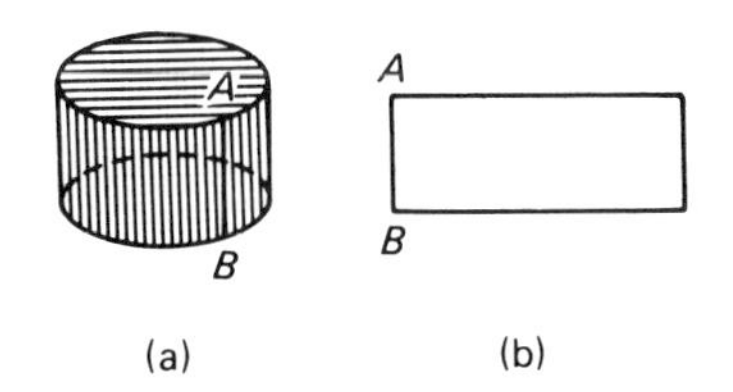

Figure 12.8. Lateral surface of cylinder

Figure 12.8b. You can then see that the lateral surface is actually a rectangular region. The height of this rectangle is the height of the cylinder, and the length is the perimeter of the base. Then the measure of area of the lateral surface is ph, where p represents the measure of perimeter of the base and h represents the measure of height of the cylinder.

Example

If a cylinder has a measure of radius of 4 and a measure of height of 7, then the measure of lateral surface area is $2\pi 4 \times 7$, or 56π.

With the understanding of lateral surface area in the last paragraph, you may now write a formula for the total measure of surface area of a right circular cylinder:

$$S = 2\pi r^2 + 2\pi rh.$$

You should explain each symbol in the formula and also indicate how the distributive property may be used to write the formula as $S = 2\pi r(r + h)$. You should recall that, for three numbers a, b, and c, the distributive property of multiplication over addition states that $a(b + c) = ab + ac$.

Example

A cylinder with a measure of radius of 3 and a measure of height of 5 has a total measure of surface area of $2\pi(3)^2 + 2\pi(3)(5) = 18\pi + 30\pi = 48\pi$. Using a rational approximation, such as 3.14, for π will give an approximate answer in decimal form if that is desired.

A person would be interested in the surface area of a water tank if it needed to be painted. If he wanted to know how much water the tank would hold, however, he would be interested in the volume of the corresponding cylindrical solid, or the interior of the cylinder.

Finding the volume of a cylindrical solid is based on an analysis very similar to that for the prismatic solid. Study Figure 12.9, which pictures a cylindrical solid with a measure of height of 3. You may think of the solid as being

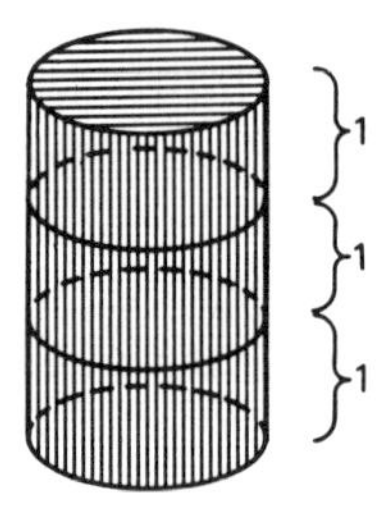

Figure 12.9. Volume of cylindrical solid

made up of three discs placed on top of each other. The measure of volume of each disc is one times the measure of area of the base, or πr^2. For this cylinder, if the measure of the radius is 2, then the measure of volume is $\pi(2)^2 3 = 12\pi$.

From the last paragraph, it is apparent that a formula for the measure of volume may be written as

$$V = \pi r^2 h \quad \text{or} \quad V = Ah,$$

where A is the measure of area of the base. In the second form, the formula is also applicable to right cylinders whose base is not a circle.

Example

Find the measure of volume of a cylindrical solid if the measure of area of the base is 7π and the measure of height is 4.

$$\begin{aligned} V &= Ah \\ &= 28\pi \end{aligned}$$

Exercise 12.2

1. Find the measure of lateral surface area and the total measure of surface area for a right circular cylinder, using the measures given. Leave the symbol π in the answers.

measure of radius	*measure of height*
4	9
15	4
6.1	7.2
$3/4$	$5/4$
$4\frac{1}{2}$	$5\frac{1}{2}$

2. Find the measure of volume for a cylindrical solid, using the measures given in Exercise 1. Leave the symbol π in the answers.
3. What is the effect on the measure of lateral surface area of a cylinder if the measure of height is doubled?
4. What is the effect on the measure of volume of a cylindrical solid if:
   (a) the measure of height is doubled?
   (b) the measure of radius of the base is doubled?

In Exercises 5–6, find the approximate number of cu in. of metal in the pipe shown in Figure 12.10.

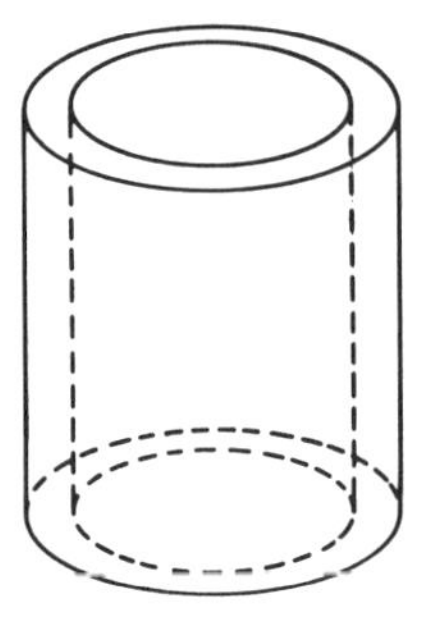

Figure 12.10

5. If the inner radius is 1 in., the outer radius is $1\frac{1}{4}$ in., and the pipe is 11 in. long. Use 22/7 for π.
6. If the inner radius is 1.4 in., the outer radius is 1.6 in., and the pipe is 12 in. long. Use 3.14 for π.

Measurements of Pyramids, Cones, and Related Solids

The measure of surface area for a pyramid may be found by adding the measure of area of the base and each of the faces. Regardless of the shape of the base, each of the faces is a triangular region; hence you can find its area if the height and base of each of the triangles are given.

Example

In Figure 12.11, the measure of surface area of the pyramid is

$$\begin{aligned} S &= 4^2 + 4\left(\frac{1}{2}\right)(4)(5) \\ &= 16 + 40 \\ &= 56. \end{aligned}$$

The measure of area of each triangular region is $\frac{1}{2}(4)(5)$, and there are four of these regions.

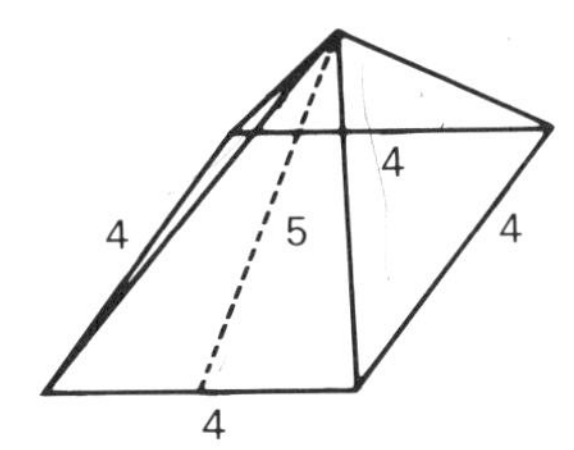

Figure 12.11. Surface area of pyramid

The rigorous establishment of a formula for finding the measure of volume of a pyramidal solid is normally a part of the study of formal solid geometry and is not included in this text. One can often arrive at the proper formula by experimenting with paper or metal models of a rectangular right prism and a pyramid with the same height with a rectangular base congruent to the base of the prism. How many times must the model of the pyramid be filled with sand to fill the model of the prism by pouring the sand from the pyramid model into the prism model? Do you agree that the answer is three?

The volume of a pyramidal solid is $^1/_3$ that of a prismatic solid with the same height if their bases are congruent. The formula for the measure of the volume of a pyramidal solid is

$$V = {}^1/_3 Ah,$$

where A is the measure of area of the base of the pyramid.

Example

A pyramidal solid with a measure of area of base of $^3/_4$ and a measure of height of $^7/_4$ has a measure of volume of $V = {}^1/_3({}^3/_4)({}^7/_4) = {}^7/_{16}$.

The surface of a right circular cone is composed of the base and the lateral surface. The measure of area of a circular base can be found by using the formula $A = \pi r^2$. At first glance, the lateral surface area does not seem to be a familiar shape. But, like the lateral surface of a cylinder, the lateral surface of a right circular cone can be thought of as being cut and laid out flat. As in Figure 12.12, the lateral surface of the cone can be drawn as a sector of a circle. The length of segment AB is called the slant height of the cone, and its measure is designated here by s. The length of the arc BC is the circumference of the base of the cone, which is $2\pi r$.

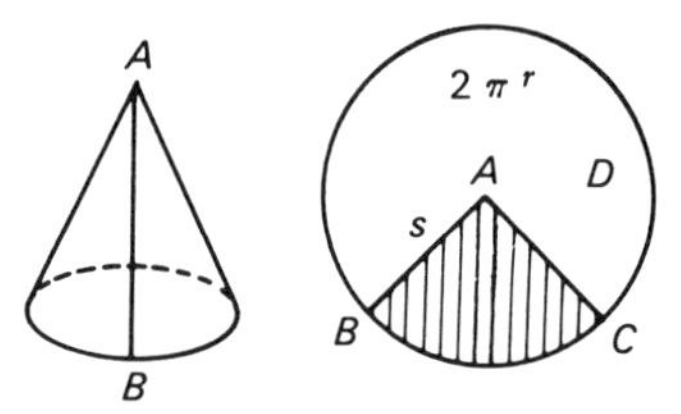

Figure 12.12. Lateral surface area of cone

The measure of area of a circular sector, indicated here by D, can be found by considering the measure of the sector as a certain fraction of the measure of area of the entire circle. The ratio of the measure of area of a sector to the measure of area of the entire circle is the same as the ratio of the measures of the arcs. Hence, $D/\pi s^2 = 2\pi r/2\pi s$ or $D/\pi s^2 = r/s$. Then

$$D = \pi r s,$$

which is the normal way of writing the formula for the measure of lateral surface of a cone. The measure of area of the lateral surface is π times the product of the measures of radius and slant height of the cone. Then the formula for the surface area of the cone is

$$S = \pi r^2 + \pi r s.$$

Example

A cone with a slant-height measure of 3 and a measure of radius of 4 has a measure of surface area of $S = \pi(4)^2 + \pi(4)(3) = 16\pi + 12\pi = 28\pi$. A formula for surface area may be written, using the distributive property of multiplication over addition, as $S = \pi r(r + s)$. Compare this with the second way the formula for the surface area of a cylinder was written.

Experimenting with paper or metal representations of a right circular cylinder and a right circular cone with the same height and with congruent bases will yield results analogous to those for the prism and the pyramid. The volume of a conical solid has a measure $^1/_3$ that of a cylindrical solid with the same height and congruent bases. The formula for the measure of the volume of a conical solid is $V = {}^1/_3 Ah$, where A is the measure of area of the base of the cone.

Example

A conical solid with a measure of area of base of $^7/_5$ and a measure of height of $^4/_5$ has a measure of volume of $V = {}^1/_3({}^7/_5)({}^4/_5) = {}^{28}/_{75}$.

Exercise 12.3

1. Find the measure of volume for pyramidal solids with the measures given.

measure of area of base	*measure of height*
9	37
$^1/_2$	$^5/_2$
6.41	7.59

2. Find the measure of surface area for a cone with the measures given. Leave the symbol for π in the answers.

measure of slant height	*measure of radius*
35	15
$^9/_4$	$^3/_4$
5.38	5.29

3. Find the measure of volume for the conical solids with the measures given.

measure of area of base	*measure of height*
6	19
$^7/_8$	$^2/_3$

4. What is the area of the base of a pyramidal solid if the measure of volume is 35 cu ft and the measure of height is 5 ft?
5. What is the effect on the measure of volume for a conical solid if the measure of height is doubled?
6. What is the effect on the measure of height of a conical solid if the measure of volume is doubled and the radius is unchanged?
7. What is the radius of a cone if the measure of slant height is 9 and the measure of lateral surface area is 18?
8. What is the radius of a cone if the measure of height is 9 and the measure of volume is 18?

Measurements of Spheres and Spherical Solids

Finding the amount of material in a ball and finding the number of square miles on the surface of the earth are examples of the need to find measures of the surface area of a spherical object. The exact formula for the measure of the surface area of a sphere cannot be determined without the aid of the limit-concept methods of calculus.

It is relatively easy to find a rough approximation for the surface area of a sphere by comparing it with the surface area of other figures that are known to have a surface area either larger or smaller than the sphere in question. For example, you may think about the sphere as being fitted into a cube of side $2r$, as in Figure 12.13. The measure of surface area for

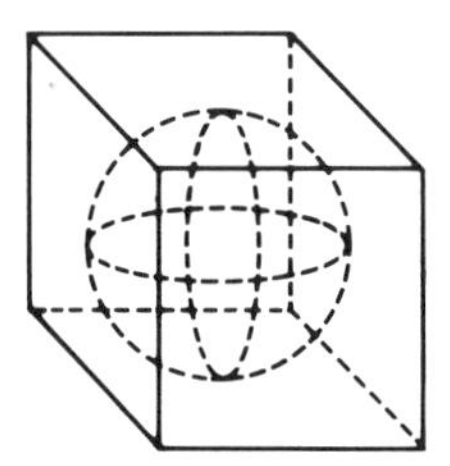

Figure 12.13. Sphere inscribed in cube

each face of the cube is $2r \times 2r = 4r^2$, and the total measure of surface area is $6 \times 4r^2$ or $24r^2$. It seems, looking at Figure 12.13, that the surface area of the cube is almost twice that of the sphere within it, so that a rough estimate of the measure of surface area for a sphere of radius r is $12r^2$. It would be misleading to claim that you could arrive at the right formula by the procedure in this paragraph, but trying several similar approaches would yield a good estimate.

The formula for the measure of surface area of a sphere with radius r is established by calculus as $S = 4\pi r^2$, which is not too far from the rough estimate.

Example

A sphere with a measure of radius of $^3/_4$ has a measure of surface area of $4\pi(^3/_4)^2 = 4\pi(^9/_{16}) = 9\pi/4$. The surface area is $9\pi/4$ square units.

A rough approximation for the volume of a spherical solid can also be found by examining Figure 12.13. The volume of the cubical solid is $(2r)^3 = 8r^3$. Since this volume seems to be about twice that of the spherical solid, you would probably estimate that the volume of the spherical solid is about $4r^3$. This estimate approximates very well the exact formula, which is

$$V = 4/3\pi r^3.$$

Example

The measure of volume of a spherical solid with a radius of $^3/_8$ is $4/3\pi(^3/_8)^3 = 9\pi/128$. The volume is $9\pi/128$ cubic units.

Example

Find the radius, correct to the nearest unit, if the volume of a sphere is 300. Use 22/7 for π.

$$V = \frac{4}{3}\pi r^3$$

$$300 = \left(\frac{4}{3} \cdot \frac{22}{7}\right) r^3$$

$$\frac{300 \cdot 21}{88} = r^3$$

$$\frac{75 \cdot 21}{22} = r^3$$

$$71\frac{13}{22} = r^3$$

By trial and error, $4^3 = 64$ and $5^3 = 125$, so that the radius is 4, to the nearest unit.

Additional insight into Cavalieri's Principle can be gained by seeing how it can be used to give a more rigorous development of the formula for the volume of a sphere. Consider the two sets of points in Figure 12.14. One is a hemisphere (half a sphere). The other is a right circular cylinder with a cone taken out as illustrated.

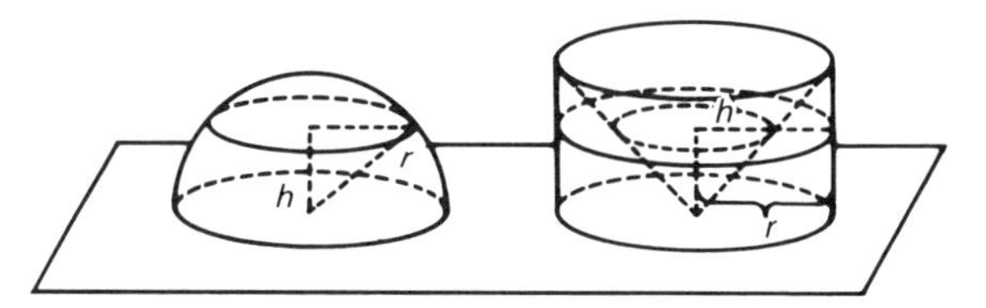

Figure 12.14. Application of Cavalieri's Principle

The cross section of the sphere is a circular region, and the cross section of the second figure is a ring. For the sphere, the area of the cross section is $\pi(r^2 - h^2)$, found by use of the Pythagorean Theorem. The cross section of the second figure is the same. This is found by subtracting the area of the inner circular region from the outer. Since both cross sections have the same area, Cavalieri's Principle states that the volumes of both solids are equal. Thus the formula for volume of a sphere can be derived from the second figure in Figure 12.14.

The total volume of the sphere is twice the volume of the hemisphere, so

$$
\begin{aligned}
V &= 2(\text{volume of cylinder minus the volume of cone}) \\
&= 2\left(\pi r^3 - \frac{\pi r^3}{3}\right) \\
&= 2\left(\frac{2\pi r^3}{3}\right) \\
&= \frac{4}{3}\pi r^3.
\end{aligned}
$$

An important problem is to find the shortest path on the sphere between two points on the surface—important because the corresponding practical problem is to find the shortest path between two points on the surface of the earth. Experiment with a globe to see if you agree with this statement: the shortest path between two points on the surface of a sphere is along a great circle. This property is illustrated for points A and B in Figure 12.15. Contrary to what one might expect, a line of latitude is not the shortest distance between two cities on that line, unless it happens to be the Equator. Also, in the Northern Hemisphere, the shortest routes seem to bend to the north. These so-called polar routes are now very common in air travel. In a sense, great circles act as "straight lines" on the surface of the sphere, since the distance measured along them is a minimum.

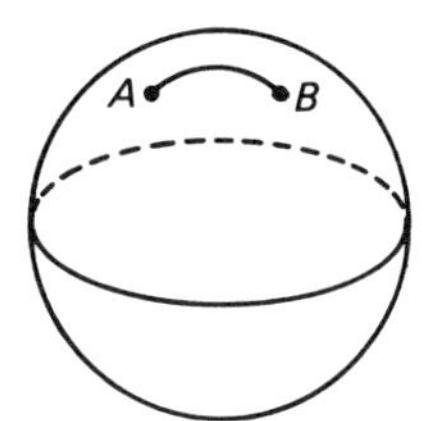

Figure 12.15. Great circle route

Although the line of latitude is not the shortest distance on the earth's surface, it is often important to find the number of miles due east or west of the starting point along a line of latitude. Mathematically, this means measuring along the circumference of a circle. The problem is an interesting application of trigonometry.

Study Figure 12.16. Suppose point C has a latitude of 30 degrees. To find the circumference of the circle of latitude through C, you can first find the radius $\overline{BC}$. Angles DAC and ACB both have a measurement of 30°. The cosine of angle BCA is the measure of $\overline{BC}$ divided by the measure of $\overline{AC}$, or a/b. Then $a = b$ cosine 30°.

The length of the 30-degree line of latitude is $2\pi a$, or $2\pi b$ cosine 30°. Since $2\pi b$ is the distance around the Equator, about 25,000 miles,

$$\text{Length of } 30° \text{ line of latitude} \approx (25{,}000)(.8660)$$
$$\cong 21{,}650.$$

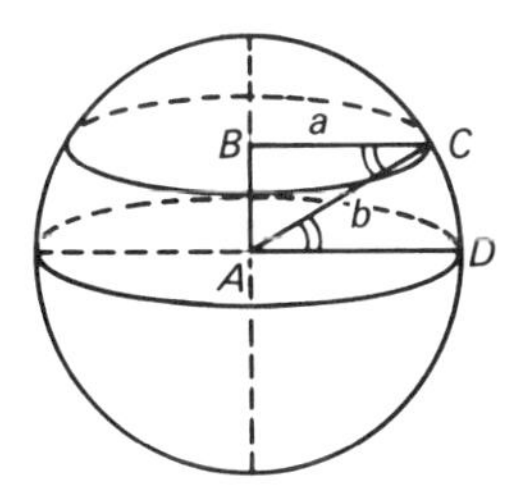

Figure 12.16. Length of line of latitude

If the total length of this line of latitude is approximately 21,650 miles, the east-west distance between two points with latitude 30°N but with longitudes of 30°W and 60°W can be found. The distance between them is $(60 - 30)/360 = {}^{1}/_{12}$ of 21,650 miles, approximately, or about 1804 miles, in an east-west direction. Remember that the great-circle distance between them is less than 1804 miles (approximately how many miles less?)

Exercise 12.4

1. Find the measure of surface area for a sphere and the measure of volume for a spherical solid, using the given measure of radius. Leave the symbol for π in the answers.
   (a) 5 (b) $^2/_3$ (c) 6.1 (d) 7.15
2. What is the effect on the measure of surface area of a sphere if the measure of radius is doubled?
3. What is the effect on the measure of volume of a spherical solid if the measure of radius is doubled?
4. Find the radius, to the nearest unit, for a sphere having a surface area of:
   (a) 75 sq in. (b) 700 sq cm
5. Find the radius, to the nearest unit, for a sphere having a volume of:
   (a) 1000 cu cm (b) 3,000,000 cu cm
6. A balloon with a radius of 3 in. is increased in size until the radius is 4 in. How much increase in surface area does this cause?
7. In Exercise 6, how much change in volume is caused?
8. Use a globe to describe briefly the great-circle routes between these pairs of cities.
   (a) San Francisco and Tokyo (b) New York and London
   (c) Los Angeles and Miami (d) New York and Sydney
9. Find the appropriate east-west mileage between two points having a latitude of 30°N and a longitude of:
   (a) 50°W and 90°W (b) 10°E and 15°E

When the theory of measurement within the mathematical model is applied to the physical world, results are always approximate. You can state the exact length of a segment, but not of a piece of rope. By using a formula, you can find exactly the measure of area of a rectangular region, but you cannot find exactly the measure of area of a garden of rectangular shape. A formula makes it possible for you to find the exact area of a circular region, but not for you to find exactly the surface area of a round table.

Among the reasons for the approximate nature of the results of applied measurement are the deficiencies of the measuring instruments used, the nature of matter itself, human limitations, and the fact that a measure may theoretically be any positive real number. Mathematicians, scientists, and technologists are all concerned with analyzing the errors that must be accepted. They want to know the greatest possible error, how this error compares with the actual measurement, and how to handle errors when computation with measures is necessary. The terminology introduced in this chapter is more widely used in mathematics than in science.

Greatest Possible Error and Precision

When the average person says that a piece of wood is 3 inches long, he usually means that, *as carefully as he cares to measure it,* the length is 3 inches. The mathematician uses a convention to standardize the meaning of a recorded measurement as the result of a physical act of measuring. The mathematician always interprets such a measurement as correct to some unit. Furthermore, he agrees that the unit will be the smallest unit stated in the measurement.

Examples

A recorded measurement of 3 in. means that 1 in. is the unit of measurement and that the measurement is correct to the nearest in. A measurement of $3\frac{1}{4}$ in. is correct to the nearest $\frac{1}{4}$ in. A measurement of 3.1 in. is correct to the nearest .1 in. A measurement of 9.26 cu in. is correct to the nearest .01 cu in.

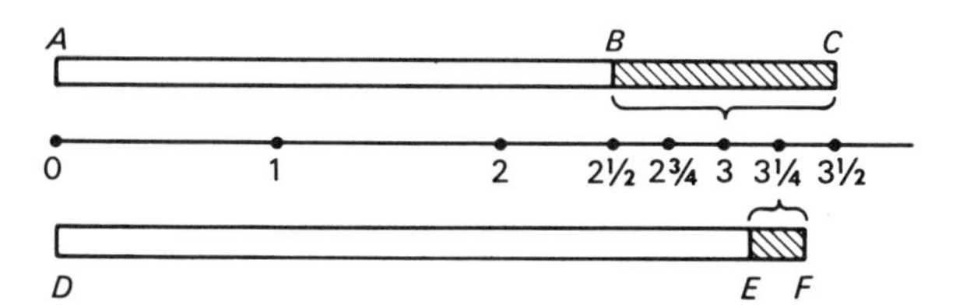

Figure 13.1. Measurement to nearest unit

Since each recorded measurement is correct to some agreed unit, it is possible, by knowing the unit, to tell just how far from the recorded measurement the actual measurement might be. For example, if the recorded measurement is 3 inches, then the actual measurement is between $2\frac{1}{2}$ and $3\frac{1}{2}$ inches. If the recorded measurement is $3\frac{1}{4}$ inches, then the actual measurement is between $3\frac{1}{8}$ and $3\frac{3}{8}$ inches. These results are summarized in Figure 13.1. The bar above the number line indicates that an object with one endpoint at A could have the other endpoint at any position between B and C and still have a recorded measurement of 3 inches, correct to the nearest inch. The bar below the number line indicates that an object could have one endpoint at D and the other endpoint at any position between E and F and have a recorded measurement of $3\frac{1}{4}$ inches, correct to the nearest $\frac{1}{4}$ inch. Similarly, a reported measurement of 9.4 square inches is within .05 square inch of the actual measurement.

When you measure correct to some unit, that unit is called the *unit of precision.* In the first example in Figure 13.1, the unit of precision is 1 inch. For the second example, the unit of precision is $^1/_4$ inch. For a measurement of 3.1 inches, the unit of precision is .1 inch. The unit of precision for the measurement 9.26 cubic inches is .01 cubic inch. For a measurement such as 11 feet, 5 inches, the unit of precision is 1 inch.

Choosing a unit of precision limits the amount of error allowed. In Figure 13.1, if you choose to measure correct to the nearest inch, then the actual measurement of the object cannot be more than $\frac{1}{2}$ inch from the stated measurement of 3 inches. If it were more than this, you would call the measurement 2 inches or 4 inches, rather than 3 inches. The *greatest possible error* in this case is $\frac{1}{2}$ inch. The greatest possible error, sometimes abbreviated g.p.e., is the greatest possible difference between the actual measurement and the recorded measurement. If the unit of precision is $\frac{1}{4}$ inch, then the greatest possible error is $\frac{1}{8}$ inch. Figure 13.2 helps to explain that, for 3.1 inches, the g.p.e. is .05 inch, since the actual measurement could be between 3.05 inches and 3.15 inches.

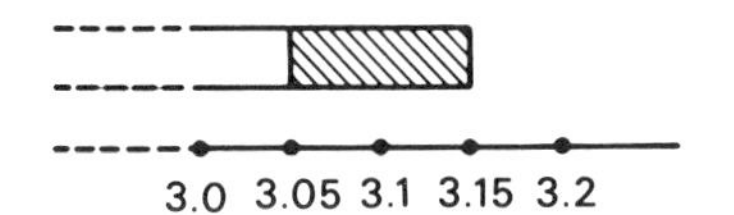

Figure 13.2. Greatest possible error

Do you see the relationship between the unit of precision and the g.p.e.? The measure of the g.p.e. is always $^1/_2$ the measure of the unit of precision. For another example, if the recorded measurement is 2.93 square meters, then the unit of precision is .01 square meter and the g.p.e. is .005 square meter.

Because of the convention of interpreting the unit of precision from the recorded measurement, it is not necessary to actually state the error in a measurement. In some cases, however, making the error explicit by showing the entire range of possible answers results in a very useful type of notation. Three examples of this notation follow: for 3 inches, you can write $(3 \pm {}^1/_2)$ inches (read 3 plus or minus $^1/_2$); for $3^1/_4$ inches, you can write $(3^1/_4 \pm {}^1/_8)$ inches; and for 3.1 inches, you can write $(3.1 \pm .05)$ inches. This notation can also be used to specify the allowable amount of error when you do not wish to use the ordinary mathematical convention. For example, $(3 \pm \frac{1}{8})$ inches means that you want to specify that the measurement is between $3^1/_8$ inches and $2^7/_8$ inches. The mathematician would normally write $3^0/_4$ inches to indicate this measurement.

How would you report a measurement, correct to the nearest inch, that seemed to be $3^1/_2$ inches? The convention often used, and the one that will be followed in this book, is to round upward in such a case. That is, the recorded measurement would be 4 inches. A second convention concerns the ambiguity when a recorded measurement ends in zeros. A measurement of 300 miles cannot be interpreted correctly unless you have more information or unless you understand some convention being used. Similarly, 8.10 feet may seem ambiguous at first. The zeros in 300 miles would not be considered significant. They should not be considered when determining the unit of precision; the unit of precision would be 100 miles, and the g.p.e. would be 50 miles. One possible way to indicate that a zero in such a position to the left of the decimal point is significant is to underline it. Thus, for $3\underline{0}0$ miles, the unit of precision is 10 miles, and for $3\underline{00}$ miles, the unit of precision is 1 mile.

Zeros following all non-zero digits to the right of the decimal point are considered significant. A measurement of 8.10 feet is correct to the nearest .01 foot. The zero would not be there if it did not indicate that a more precise measurement had been made. Similarly, a measurement of 4.120 miles is correct to the nearest .001 mile.

To say that one measurement is more precise than another means that the more precise one has a smaller g.p.e.; the more precise the measurement, the smaller the g.p.e. The measurement $3^1/_4$ inches is more precise than the measurement 3 inches, and the measurement 3.1 inches is more precise than either of the others. It is not uncommon today to need to measure correctly to the nearest .001 inch, and measurements much more precise than this are sometimes necessary in science and technology.

Exercise 13.1

1. Write the measurement of length of each bar shown in Figure 13.3 correct to the nearest:
   (a) $^1/_2$ in. (b) $^1/_4$ in.

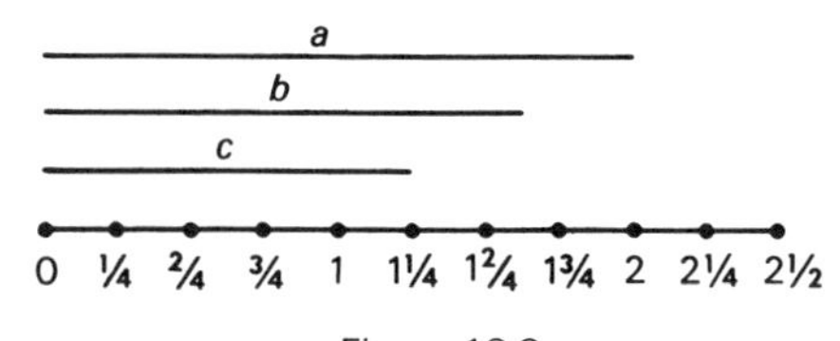

Figure 13.3

In Exercises 2–16, complete the table.

	measurement	*unit of precision*	*g.p.e.*	*notation with* ±
2.	5 ft	____	____	____
3.	6.2 in.	____	____	____
4.	$7^1/_2$ in.	____	____	____
5.	3.04 m	____	____	____
6.	354 sq ft	____	____	____
7.	$6^1/_4$ sq cm	____	____	____
8.	521 cu cm	____	____	____
9.	3.2 cu in.	____	____	____
10.	50 yd	____	____	____
11.	5$\underline{0}$ yd	____	____	____
12.	1$\underline{0}$00 mi	____	____	____
13.	7.0 in.	____	____	____
14.	4.20 ft	____	____	____
15.	3000 sq ft	____	____	____
16.	1.00 cu in.	____	____	____

In Exercises 17–22, tell which of each pair of measurements is more precise.

17. 14 ft or 14 in.
18. 5 m or 5.2 m
19. 305 km or 305 mi
20. 42 yd or 3652 in.
21. 1.2 sq km or 3.16 sq km
22. 17 cu in. or 18 cu in.

Relative Error and Accuracy

The greatest possible error is the maximum amount of error that may be involved in a recorded measurement, but the g.p.e. does not always indicate just how serious this error is. An error of $^1/_2$ inch for a measurement of 1 inch seems much more important than an error of $^1/_2$ inch for a measurement of 39 inches, although the unit of precision is 1 inch in both cases. An error of 1 million miles in a measurement in astronomy might be less significant than an error of one-millionth of an inch in atomic measurements.

Comparing the g.p.e. with the recorded measurement will provide a numerical way of determining the seriousness of an error. By expressing the ratio of the measure of the g.p.e. to the measure of the recorded measurement, a number called the *relative error* is obtained.

Examples

The relative error for a measurement of 1 in. is ${}^{1}/_{2}/1 = {}^{1}/_{2}$. For a measurement of 39 in., the relative error is ${}^{1}/_{2}/39 = 1/78$. An error of 1 in 2 is much more serious than an error of 1 in 78. The relative error for 3.2 m is $.05/3.2 \approx .016$. The relative error for 853 sq mi is .5/853, and the relative error for .1257 cu mm is .00005/.1257.

The smaller the relative error, the more *accurate* the measurement; the more accurate a measurement, the less significant the error. Notice that relative error is expressed without units of measurements, since it is the ratio of two measures. Often, for convenience in interpretation, the relative error is written as a percent. When this is done, the answer is called a percent of error. For 1 inch, the percent of error is 50%, but for 39 inches, it is only 1.3%, approximately. For 3.2 meters, the percent of error is approximately 1.6%.

Sometimes the word *tolerance* is used to mean either the g.p.e. or the relative error. The tolerance for a measurement of 5 inches might be expressed as $(5 \pm {}^{1}/_{2})$ inches, or 5 inches $\pm$ 10%. In its popular usage, *tolerance* refers in one way or another to the amount of error that can be accepted.

Example

The tolerance for a measurement of 96 sq in. is expressed as 96 sq in. $\pm$ 5%. Find the greatest amount of error, in terms of sq in., that can be allowed.

Since 96(.05) = 4.80, the greatest amount of error that can be allowed is 4.80 sq in.

Exercise 13.2

For Exercises 1–9, express the relative error for each measurement as a fraction and then as a percent of error to the nearest ${}^{1}/_{10}$ of 1%. Use the symbol for "approximately equal" when necessary.

1. 5 ft
2. 6.2 in.
3. $7\frac{1}{2}$ sq in.
4. 3.04 m
5. $5\underline{0}$ yd
6. 50 cu yd
7. $1\underline{0}00$ mi
8. 7.0 in.
9. 4.20 sq ft

In Exercises 10–15, tell which of each pair of measurements is more accurate.

10. 14 ft or 14 in.
11. 5 m or 5.2 m
12. 305 km or 305 mi
13. 42 yd or 3652 in.
14. 50 sq ft or 50 sq mi
15. 80 cu in. or 2 cu mi

Operations with Measures

Since measurements that result from a physical act of measuring are approximate, great care must be employed in operating with them. When the measures are added or multiplied, for example, the effect on the error must be analyzed. Several examples will help to explain how to interpret the results of operations performed on measures.

If two boards, $^3/_8$ inch thick and $^5/_8$ inch thick, are nailed together, what is their combined thickness? The combined thickness is *approximately* $^8/_8$ or 1 inch. The first way of interpreting the results of operations on measures is simply to report that the results are approximate.

A second example of approximate results is found in this problem: "What is the area of a plot of grass 12 feet by 16 feet, in the shape of a rectangle?" Again, an acceptable answer is "approximately 192 square feet." Operating with the measures of physical measurements yields approximate results, and recognizing this is often enough for practical applications.

A third example of approximate measure is attempting to find the volume of a spherical container if the radius is measured as 5 inches, correct to the nearest inch.

In the first example, you may arrive at a range of possible answers by remembering that each of the given measurements actually implies a possible range of values, because of the g.p.e. Since $^3/_8$ inch may apply to a board that is actually between $^5/_{16}$ and $^7/_{16}$ inch thick, and $^5/_8$ inch may apply to a board actually between $^9/_{16}$ and $^{11}/_{16}$ inch thick, the sum of the measures might really be any number between $^5/_{16} + {}^9/_{16} = {}^{14}/_{16} = {}^7/_8$ and $^7/_{16} + {}^{11}/_{16} = {}^{18}/_{16} = {}^9/_8$.

A more concise way of indicating the resulting measure in this first example is $^7/_8 < m < {}^9/_8$ (read "$^7/_8$ is less than the measure, and the measure is less than $^9/_8$"). In other words, the actual combined thickness may be any thickness between $^7/_8$ inch and $^9/_8$ inch. Notice that the g.p.e. of the combined measure ($^1/_8$) is the sum of the g.p.e.s. ($^1/_{16} + {}^1/_{16}$) of the two original measures.

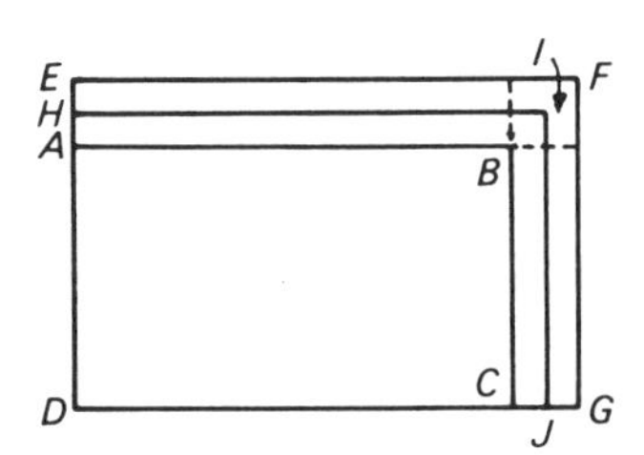

Figure 13.4. Error in area measurement

In Figure 13.4, the garden with the measurement of area of approximately 192 square feet could be as small as indicated by rectangle $ABCD$, which is $11^1/_2$ feet by $15^1/_2$ feet. It could be as large as indicated by rectangle $EFGD$, which is $12^1/_2$ feet by $16^1/_2$ feet. Then the measure of area in square feet is between 178.25 and 206.25; this may be written as $178.25 < m < 206.25$. A g.p.e. of only $^1/_2$ foot in the measurements of the sides of the garden can make a difference of many square feet in the possible results for the measurement of area. Notice that 192, probably the best single answer, is not the average of 178.25 and 206.25. This fact can be understood after a study of the shape of $EFGCBA$ (Fig. 13.4). Note that less than half of the dotted rectangular region at the upper right is included in the rectangular region $DHIJ$.

The measurements for the third example are illustrated in Figure 13.5. The actual volume V is between $\pi(4\frac{1}{2})^3$ and $\pi(5\frac{1}{2})^3$ cubic inches, or $91\frac{1}{8}\pi < V < 166\frac{3}{8}\pi$.

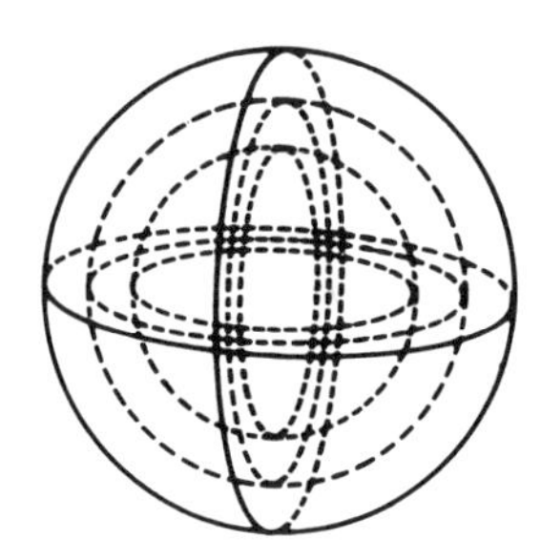

Figure 13.5. Error in volume measurement

The mathematician sometimes uses arbitrary working rules for selecting one particular number as the best possible answer for an operation on measures. His concern is that he not claim to have a better answer than the original data will allow, considering the errors. The operators of modern computers must also consider the effect of errors when thousands of operations are performed involving numbers expressed to several decimal places. As you can see, errors accumulate to reduce the usefulness of the results, even when you start with a relatively small possible error.

The mathematician must also establish agreements when the degrees of precision differ for the initial recorded measurements. For example, if 3.24 inches and 9.1 inches were given as recorded measurements, the mathematician might decide to round the first to 3.2 inches before using it in computation.

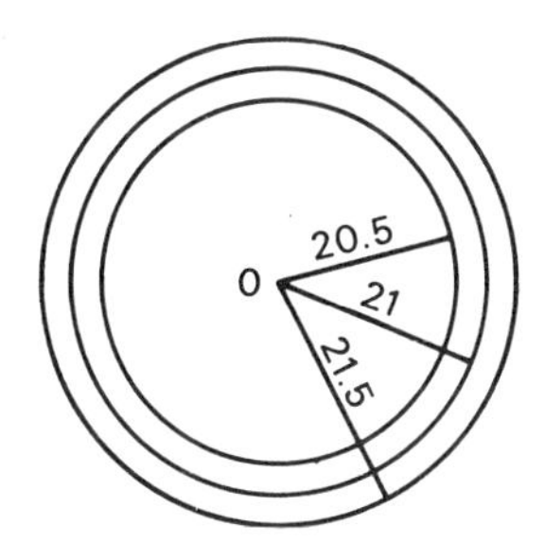

Figure 13.6. Application of error analysis

A final example will illustrate again how a small error in the given measurement affects the result. In the design of helicopters, determining the area of the circle swept out by thc rotor blade is very important. The radius of the circle (the length of the blade) must be measured very carefully if the error in the area is to be small. Suppose that, for a particular helicopter, the radius was stated as 21 feet. Study Figure 13.6. The smallest possible measure of area is $\pi(20.5)^2 = 420.25\pi$ square feet, and the greatest possible measure of area is $\pi(21.5)^2 = 462.25\pi$ square feet. If you claimed that the area was $\pi(21)^2 = 441\pi$ square feet, then you might possibly have made an error of, say, 20π square feet, or approximately 62.8 square feet. This error might make a great deal of difference in the performance of the craft, and, of course, it is necessary to measure the radius more precisely. The difficulty of the problem, however, is further illustrated by the fact that you would still have a possible error of approximately 6 square feet in area if you measured the radius correct to the nearest tenth of a foot.

Exercise 13.3

For each exercise, use the least possible and greatest possible measures to write two measurements between which the answer must lie.

1. The Allen family traveled 374 mi on the first day of their vacation and 346 mi the second day. How far did they travel on the two days?
2. What is the perimeter of a post in the shape of a regular hexagon, if each side is 9 in. across?
3. How many sq in. are in a sheet of paper 9 by 11 in.?
4. What is the area of a field in the shape of a parallelogram, if the base is 230 ft and the altitude is 120 ft?
5. How many sq ft of canvas are in a piece shaped like a triangle with a base of 5 ft and an altitude of 7 ft?
6. How far does a wheel roll during one revolution, if the radius is 8 in.?
7. How many sq in. are in a circular stained-glass window with a radius of 14 in.?
8. How many cu in. of wood are contained in a spherical ball with a radius of 2 in.?
9. How many cu ft of water can be held in a cylindrical tank with a radius of 3 ft and a height of 5 ft?

Our intuitive study of geometry up to this point has been based on the same foundation as that of ordinary school geometry. It has been, in other words, a modern, intuitive presentation of Euclidean geometry. In this final chapter, you will have a chance to enlarge your ideas of what geometry actually is by considering briefly four geometries that are different from Euclidean geometry. These four are *projective geometry, topology, finite geometry*, and *non-Euclidean geometry.* All of these are what are called *modern* geometries, since they are rather recent additions to mathematics. In this informal approach, we will explain some of the basic ideas of each geometry and then list a few significant theorems not found in Euclidean geometry. You should note some of the differences among the various geometries. You should also be aware of the great variety of interesting concepts that are studied under the general heading of "geometry." Often, various kinds of geometry are encountered as separate college courses.

Introduction to Projective Geometry

Projective geometry was invented early in the nineteenth century by the Frenchman Poncelet, while he was in a Russian prison during the wars of

Napoleon. Some basic ideas of projective geometry, however, go back as far as the ancient Greeks and also include the work of Desargues and

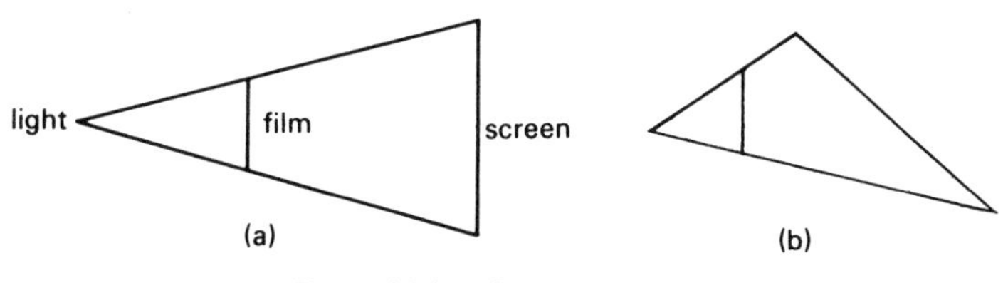

Figure 14.1. Central projection

Pascal. The idea of a *central projection* is familiar to everyone through the motion-picture projector, which projects a picture onto a screen, as diagramed in Figure 14.1a. Because the screen and the film are parallel, the image on the screen is similar to that on the original film. But what happens if the screen is not parallel to the film, as in Figure 14.1b? You certainly get distortion; yet some properties remain the same.

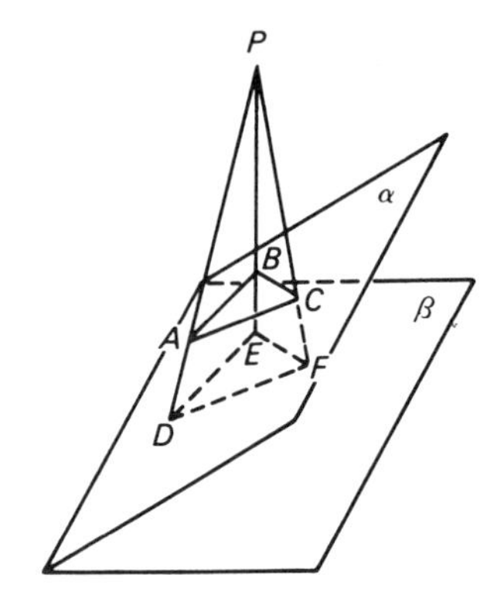

Figure 14.2. Projection of a triangle

Projective geometry is concerned with those properties of a set of figures that do not change while undergoing central projection. A central projection illustrating that the property of being a triangle is an invariant (something that does not change) is illustrated in Figure 14.2. In this figure, P is the point of projection. You may think of the two triangles on planes *alpha* and *beta* as being projected into each other in either direction. There is a one-to-one correspondence between the points of the two triangles. Points A and D, B and E, and C and F correspond, for example. Projection is a more general type of transformation that includes the motions of Euclidean geometry as a special case. Other properties that do not change under projection are those having to do with intersection. For example, if two lines meet, then their projections will meet also.

A basic idea in projective geometry is that of *perspective* and *projective forms.* Figure 14.3a shows two sets of collinear points perspective from

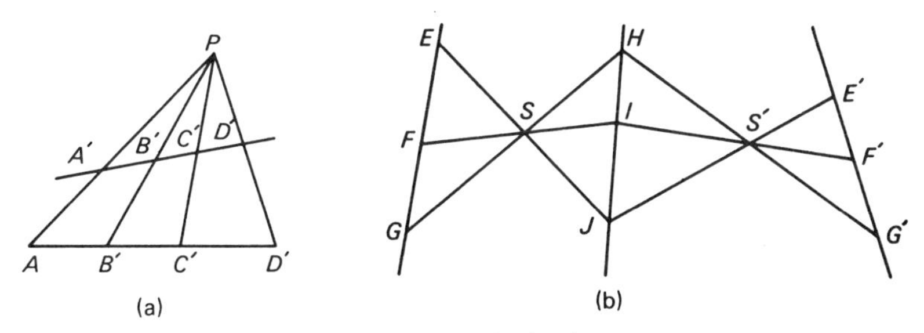

Figure 14.3. Projective forms

point P. Pairs of corresponding points, such as A and A', lie on a line through P. For two sets of collinear points to be perspective from a point simply means that pairs of corresponding points, one on each of the lines, are collinear with the points from which the sets are perspective. In Figure 14.3b, points E, F, and G are said to be projected into points E', F', and G'. A projectivity is a chain of perspectivities. That is, S is the first center of perspectivity, and S' is the second center of perspectivity. Points E, F, G and J, I, H are perspective, and, in turn, J, I, H and E', F', G' are perspective; thus, E, F, G and E', F', G' are projective.

The axiomatic foundation of projective geometry is very similar to that of Euclidean geometry. Technically, Euclidean geometry is really a special case of the more general projective geometry; it might be called a subgeometry of projective geometry. The axiom that distinguishes projective geometry from Euclidean geometry is the following. *In projective geometry, any two lines in the same plane meet at a point.* This axiom is also true in Euclidean geometry for many pairs of lines, but there is an important exception. In Euclidean geometry, parallel lines do not meet.

If two lines are parallel in the ordinary sense, then they are said to meet at an *ideal point* in projective geometry. The concept of an ideal point helps to explain that the lines do have something in common: a common direction. Figure 14.4a shows a set of parallel lines all passing through the same ideal point. Since exactly one ideal point is on a line, that ideal point can be approached by going in either direction along the line. Technically, the

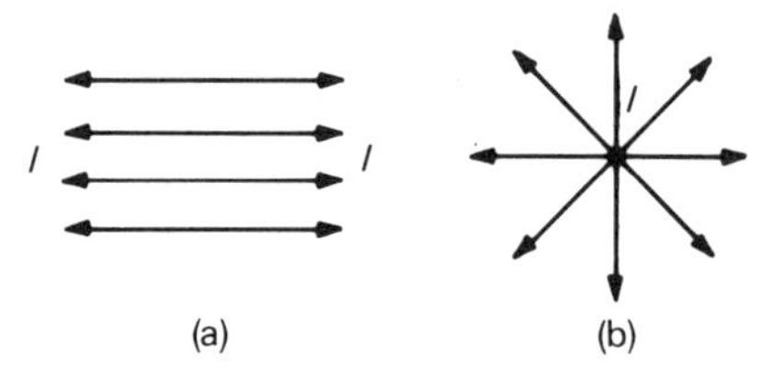

Figure 14.4. Lines meeting at ideal point

use of an ideal point in this way to extend the Euclidean line results in the study of *affine* geometry, in which ideal points are special cases. In projective geometry, considered from a more formal point of view, the ideal points are not special cases at all but are dealt with as regular points of a projective plane. This point of view is illustrated in Figure 14.4b. If *I* is an ideal point, then all the lines shown can be considered as parallel, whether they look parallel or not.

The introduction of ideal points makes possible a remarkable symmetry of statement in projective geometry known as *duality*. In *plane duality*, the words *point* and *line* may be interchanged in a valid statement or theorem, with appropriate changes in connecting words, to form another statement, which is also a valid statement or theorem.

Example

Statement: Any two points of a plane determine a line.
Plane Dual: Any two lines of a plane determine a point.

The plane dual is not always true in ordinary geometry because there is an exceptional case: parallel lines. For three dimensions, *space duality* results from the interchange of the words *point* and *plane*, along with other necessary changes.

Example

Statement: Any two distinct planes lie on a line.
Space Dual: Any two distinct points lie on a line.

One of the basic sets of points studied in projective geometry is a *complete quadrangle,* pictured in Figure 14.5. This complete quadrangle consists of points *A*, *B*, *C*, *D* and the six lines determined by joining these points in all

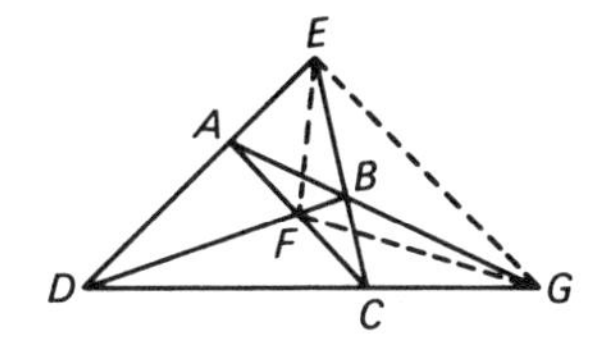

Figure 14.5. Complete quadrangle

possible pairings, extending each pair of opposite sides if necessary so that they meet at points E, F, and G. The four points and the six lines they determine may be projected into another complete quadrangle; hence the property of being a complete quadrangle is invariant in projective geometry. Triangle EFG, shown by the dotted lines in Figure 14.5, is called the *diagonal triangle* of the complete quadrangle, since its three vertices are the three *diagonal points,* which are the points of intersection of the three pairs of opposite sides of the quadrangle. The plane dual of a complete quadrangle is called a *complete quadrilateral.*

Although the complete quadrangle is interesting in its own right, it is significant in projective geometry partly because of the fact that it is used in the theory of constructing a coordinate system not dependent on distance, as is the coordinate system in ordinary Euclidean geometry. The basic idea used is to *define* what is meant by the midpoint of a segment in a different way. Thus, in Figure 14.6, it would be possible to construct a

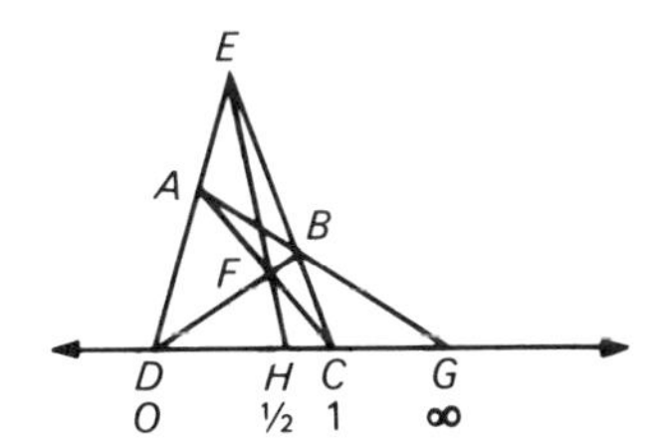

Figure 14.6. Projective coordinates

projective coordinate system on line DG by defining the point H where $\overleftrightarrow{EF}$ meets $\overleftrightarrow{DG}$ as the midpoint of $\overline{DC}$. If the coordinates of D, C, G are 0, 1, ∞ (infinity), respectively, then the coordinate of H is defined as $^1/_2$. In general, this would not be the ordinary midpoint, and we would have a coordinate system in which the distance from 0 to ½ was not really the same as the distance from ½ to 1. Since distances do change in projective geometry, it is good to have a coordinate system based on a complete quadrangle, which is an invariant in projective geometry.

One interesting property that is invariant under projection (preserved under projective transformations) is the *cross ratio* of four points on a line. Cross ratios are also involved in the theory of developing projective

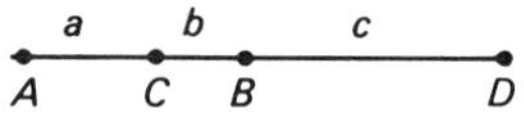

a b c
A C B D

Figure 14.7. Cross ratio

geometry from an analytic point of view, using coordinates. In Figure 14.7, the cross ratio (AB, CD) is defined as

$$\frac{\dfrac{a}{b}}{\dfrac{a+b+c}{-c}} \quad \text{or} \quad \frac{a}{b} \cdot \frac{-c}{a+b+c}.$$

Four points on a line, when projected into four other points on another line or on the same line, still have the same cross ratio, although the relative positions and the ratios of measures of distances may change. The cross ratio is actually a ratio of ratios. It compares the ratio of internal division of segment AB by point C and the ratio of division of this same segment by the point D. In other words, it compares the ratio of measures of $\overline{AC}$ and $\overline{CB}$ with the ratio of measures of $\overline{AD}$ and $\overline{DB}$. To find the correct value for this ratio of ratios, one direction along the line is considered as positive, which explains why a $-c$ was used, since it was measured from D back to B. D is said to divide segment AB externally, since it does not lie between them. Thus the cross ratio compares the ratio of internal division and external division of a given segment.

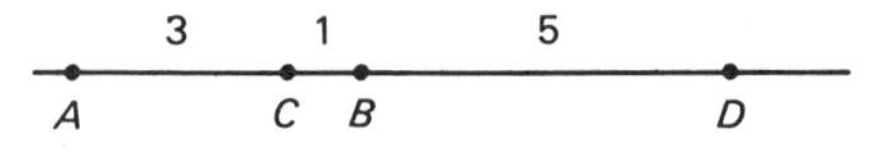

Figure 14.8. Example of cross ratio

Example

Here is a numerical example of finding the cross ratio for four points, if the measures of the segments are as given in Figure 14.8. The cross ratio (AB, CD) is

$$\frac{3/1}{9/-5} = \frac{3}{1} \cdot \frac{-5}{9} = \frac{-5}{3}.$$

One special case of the cross ratio, when it equals negative 1, is of great importance. In this case, illustrated by a particular example in Figure 14.9,

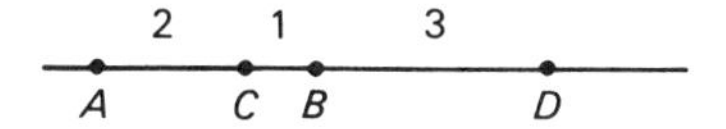

Figure 14.9. Harmonic set of points

points C and D divided the segment AB internally and externally in ratios with the same absolute value. Here the cross ratio (AB, CD) is $^{2}/_{1}/^{6}/_{-3}$ $= {}^{2}/_{1} \cdot {}^{-3}/_{6} = -1$. In this special case, the four points constitute a *harmonic set* of points. This property of being a harmonic set is an invariant under projection.

Several other significant theorems in projective geometry are illustrated briefly:

1. *Desargues' Theorem* states that if two triangles are perspective from a point, they are perspective from a line. Study Figure 14.10. For two triangles to be perspective from a line means that the corresponding sides of

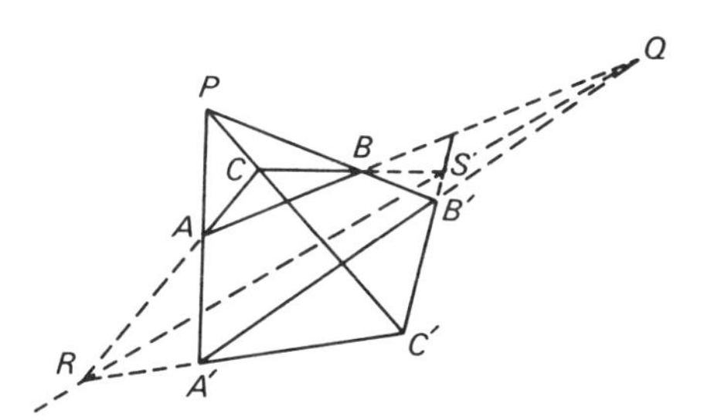

Figure 14.10. Desargues' Theorem

the two triangles meet at collinear points. In the figure, triangles ABC and $A'B'C'$ are perspective from point P. They are also perspective from the line RS.

2. A projectivity between two sets of collinear points is uniquely determined when three pairs of corresponding points are known. This minimum of information is enough to find the corresponding point from some fourth point given on one of the lines. For example, the seven points shown in Figure 14.11 can be chosen at random to be pairs of corresponding points in a projectivity, but then the image of D will be uniquely determined by the given information.

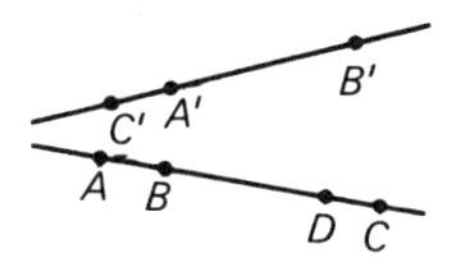

Figure 14.11. Corresponding points in projectivity

The idea is formalized in what is known as the *Fundamental Theorem of Projective Geometry.*

Theorem

If A, B, C, and D are any four points on a line, and if A', B', and C' are any three points on another or the same line, and if there are two projectivities such that the following pairs of points correspond,

$$\begin{array}{ll} A \leftrightarrow A' & A \leftrightarrow A' \\ B \leftrightarrow B' & B \leftrightarrow B' \\ C \leftrightarrow C' & C \leftrightarrow C' \\ D \leftrightarrow D' & D \leftrightarrow D'' \end{array}$$

then $D' = D''$.

3. The property of being a conic is invariant under projection. For example, a circle may be projected into an ellipse, as illustrated in Figure 14.12,

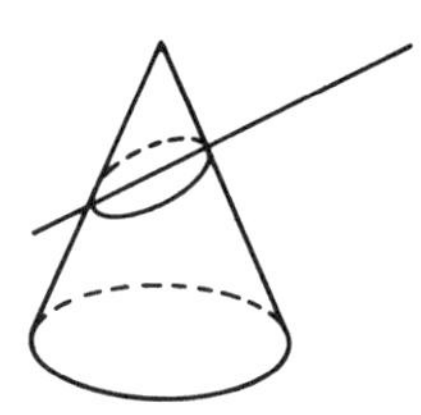

Figure 14.12. Circle projected into ellipse

or into some other conic section, but one conic section is always projected into another conic section.

4. The following theorem, which is stated in a slightly simplified form, is known as *Pascal's Theorem*: If a hexagon is inscribed in a conic, then the three intersections of pairs of opposite sides are collinear. The line of

collinearity is designated as the *Pascal line* of the hexagon. Pascal's Theorem is illustrated for a circle in Figure 14.13. The sides must be extended to meet.

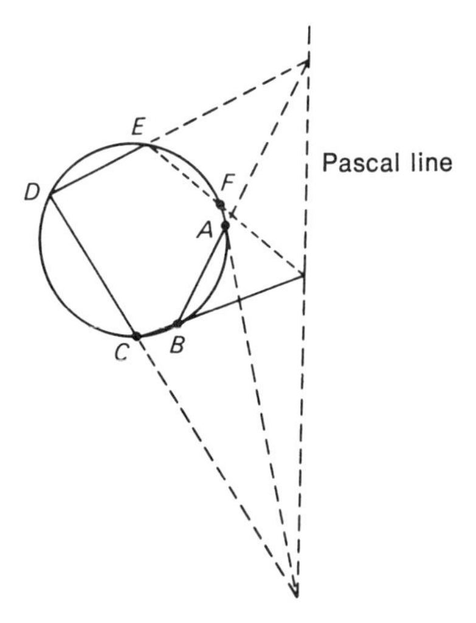

Figure 14.13. Pascal line for *ABCDEF*

In projective geometry, any specified order for the six points on a conic results in a hexagon, even though the sides may cross. For example, Figure 14.14 shows the hexagon *ADBECF*, with vertices in this order.

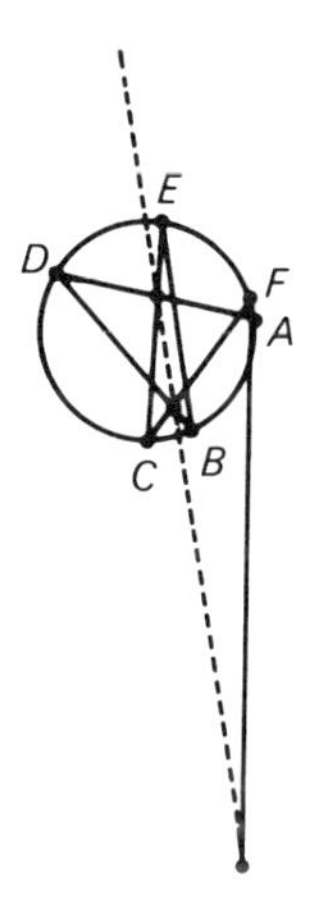

Figure 14.14. Pascal line for *ADBECF*

Pairs of opposite sides are $\overleftrightarrow{AD}$ and $\overleftrightarrow{EC}$, $\overleftrightarrow{DB}$ and $\overleftrightarrow{CF}$, and $\overleftrightarrow{BE}$ and $\overleftrightarrow{AF}$. The three points of intersection determine a different Pascal line. In fact, each of the 60 different hexagons has a distinct Pascal line.

Exercise 14.1

1. Tell whether the following seem to be invariants for a set of points under any projection.
   (a) Measure of angles (b) Measure of area
   (c) Measure of length (d) Collinearity of points
   (e) Shape of a quadrilateral (f) Size of a triangle

In Exercises 2–3, draw a picture like Figure 14.2 (p.276), showing:

2. Two intersecting lines on one plane projected into two intersecting lines on the other.
3. A quadrangle on one plane projected into a parallelogram on the other.
4. In Figure 14.3b (p. 277), suppose a fourth point, x, is given on $\overleftrightarrow{EG}$. Explain how to find its image, x', on $\overleftrightarrow{E'G'}$.
5. Write the plane dual of Desargues' Theorem.

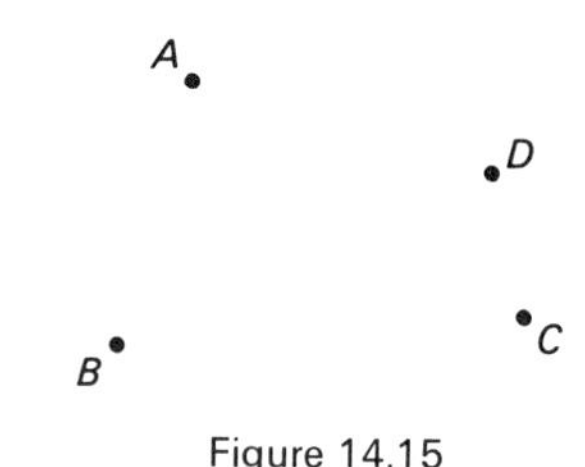

Figure 14.15

6. Copy Figure 14.15 and use the four points as vertices to draw a complete quadrangle and its diagonal triangle.
7. Write the plane dual of the definition of a complete quadrangle.
8. In Figure 14.6 (p. 279), explain what would happen to point H if $\overleftrightarrow{AB}$ were parallel to $\overleftrightarrow{DC}$, in the Euclidean sense.
9. In Figure 14.7 (p. 280), find the cross ratio (AB, CD) for the given values of a, b, and c.
   (a) $a = 3, b = 4, c = 5$ (b) $a = 1, b = 3, c = 2$
10. Copy Figure 14.16 and locate point D so that the four points (AB, CD) will be a harmonic set of points.

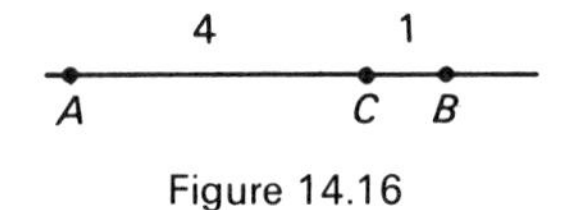

Figure 14.16

11. Copy Figure 14.17 and locate the Pascal line for hexagon *ABCDEF*.

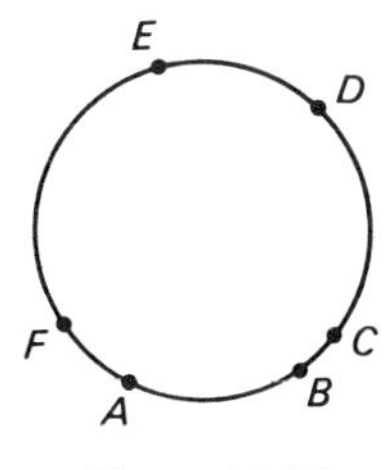

Figure 14.17

12. In Figure 14.17, locate the Pascal line for *ACBDEF*.

Introduction to Topology

Topology is an even more general kind of geometry than is projective geometry. The development of topology as a subject in its own right has occurred within the last hundred years. No single person is given credit for the invention or discovery of topology, although the names of Moebius, Listing, and Riemann must be given prominent mention. Euler's formula $V - E + F = 2$, studied earlier in this text, is essentially a topological formula.

Topology has sometimes been characterized loosely as "rubber-sheet" geometry, because the transformations of topology are those that could be performed if the plane were treated as a rubber sheet that could be pulled and twisted but not torn. For example, angles, distances, and areas are not invariants. A triangle can be deformed into a circle or a square. Figure 14.18 shows intuitively one way in which figures can be changed in topology. Such Euclidean invariants as convexity and number of sides of a polygon are not invariant under the set of topological transformations.

The *Jordan Curve Theorem,* which can be proved using the axiomatic system of topology, asserts that a simple closed curve in a plane partitions

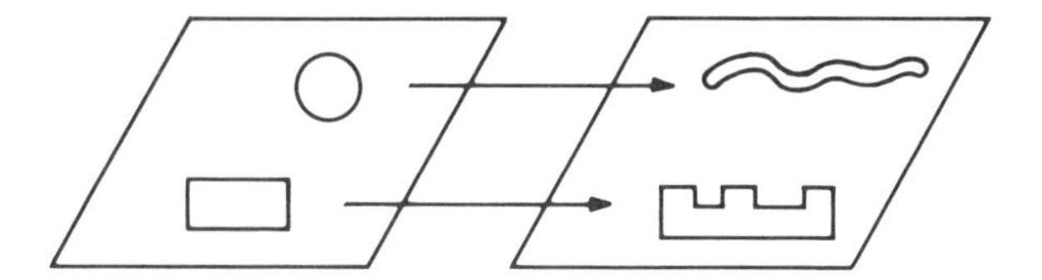

Figure 14.18. Topological transformations

the plane into the set of points on the curve, the set of points in the interior, and the set of points in the exterior. All simple closed curves have the same topological properties, since one can be transformed into another. In ordinary Euclidean geometry, on the other hand, it is necessary to distinguish among many different simple closed curves that have different properties.

With such a general type of transformation allowed, it may be hard to think of a topological invariant. One example is what is known as *degree of connectivity*. A set is *connected* if any two of its points can be joined by some curve lying wholely in the set. For example, the first set in Figure 14.19 is connected, but the second is not.

Figure 14.19. Connected and not-connected sets

The interior of a simple closed curve is said to be *simply* connected and has a degree of connectivity of 1.

The curves in Figure 14.20 are *multiply* connected. The first has a degree of connectivity of 2, and the second has a degree of connectivity of 3. Intuitively, it can be seen that the degree of connectivity is one more than the

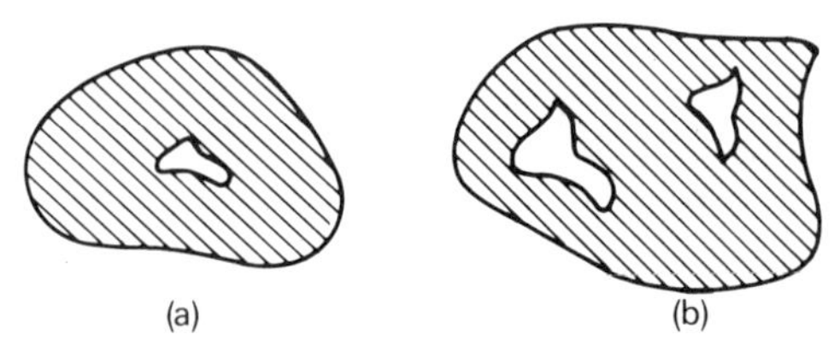

Figure 14.20. Degree of connectivity

number of holes. Any two sets with the same degree of connectivity can correspond in a topological transformation.

Three other topics from topology will be discussed briefly: the *four-color map problem,* the *Brouwer fixed-point theorem,* and the idea of a *one-sided surface.* The four-color map problem is an example of a statement in mathematics that has not yet been proved. It states that any map, no matter how many countries it contains or how they are situated, may be colored, using only four different colors, so that any two countries having a common border are differently colored. It is understood that countries must touch at more than one point in order to have a common boundary.

Figure 14.21 illustrates two simple maps that may be colored by four colors as indicated by the four numbers 1, 2, 3, 4. Try other maps to see

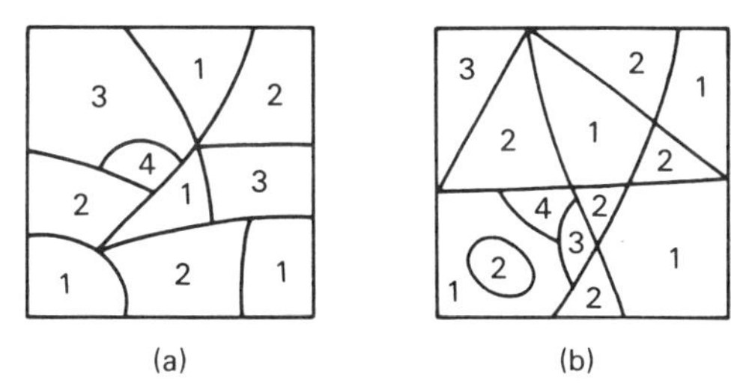

Figure 14.21. Map-coloring problem

that each can be colored using only four colors. Although the general problem has remained unsolved, mathematicians have proved that five colors are always enough. They have also shown that four colors are sufficient for maps of not more than 36 countries.

A *topological transformation* is a mapping in which each point of a set has an image and which leaves topological properties invariant. The Brouwer fixed-point theorem states that if all the points in a circular region are subjected to some topological transformation, then at least one of the points remains fixed. It is necessary to specify that the points are transformed into the other points of the same set and do not go outside the circle. A very simple example of what might be done could be illustrated by thinking of rotating a paper disc. In this case, the center is the fixed point, since it would not change position.

The final topic to be considered in topology is the idea of a surface with only one side. The classic example of such a surface is the *Moebius strip,* illustrated in Figure 14.22. It may be helpful to consider that a fly crawling

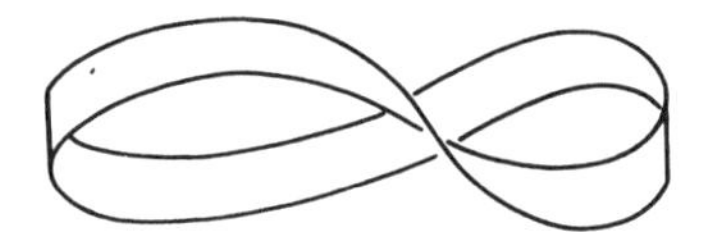

Figure 14.22. Moebius strip

on such a surface could crawl from one point to any other point without ever having to crawl over the edge of the strip.

To illustrate a Moebius strip, simply take a strip of paper about 18 inches long and 1 inch wide, give it a half twist, and glue the ends together. The result is a one-sided surface similar to Figure 14.22. The property of being a one-sided surface is a topological property: a one-sided surface cannot be changed into a two-sided surface by a topological transformation. Figure 14.23 shows two other examples of one-sided surfaces. Figure 14.23a is called a Klein bottle, named after the mathematician Felix Klein.

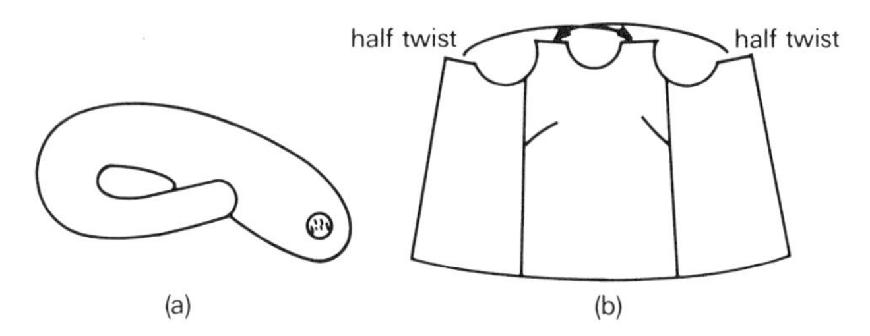

Figure 14.23. One-sided figures

Figure 14.23b is a one-sided dress. (See "Dressing up Mathematics," by Jean J. Pedersen, *The Mathematics Teacher,* Vol. LXI, No. 2, February 1968.)

Obviously, topology includes many ideas that have not been mentioned. It is hoped, however, that you will notice how different some of the topics investigated in topology seem from the topics you have studied in conventional geometry.

Exercise 14.2

1. Do the following properties seem to be invariant in topology?
   (a) Property of being a straight line

(b) Property of being a polygon
(c) Intersection of two curves
(d) Cross ratio of four points on a line

2. Give the degree of connectivity for the sets shown in Figure 14.24.

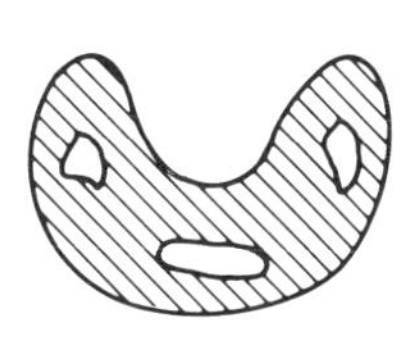
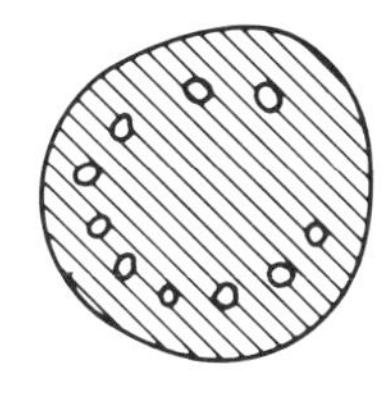

Figure 14.24

3. Copy the maps in Figure 14.25 and color them, using not more than four colors.

(a)

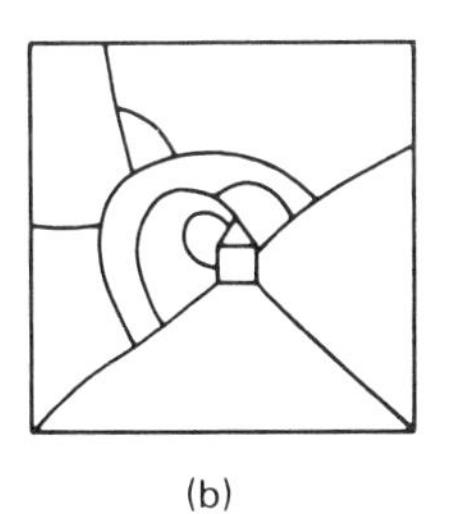
(b)

Figure 14.25

4. Draw a map with 12 countries that can be colored using only three different colors.
5. Draw a map with 12 countries that can be colored using only two different colors.
6. Make a Mocbius strip out of paper and cut it down the middle lengthwise. What is the result?
7. Experiment to see if the same thing as in Exercise 6 happens if you first give the paper a full twist rather than a half twist before gluing.
8. What is the result if a Moebius strip is cut lengthwise beginning $^{1}/_{3}$ of the width from one edge?

Introduction to Finite Geometry

A *finite geometry* is one that contains only a finite number of elements, such as points, lines, and planes. For example, a finite geometry might contain only seven points and seven lines, all on the same plane. Such geometries seem simple, because there are only a few elements to consider; yet, oddly enough, finite geometry has been studied only recently. In fact, it is a topic in mathematics that has developed largely during the twentieth century and is in every sense of the word a part of modern mathematics.

One example of a finite geometry is the geometry of the complete quadrangle. This geometry has points and lines as undefined terms and includes only three axioms:

1. There are exactly four lines in the geometry.
2. Each pair of distinct lines intersects in exactly one point.
3. Each point is on exactly one pair of distinct lines.

On the basis of the three axioms, a few theorems can be proved. For example, it is possible to show that there are only six points in the geometry. Many different pictorial representations of this finite geometry are possible, and it is obvious that lines in finite geometry are not the same as lines in ordinary Euclidean geometry. Two possible pictures of the geometry appear in Figure 14.26.

Figure 14.26. Finite geometry of the complete quadrangle

To emphasize the abstractness and generality of finite geometries, it is helpful to write various other representations using words other than *point* and *line*. Here is an interpretation of the finite geometry using *student* for point and *class* for line.

1. There are exactly four classes.
2. Each pair of distinct classes has exactly one student in common.
3. Each student belongs to exactly one pair of distinct classes.

It can be shown that there are six students and that each class has exactly three students.

The second example of a finite geometry is a seven-point projective geometry, sometimes called *Fano's geometry*. There are only five axioms for this geometry:

1. There exists at least one line.
2. Exactly three points lie on every line.
3. Not all the points lie on the same line.
4. Any two distinct points determine exactly one line.
5. Any two distinct lines determine at least one point.

Although the axioms do not clearly indicate seven points, you should be able to show that the axioms do result in seven points. Study Figure 14.27.

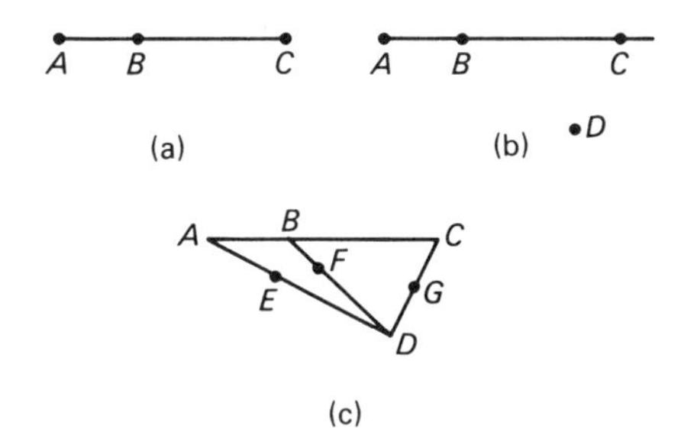

Figure 14.27. Development of Fano's geometry

Axioms 1 and 2 show the need for one line, but this line contains three points, labeled A, B, and C in Figure 14.27a. Axiom 3 means that a fourth point is needed, which is point D in Figure 14.27b. But, by axiom 4, this implies the existence of three more lines, each of which has a third point on it, making seven points in all.

It would seem at first that this procedure of creating more points and lines might be continued indefinitely, instead of producing only seven points. The reason this does not happen is that axiom 2 specifies that each line has only three points. In Figure 14.27c, the line through B and E does not meet the line through CD in still another point but must meet it at G, its third point. Similarly, the line through F and G must meet line BC at its third point, A. It is obviously necessary to modify your intuitive ideas of what is meant by a straight line to accept this representation of finite geometry. Study Figure 14.28, which shows the seven points numbered. In this picture, points 2, 7, 6, must be considered as on one of the seven lines, represented by the curve in the drawing.

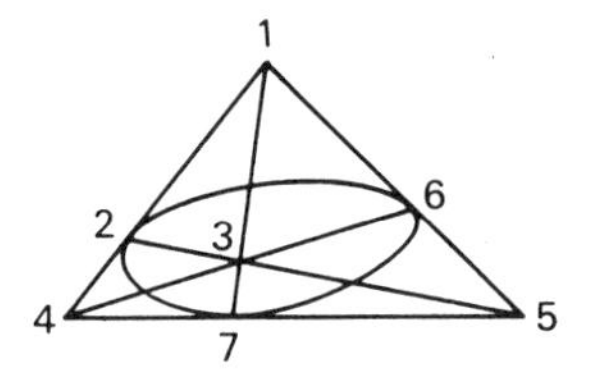

Figure 14.28. Representation of Fano's geometry

Although there are many finite geometries, it is customary to classify them by the number of points on a line. Finite projective geometries exist for three, four, five, and six points on a line, but after that the pattern is not so regular. A geometry rather similar to the one just investigated has 13 points and 13 lines, but there are exactly four points on each line. The ten labeled points and ten lines in Figure 14.10 (p. 281) constitute the elements in the Desargues finite geometry, which has three points on a line and three lines on a point.

Exercise 14.3

1. Is a geometry with a finite number of axioms a finite geometry?

Answer Exercises 2–5 for the geometry of the complete quadrangle.

2. In this geometry, how many points are there on each line?
3. Are each two points of the geometry connected by a line?
4. Are there pairs of lines that have no points in common?
5. Write an interpretation for this geometry using the word *book* for point and *library* for line.

For Exercises 6–9, answer yes or no, or not enough information to tell, for the seven-point geometry studied:

6. No four points lie on the same line.
7. All seven points may lie on the same line.
8. Some of the points lie on three of the lines.
9. A line is infinite in length.
10. Suppose that, in Figure 14.28, two lines that meet on line 1, 6, 5 are called parallel. Which of these pairs of lines are parallel?
    (a) 4, 7, 5 and 5, 3, 2
    (b) 6, 3, 4 and 1, 2, 4
    (c) 4, 2, 1 and 7, 3, 1
    (d) 2, 7, 6 and 4, 3, 6

Answer Exercises 11–15 for the star finite geometry shown in Figure 14.29.

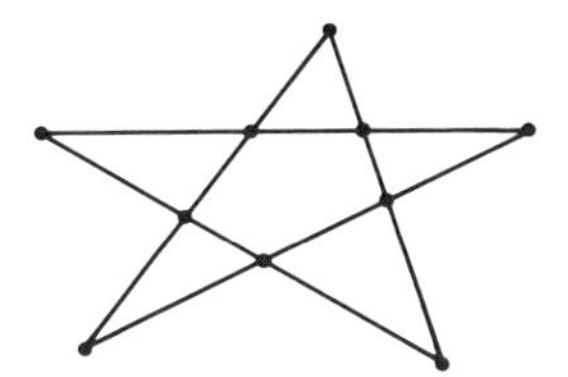

Figure 14.29

11. There are how many lines in the geometry?
12. There are how many points in the geometry?
13. Are each of the points connected by a line?
14. How many lines pass through each point?
15. How many points lie on each line?

Introduction to Non-Euclidean Geometry

The expression *non-Euclidean geometry* normally refers to a geometry that substitutes some other axiom for the last one of Euclid. Euclid assumed, you will recall, that through a point not on a line, exactly one line in the plane may be drawn parallel to a given line in the plane. The same parallel axiom has been included in the set of axioms for modern elementary Euclidean geometry. What are the alternatives? You could assume that more than one parallel could be drawn or that no parallels could be drawn. The first of these assumptions results in what is called *hyperbolic geometry*; the second results in *elliptic geometry*. Both of these non-Euclidean geometries follow just as logically from their axioms as Euclidean geometry does from its axioms.

In this section, some of the consequences of the first alternative, hyperbolic geometry, are presented briefly. The credit for the discovery of hyperbolic geometry and publication of the discoveries is usually given to Johannes Bolyai, a Hungarian, and to Nicolai Lobachevsky, a Russian. Early in the nineteenth century, both explored independently many of the basic ideas introduced here. The entire subject of non-Euclidean geometry is a part of what must be included under the general term *modern mathematics*.

Figure 14.30 illustrates the basic postulate in hyperbolic geometry—that more than one line in the plane can be drawn through P and parallel to $\overleftrightarrow{AB}$. If there are two distinct lines parallel to $\overleftrightarrow{AB}$, then all the lines through P

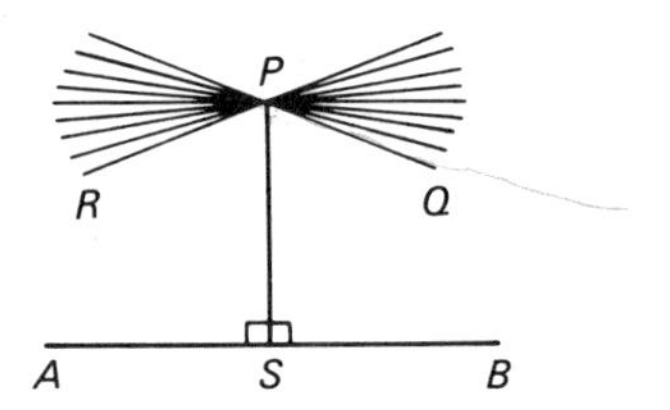

Figure 14.30. Lines in hyperbolic geometry

lying between these two also do not meet $\overleftrightarrow{AB}$, hence there is an infinite number of such lines. The first lines on either side of P that do not meet $\overleftrightarrow{AB}$, which are $\overleftrightarrow{PQ}$ and $\overleftrightarrow{PR}$ in Figure 14.30, are called *parallels* to $\overleftrightarrow{AB}$; all the other lines through P that do not meet $\overleftrightarrow{AB}$ are called *non-intersecting* lines. The idea that parallel lines are everywhere the same distance apart is only correct in Euclidean geometry. By definition, parallel lines simply do not have an ordinary point in common.

In Figure 14.30, angles SPQ and SPR are congruent acute angles. These two angles are called *angles of parallelism.* Oddly enough, their measure depends on the length of $\overline{PS}$; the farther the point P from $\overleftrightarrow{AB}$, the smaller the measure of the angle of parallelism.

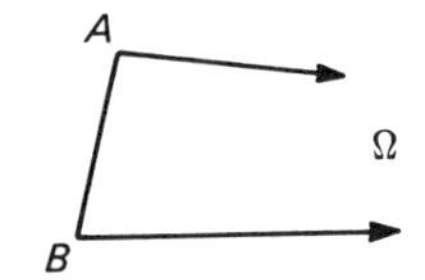

Figure 14.31. Omega triangle

One of the interesting concepts introduced in hyperbolic geometry is that two parallel lines are said to meet in an *ideal point.* The parallel lines do have something in common; they point in the same direction and do not meet in an ordinary point. In non-Euclidean geometry, the usual symbol for an ideal point is omega (Ω); a "triangle" with an ideal point as a vertex, as shown in Figure 14.31, is called an *omega triangle.* In hyperbolic geometry, there are two distinct ideal points on each line, as shown in Figure 14.32.

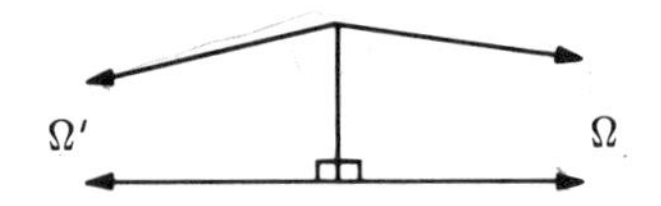

Figure 14.32. Ideal points on line

Two of the basic figures investigated in hyperbolic geometry are illustrated in Figure 14.33. The quadrilateral $ABCD$, with two right angles at C and D, and with $\overline{BC} \cong \overline{AD}$, is called a *Saccheri quadrilateral*, after one of the early investigators of hyperbolic geometry. In this quadrilateral, the angles at A and B are *not* right angles, and $\overline{AB}$ and $\overline{CD}$ are *not* congruent. The angles at A and B are acute, and $AB > CD$.

In Figure 14.33, quadrilateral $FBCE$ is called a *Lambert quadrilateral.* It has three right angles and one acute angle. This interesting property of a Lambert quadrilateral—that the sum of the measures of the four angles is

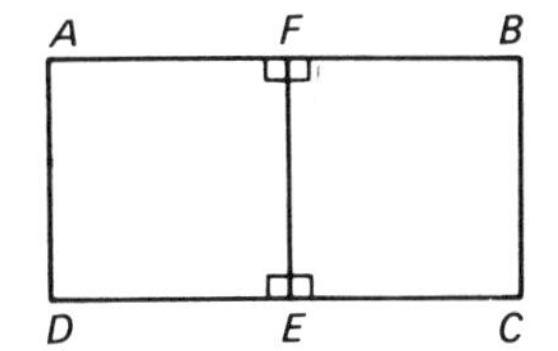

Figure 14.33. Saccheri quadrilateral

less than 360 degrees—is used to establish one of the distinctive theorems of hyperbolic geometry: The sum of the measures of the angles of a triangle is less than 180 degrees. Furthermore, the sum of the measures of the angles of different triangles may be different.

Here is a compilation of some other interesting theorems of hyperbolic geometry, not valid in Euclidean geometry:

1. The sum of the measures of the angles of every quadrilateral is less than 360 degrees.
2. The measure of the segment joining the midpoints of two sides of a triangle is less than half the measure of the third side.
3. Similar triangles that are not congruent do not exist in hyperbolic geometry. If the three angles of a triangle are congruent to the corresponding angles of a second triangle, then the two triangles are congruent.
4. Two parallel lines converge in the direction of parallelism.
5. There are no squares in hyperbolic geometry.
6. An angle inscribed in a semicircle does not have a measure of 90 degrees.
7. There exists one particular triangle with a maximum area. This triangle has three ideal vertices.

Whether the space of our experience is Euclidean or non-Euclidean is not known. At any rate, non-Euclidean geometry has provided a good model for recent investigations of the nature of the universe. A real understanding of what is meant by a set of axioms often comes for the first time in non-Euclidean geometry, since the student reaches logical conclusions that may seem very strange to him but that are the result of axioms he has accepted. Non-Euclidean geometry is consistent if Euclidean geometry is. Does it also agree with your intuitive ideas of space?

Exercise 14.4

1. Make an appropriate drawing for hyperbolic geometry showing four lines meeting at the same ideal point.
2. In Figure 14.33, are $\overleftrightarrow{AB}$ and $\overleftrightarrow{CD}$ parallel or non-intersecting?
3. What is one good reason why the development of the concept of area is different in non-Euclidean and Euclidean geometry?
4. Why can measurement not be used to decide the sum of the measures of the three angles of a triangle?
5. Which theorems of Euclidean geometry are also valid in hyperbolic geometry?

In Exercises 6–8, refer to Figure 14.34.

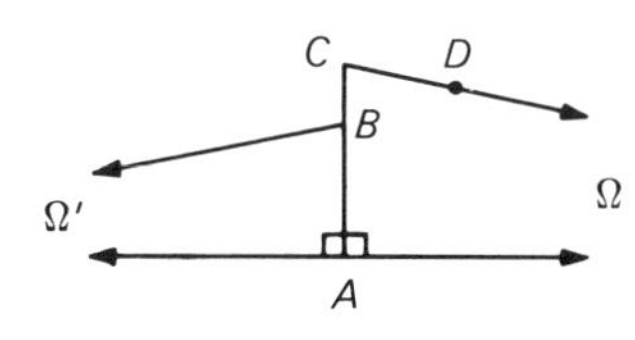

Figure 14.34

6. Which angle is greater, $\angle AB\Omega'$ or $\angle AC\Omega$?
7. What can you say about the sum of the angles at ordinary vertices for the omega triangles shown?
8. Which is closer to line $\Omega\Omega'$—point C or point D?

Adler, Claire Fisher, *Modern Geometry.* New York: McGraw-Hill, 1958.

Banks, J. Houston, *Elements of Mathematics,* 2nd Edition. Boston: Allyn and Bacon, 1961.

Barry, Edward H., *Introduction to Geometrical Transformations.* Boston: Prindle, Weber and Schmidt, 1966.

Courant, Richard, and Herbert Robbins, *What Is Mathematics?* New York: Oxford University Press, 1941.

Coxeter, H. S. M., *Introduction to Geometry.* New York: Wiley, 1961.

Coxeter, H. S. M., *Projective Geometry.* New York: Blaisdell, 1964.

Davis, David R., *Modern College Geometry.* Reading, Mass.: Addison-Wesley, 1954.

Eves, Howard, *A Survey of Geometry*, Volume 1. Boston: Allyn and Bacon, 1963.

Gans, David, *Transformations and Geometries.* New York: Appleton-Century-Croft, 1969.

Heddens, James W., *Today's Mathematics.* Chicago: Science Research Associates, 1964.

Hilbert, David, *The Foundations of Geometry.* La Salle, Ill.: The Open Court, 1950.

Johnson, Paul B., and Carol H. Kipps, *Geometry for Teachers.* Belmont, Calif.: Brooks/Cole, 1970.

Keedy, Mervin L., and Charles W. Nelson, *Geometry: A Modern Introduction.* Reading, Mass.: Addison-Wesley, 1965.

Levi, Howard, *Topics in Geometry.* Boston: Prindle, Weber and Schmidt, 1968.

Meserve, Bruce E., and Joseph A. Izzo, *Fundamentals of Geometry.* Reading, Mass.: Addison-Wesley, 1969.

Middlemiss, Ross R., John L. Marks, and James R. Smart, *Analytic Geometry,* 3rd Edition. New York: McGraw-Hill, 1968.

Moise, Edwin, *Elementary Geometry from an Advanced Standpoint.* Reading, Mass.: Addison-Wesley, 1963.

Norton, M. Scott, *Geometric Constructions.* St. Louis: Webster, 1963.

Pearson, Helen, and James Smart, *Geometry.* Boston: Ginn, 1971.

Perfect, Hazel, *Topics in Geometry.* Oxford, England: Pergamon, 1963.

Ruchlis, Hy, and Jack Englehardt, *The Story of Mathematics.* New York: Harvey House, 1958.

School Mathematics Study Group, *Studies in Mathematics,* Vol. VII. Author, 1961.

School Mathematics Study Group, *Studies in Mathematics,* Vol. IX. Author, 1963.

Tuller, Annita, *A Modern Introduction to Geometries.* Princeton, N.J.: Van Nostrand, 1967.

von Baravalle, Hermann, *Geometrie als Sprache der Formen.* Stuttgart: Verlag Freies Geistes Leben, 1963.

Wolfe, Harold E., *Introduction to Non-Euclidean Geometry.* New York: Dryden, 1945.

Wren, F. Lynwood, *Basic Mathematical Concepts.* New York: McGraw-Hill, 1965.

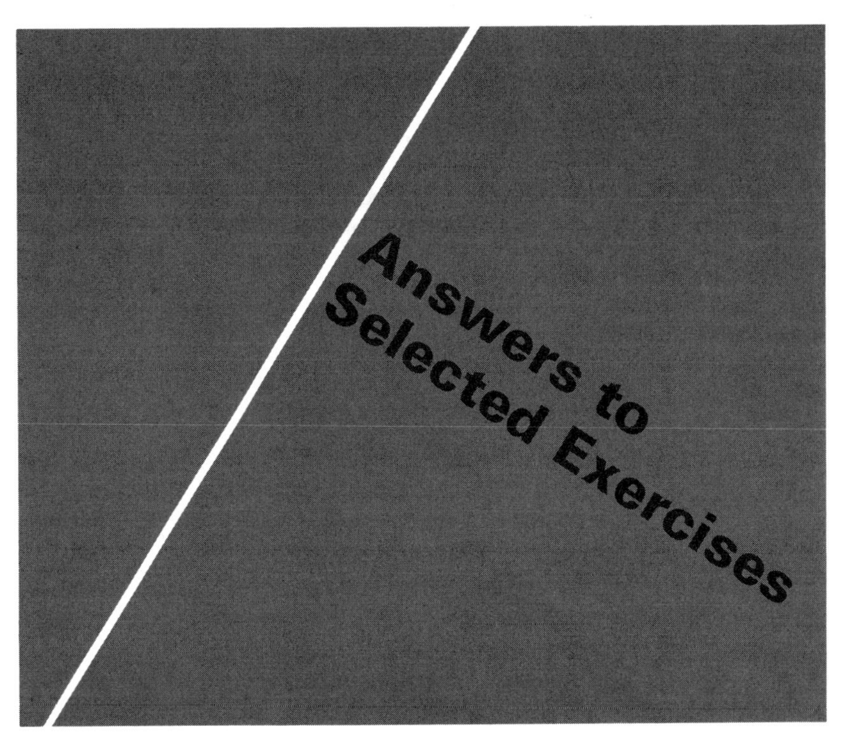

Exercise 1.1

1. No **3.** (a) Indivisible (b) Magnitude **5.** (a) $\{A, D\}$ (b) $\{E\}$

Exercise 1.2

3. Not independent **5.** No **7.** No **9.** The sum of the measures of the three angles of a triangle is 180 in degrees. The sum of the measures of the three angles of a triangle is less than 180 in degrees. The sum of the measures of the three angles of a triangle is greater than 180 in degrees. **11.** The base angles of an isosceles triangle have the same measure. The base angles of an isosceles triangle do not have the same measure.

Exercise 1.3

1. (a) A sphere is not a three-dimensional figure. (b) A disjunction is not a compound statement. **3.** A right triangle has one right angle if and only if an acute triangle has three acute angles. **5.** A right triangle has one right angle and an acute triangle has three acute angles. **7.** True **9.** True **11.** False **13.** True

15. Converse: If all the points of the line are in the plane, then two points of a line are in the plane. Inverse: If two points of a line are not in a plane, then all the points of the line are not in the plane. Contrapositive: If all the points of the line are not in the plane, then two points of the line are not in the plane.

Exercise 2.1

1. Yes **3.** Yes **5.** Yes **7.** Yes **9.** Yes **11.** No **13.** Yes **19.** Always true **21.** (a) $\overline{EF}$ (b) $\overline{AD}$ (c) half-plane above $\overleftrightarrow{CD}$ (d) half-plane below $\overleftrightarrow{AB}$

Exercise 2.2

1. Two **3.** Convex **5.** Convex **7.** Convex **9.** Non-convex **11.** Non-convex **13.** Convex

Exercise 2.3

1. Sample answers: roof of church, foul lines in baseball diamond, open door **3.** (a) 3 (b) 6 (c) 10 **5.** $n(n + 1)/2$ **7.** (a) 6 (b) 12 (c) 20 **9.** No **11.** A line in the same plane parallel to one side of an angle intersects the line containing the second side.

Exercise 2.4

1. (a) Equal (b) Equal (c) Not equal (d) Not equal **5.** (a) Not equal (b) Not equal (c) Not equal (d) Not equal **7.** (a) 8 segments (b) 16 segments **9.** Empty set, 1 point, open segment **11.** The intersection of two convex sets is a convex set.

Exercise 3.1

1. a, b, c, d **3.** (a) Approximate (b) Approximate (c) Exact (d) Approximate **5.** Yes **7.** Yes **9.** No

Exercise 3.2

1. (a) Yes (b) Yes (c) No **3.** 7

5.

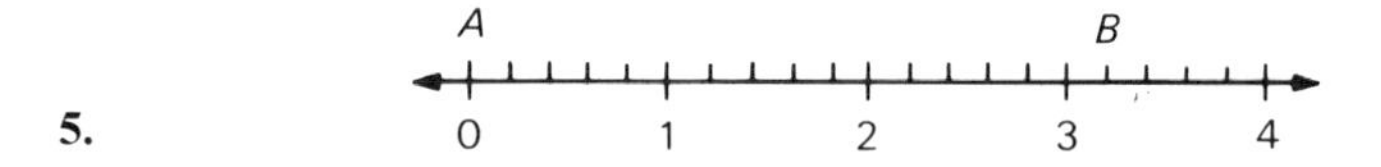

7. Pace, digit **9.** (a) 1760 yd (b) 320 rd (c) 198 in. (d) 6.125 mi **11.** 10,000
13. (a) No (b) Yes (c) No (d) Yes (e) Yes (f) Yes **15.** ≈ 219 and ≈ 1640
17. ≈ 19 cm **19.** ≈ 2.16 dec by ≈ 2.79 dec

Exercise 3.3

1. (a) 5 (b) 3 (c) 8 **3.** (a) 75° (b) 45° (c) 120° **5.** All acute except c, which is obtuse **7.** (a) 55 (b) 27 (c) 74.6 (d) $40\frac{1}{4}$ (e) .01 **9.** (a) 63° 30′ (b) 75° 6′ (c) 12° 54′ **11.** $\angle FCD$, $\angle BOG$, $\angle HOA$ **13.** The sum is 180.
15. 45°, $22\frac{1}{2}$°, $11\frac{1}{4}$°

Exercise 4.1

1. a, b, c, d, e, f, g, h, j, k, l **3.** a, b, d, e, f **5.** Yes **7.** Yes **9.** No **11.** No
13. Curve **15.** Closed curve **17.** Not a curve

Exercise 4.2

1. All correct except g **3.** (a) 2, 1, or 0 points or a segment (b) segment or 2, 1, or 0 points (c) Broken line, 4, 3, 2, 1, or 0 points, segment, segment and point, segment and 2 points (d) segment, broken line, 6, 5, 4, 3, 2, 1, or 0 points, 1 segment and 1 point, 1 segment and 2 points, 1 segment and 3 points **5.** Yes
7. (a) <40 (b) <63 (c) <75.7 **9.** Each has a measure of 60 in degrees.
11. (a) True (b) True **13.** That two corresponding sides are congruent

Exercise 4.3

1. Yes **3.** No **5.** 1 **7.** Parallelogram **9.** Measure of first is half the sum of the other two. **11.** Always congruent **13.** Not always congruent **15.** Always congruent **17.** 5.0 m **21.** Rhombuses

Exercise 4.4

1. Some street intersections, some containers for cosmetics, some lamp bases, some schoolrooms **3.** (a) 4 (b) 3 (c) 2 (d) 1 (e) 0 (f) 0 **5.** 2, 5, 9, 14, 20
7. 104 **9.** 1440° **11.** Square **13.** No **15.** 60°

Exercise 4.5

1. 10 **3.** No **5.** False **7.** (a) P is multiplied by 4 (b) Not changed (c) n is halved (d) n is mutiplied by 4 (e) Not changed **9.** 13 **11.** $\sqrt{149}$ m
13. $\sqrt{761}$ m **15.** 22 ft **17.** 8 in. **19.** 8 ft

Exercise 5.1

1. (a) No (b) It is not made up of line segments. **3.** Diameter **5.** 4, 3, 2, 1, or no points **7.** 3 **9.** Yes **11.** No **13.** (a) Yes (b) No (c) No

Exercise 5.2

1. (a) Yes (b) No (c) No **3.** A, B, C, D, E, F **5.** $\overline{BF}$ is a diameter **7.** (a) ≈ 50 (b) ≈ 165 (c) ≈ 65 (d) ≈ 155 **9.** (a) Have same measure (b) Have same measure **11.** 6 **13.** $\overline{CD}$ and $\overline{EF}$ are the same perpendicular distance from O and hence are congruent chords **15.** The point of tangency is on the line of centers. **17.** The measure of an inscribed angle is half the measure of its intercepted arc. **21.** Yes

Exercise 5.3

1. Yes, congruent **3.** 6 in. **5.** The measure of the first is twice the measure of the second. **7.** 1, 22/7, 2/3, 88/21, 399/176, 399/88, 5, 110/7 **9.** $\pi/6$, 90, $2\pi/3$, 225, $\pi/20$, 15 **11.** 6 **13.** 7 **15.** 6 **17.** 3 **19.** 3 **21.** 2 **23.** 3 **25.** 3 **27.** 7

Exercise 5.4

1. Tree, rainbow **3.** (a) No (b) Yes (c) No (d) Yes (e) No (f) Yes **5.** (a) No (b) No (c) No (d) No (e) No (f) No (g) No (h) Yes (i) No **7.** (a) No (b) Yes (c) Yes (d) Yes (e) Yes (f) No **9.** (a) Yes (b) No (c) No (d) No (e) No (f) No (g) No **15.** D **17.** A **19.** C **21.** A

Exercise 6.1

13. Use the entire length of the segment for the radius of the arcs.

Exercise 6.2

7. 6 **9.** 9

Exercise 6.3

3. It is outside the triangle. **5.** 130, 105 **9.** Same as centroid and incenter **11.** The angle in a triangle opposite the longer side has a larger measure. **13.** The product of the measures of $\overline{AF}$, $\overline{BD}$, and $\overline{CE}$ equals the product of the measures of $\overline{FB}$, $\overline{DC}$, and $\overline{EA}$. **15.** (a) 1 (b) 1 (c) 1 **19.** 120 and 130 in degrees **21.** $BFEC$ and $AFDC$

Exercise 7.1

1. (a) 1296 (b) 1/640 (c) 10,000 (d) 10 billion (e) 102,400 (f) 10,000
3. (a) 256 (b) $6\frac{1}{4}$ (c) .2116 (d) $60\frac{1}{16}$

Exercise 7.2

3. $45\frac{1}{2}$ sq in., 30.24 sq ft, 11.315 sq mm, .8 mi, ≈ 39.6 cm, $14\frac{2}{9}$ yd **5.** Multiplied by 1/4 **7.** Unchanged **9.** Multiplied by 1/4 **11.** Unchanged

Exercise 7.3

1. 78, 18.755, $13\frac{3}{4}$, $7\frac{1}{17}$, 64

Exercise 7.4

3. 154, 490.6, 1.5, 9 **5.** $3\frac{13}{81}$ **7.** (a) $\approx \frac{5}{2}$ (b) ≈ 2 (c) ≈ 3 **9.** $\approx \frac{11}{4}$

Exercise 8.1

1. Airplanes and model airplanes, paper clips of different sizes **3.** 1 to 1 **5.** No
7. No **9.** $\frac{3}{4}$ **11.** $\frac{14}{3}$ **13.** 4.5 **15.** $\frac{49}{4}$ **17.** $\triangle ADB \sim \triangle ABC \sim \triangle BDC$

Exercise 8.2

1. (a) 1 in.: 10 ft (b) 1 in.: 20 yd **3.** (a) $\frac{1}{2}$ in. (b) $\frac{7}{4}$ in. (c) $\frac{4}{5}$ ft (d) $\frac{19}{20}$ m
5. (a) ≈ 37 mi (b) 20 mi (c) ≈ 28 mi (d) 45 mi **7.** c and d

Exercise 8.3

1. Finding distance to a mountain without going there, finding distance from observer to an airplane **3.** (a) 400 ft (b) 375 ft

Exercise 8.4

1. Reproduce the given angle. On the sides of this angle, lay off segments with twice the measures of the two sides of the given triangle. Connect the endpoints to form the third side **3.** $\frac{16}{49}$ **5.** $\frac{4}{14}$

Exercise 9.1

3. (a) Third (b) Second (c) First (d) Fourth **5.** $(\frac{7}{2}, 2)$ **7.** $(\frac{5}{2}, -2)$
9. $\sqrt{37}$ **11.** $\sqrt{85}$ **13.** 6 **15.** $-\frac{9}{2}$

Exercise 9.2

1. $\{(x, y)|3y = x\}$ **3.** $\{(x, y)|y = 3x - 2\}$ **5.** a and b **7.** $2/3$ **9.** $y - 3 = 1/2(x - 2)$ **11.** $y = 3x + 4$ **13.** $(2\frac{1}{2}, 2\frac{1}{2})$ **15.** c and d

Exercise 9.3

1. Increase **3.** 20° **5.** 45° **7.** If the coordinates are (a, b), the tangent of the angle formed by the positive end of the x-axis and the ray from the origin through (a, b) is b/a. **9.** $\approx$ 119 ft **11.** $\approx$ 170 ft **13.** $\approx$ .9 mi

Exercise 9.4

3. (a) $\sqrt{58}$ (b) $\sqrt{17}$ (c) $\sqrt{29}$ (d) $\sqrt{20}$ **11.** (1, 6) **15.** (7, 3) **19.** $\approx$37 mi, $\approx$70° west of south **21.** $\approx$113 mph, $\approx$27° east of north

Exercise 10.1

1. b and d **3.** (a) $\overleftrightarrow{CD}$ (b) B (c) A (d) { } (e) $\overleftrightarrow{AC}$ (f) C **9.** $\overleftrightarrow{EF}$, $\overleftrightarrow{BF}$, $\overleftrightarrow{GH}$, $\overleftrightarrow{CG}$

Exercise 10.2

1. No **3.** Yes **5.** Yes **7.** Yes

9.

V	E	F	$V + F - E$
8	12	6	2
6	9	5	2
10	15	7	2
12	18	8	2
6	12	8	2

11. 6, 12, 12, 30, 30 **13.** Square and its interior, empty set, point, rectangle, square, triangle, segment

Exercise 10.3

1. It has one point in common with the surface. **3.** Ice-cream cone, clown's hat **5.** Yes **7.** Yes **9.** Segment **11.** Rectangular region **13.** A polygonal region similar to the base

Exercise 11.1

5. (a) (0, −2, 3) (0, 5, 18) (0, 5, 2) (b) (1, 0, −4) **7.** (a) $y = 2$ (b) $x = 1$ (c) $z = 3/4$ **9.** (a), (c), (d) **11.** $x^2 + y^2 + z^2 = 16$

Exercise 11.2

1. 31°E, 30°E, 4°W, 74°W, 77°E **3.** Northern Temperate zone **5.** (a) 38°N, 122°W (b) 10°N, 67°W (c) 34°S, 18°E (d) 60°N, 11°E **7.** (a) Water (b) Water (c) Land (d) Land

Exercise 12.1

1. (a) 104 (b) 280 (c) $9\frac{7}{24}$ (d) $4\frac{28}{75}$ (e) $92\frac{1}{2}$ (f) $108\frac{7}{8}$ **3.** (a) 60 (b) 288 (c) $\frac{7}{4}$ (d) $\frac{12}{25}$ (e) $56\frac{1}{4}$ (f) $73\frac{5}{16}$ **5.** (a) 512 (b) $\frac{27}{64}$ **7.** 46,656 **9.** $\frac{1}{2}$ in., 3 m, $\frac{2}{27}$ ft **11.** Multiplied by 27 **13.** 1520 sq ft **15.** $\sqrt{14}$

Exercise 12.2

1. 72π, 104π
120π, 570π
87.84π, 162.26π
$(15/8)\pi$, 3π
$49\frac{1}{2}\pi$, 90π
3. Doubled **5.** $\approx 19\frac{25}{56}$ cu in.

Exercise 12.3

1. 111, $\frac{5}{12}$, 16.2173 **3.** 38, $\frac{14}{72}$ **5.** Doubled **7.** $2/\pi$

Exercise 12.4

1. (a) 100π, $(500/3)\pi$ (b) $(16/9)\pi$, $(32/81)\pi$ (c) 148.84π, $\approx 302.64\pi$ (d) $\approx 204.4\pi$, $\approx 485.9\pi$ **3.** Multiplied by 8 **5.** (a) 6 cm (b) 90 cm **7.** $(49\frac{1}{3})\pi$ cu in. **9.** (a) 2406 mi (b) 301 mi

Exercise 13.1

1. (a) $2\frac{0}{2}$ in., $1\frac{1}{2}$ in., $1\frac{1}{2}$ in. (b) $2\frac{0}{4}$ in., $1\frac{2}{4}$ in., $1\frac{1}{4}$ in. **3.** .1 in. .05 in. (6.2 + .05) in. **5.** .01 m .005 m (3.04 + .005) m **7.** $\frac{1}{4}$ sq cm $\frac{1}{8}$ sq cm $(6\frac{1}{4} + \frac{1}{8})$ sq cm **9.** .1 cu in. .05 cu in. $(3.2 \pm .05)$ cu in. **11.** 1 yd .5 yd $(50 \pm .5)$ yd **13.** .1 in. .05 in. $(7.0 \pm .05)$ in. **15.** 1000 sq ft 500 sq ft (3000 ± 500) sq ft **17.** 14 in. **19.** 305 km **21.** 3.16 sq km

Exercise 13.2

1. $\frac{1}{10}$, 10.0% **3.** $\frac{1}{30}$, $\approx 3.3\%$ **5.** $\frac{1}{10}$, 10.0% **7.** $\frac{1}{20}$, 5% **9.** $\frac{1}{840}$, $\approx .1\%$ **11.** 5.2 m **13.** 3652 in. **15.** 80 cu in.

Exercise 13.3

1. 719 mi, 721 mi **3.** $89\frac{1}{4}$ sq in., $109\frac{1}{4}$ sq in. **5.** $14\frac{5}{8}$ sq ft, $20\frac{5}{8}$ sq ft **7.** 182.25π sq in., 210.25π sq in. **9.** 67.375π cu ft, 28.125π cu ft

Exercise 14.1

1. (a) No (b) No (c) No (d) Yes (e) No (f) No **5.** If two triangles are perspective from a line, they are perspective from a point. **7.** Four lines and the six points they determine in pairs **9.** (a) $-\frac{5}{16}$ (b) $-\frac{1}{9}$

Exercise 14.2

1. (a) No (b) No (c) Yes (d) No **7.** Two interlocking twisted bands

Exercise 14.3

1. Not necessarily **3.** No
5. There are exactly four libraries.
Each pair of distinct libraries has exactly one book in common.
Each book is in exactly one pair of distinct libraries. **7.** No **9.** Not enough information to tell **11.** 5 **13.** No **15.** 4

Exercise 14.4

3. There is no square in hyperbolic geometry. **5.** Those that do not depend on Euclid's axiom about parallels **7.** Less than 180 in degrees

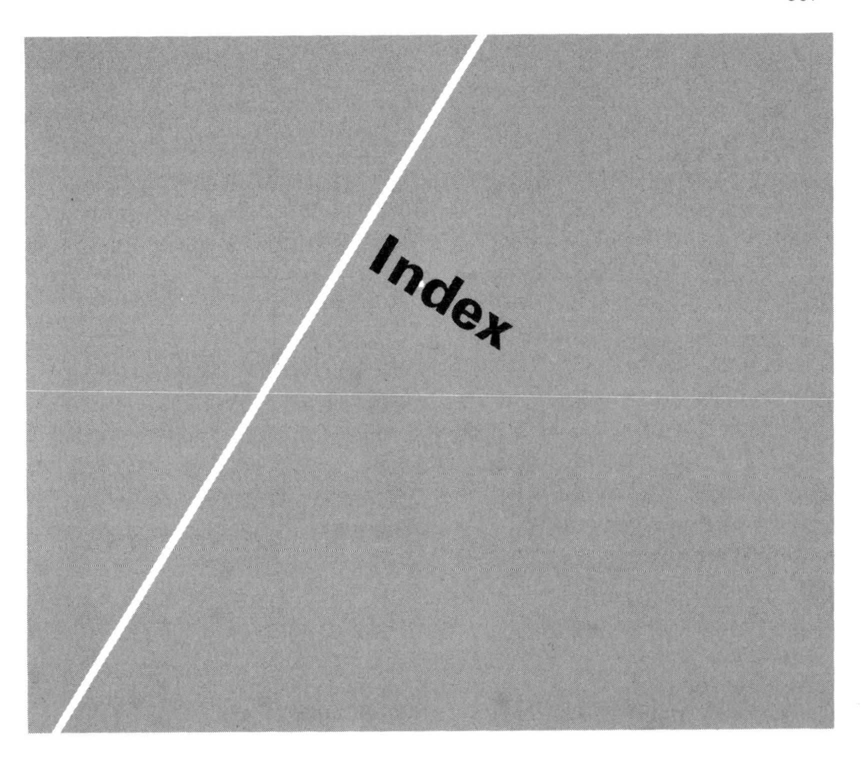

Absolute value, 48–49, 190
Accuracy, 269–270
Acre, 147–148
Acute angles, 58
Acute triangle, 71
Addition of vectors, 211–212
Adjacent angles, 32, 59
Adjacent side, 204
Affine geometry, 278
Alternate interior angles, 61
Altitude:
  of parallelogram, 151
  of trapezoid, 155–156
  of triangle, 136–137, 152
Analytic geometry, 189, 194–201
Angle(s):
  acute, 58
  adjacent, 32, 59
  alternate interior, 61
  associated with parallel lines, 60–61
  bisector of, 127, 136–137
  central, 98, 101–109
  complementary, 59
  concept of, 30–34
Angle(s)—(continued)
  congruent, 54
  corresponding, 60–61
  definition of, 31
  dihedral, 220–221
  of elevation, 178–179
  exterior, 72, 140
  exterior of, 31–32
  included, 73
  inscribed, 98
  interior of, 31–32
  measurement of, 54–62
  obtuse, 58
  of parallelism, 294
  reflex, 77
  reproducing, 128
  right, 58
  sides of, 31
  straight, 59
  sum of measures in hyperbolic
    geometry, 295
  sum of measures in polygons, 85
  supplementary, 59
  of a triangle, 70

Angle(s)—(continued)
  trigonometric ratios of, 203–207
  trisecting, 134
  unit, 54–55
  vertex of, 31
  vertical, 32, 59–60
Antarctic Circle, 243
Apex, 225
Approximately equal to, 111
Approximations in measurement, 265–274
Arc, 101
Arc degree, 101–103
Archimedes, 161
Arctic Circle, 243
Area:
  approximate measure, 162–163
  in calculus, 163
  circular regions, 159–161
  concept of, 146–150
  lateral surface of prism, 246–247
  polygonal regions, 155–159
  rectangular region, 147–150
  region with parallelogram, 151–152
  square, 149
  standard units of, 147–148
  surface of prism, 245–247
  trapezoidal region, 155–156
  triangular region, 152–154
Aristotle, 11
Assumptions, 4
Axes:
  in plane, 188
  in space, 236
  of symmetry, 116–117
Axioms:
  consistent, 8
  definition of, 6
  of Euclid, 7
  for Euclidean geometry, 25–30, 32, 48, 56, 58, 73, 103, 148, 152, 156–157, 188, 219, 220, 251–252
  existence, 26
  of Hilbert, 7–8
  independent, 8
  of Pearson and Smart, 8
  for projective geometry, 277

Base(s):
  cone, 230
  non-circular cylinders, 229–230
Base(s)—(continued)
  parallelogram, 151
  pyramid, 225
  right circular cylinder, 229
  right prism, 223
  trapezoid, 155–156
  triangle, 152
Between:
  on arc, 101
  undefined term, 21
Bisecting segment, 126
Bisector:
  angle, 76, 127, 136–137
  external angle, 140
Bolyai, Johannes, 293
Broken line, 36
Brouwer fixed-point theorem, 287

Calculus:
  area in, 163
  derivative, 207–209
Cardinal number, 19
Caroms, 185
Cavalieri's Principle, 250, 261–262
Centare, 147
Center:
  of gravity, 137
  of sphere, 230
Centimeter, 51
Central angle, 98, 101–109
Central projection, 276
Centroid, 137
Ceva, 139
Ceva's Theorem, 139
Cevian, 139–141
Chord(s):
  concept of, 97
  congruent, 104
  products of, 170
Christmas star, 134
Circle(s):
  basic concepts, 95–100
  circumference, 109–115
  concentric, 97
  equation, 199–200
  great, 232, 241
  as mathematical model, 3
  small, 232, 241
  squaring, 134
  unit, 203

Circular number line, 113
Circular regions, 159–161
Circumcenter, 142
Circumcircle, 142–143
Circumference, 109–115
Circumscribed polygon, 97
Closed half-plane, 23
Closed segment, 22
Common notions of Euclid, 7
Compass, 125, 135
Complementary angles, 59
Complete quadrangle, 278–279, 290
Complete quadrilateral, 279
Composite regions, 161
Compound statements, 11–16
Comte de Buffon, 111
Concentric circles, 97
Conclusion, 9
Cones, 229–230
Congruence:
  axioms of, 8
  conditions for triangles, 73–75
  equivalence relation, 46
  in measurement, 44–45
  quadrilaterals, 81
  and symmetry, 118
Congruent angles, 54
Congruent chords, 104
Congruent polygons, 86
Congruent segments, 47
Congruent triangles, 72–76
Congruent unit regions, 150
Conic, 282
Conical solid, 232, 258–259
Conic sections, 231–232
Conjunction, 12–13
Connected sct, 286
Connection, axioms of, 8
Connectives, 11–16
Connectivity, degree of, 286
Consistency:
  non-Euclidean geometry, 296
  sets of axioms, 8
Constant width, 99
Construction(s):
  dividing a segment in a given ratio, 182–183
  instruments for, 124–131
  mathematical laboratory experiences, 136–145
  one instrument, 135
Construction(s)—(continued)
  problems, 131–135
  proof of, 126–130
  similar figures, 181–184
  square root, 184
  triangle, 128–130
Continuity, 8
Continuous variable, 149
Contrapositive, 15–16, 28
Converse, 15–16
Convex sets:
  concept of, 25–27, 67
  polygonal region, 78, 156
  polyhedrons, 226–227
  quadrilaterals, 77
Coordinates:
  midpoint of segment, 189–190
  points in plane, 187–194
  points in space, 234–241
  projective geometry, 279
  vectors, 212–213
Corresponding angles, 60–61
Cosine, 203–205
Counting numbers, 48
Cross ratio, 280
Cross section, 231
Cube, 134, 224, 226
Cubical solid, 248
Cubic centimeter, 252
Cubic foot, 251
Cubic inch, 251
Cubic meter, 252
Cubic millimeter, 252
Cubic unit of volume, 247–248
Cubic yard, 251
Curve, 63–68
Curved surface, 229
Cyclic quadrilaterals, 143–144
Cylinders, 229–230, 252–254
Cylindrical solid, 232, 254–255

Decimal expressions for pi, 110
Decimals in angle measurement, 56
Decimeter, 51
Defined terms, 4, 17–25
Degree, 55
Delta notation, 207
Derivative, 207–209
Desargues, 276
Desargues' finite geometry, 292

Desargues' Theorem, 281
Descartes, René, 189
Designs with constructions, 131–135
Diagonal, 79
Diagonal points, 279
Diagonal triangle, 279
Diameter, 96
Dihedral angle, 220–221
Dimensions, 234–235, 240
Directed segment, 210
Direct motions, 120
Discrete variable, 149
Disjoint sets, 20
Disjunction, 12–13
Distance:
  between parallel planes, 221
  between points in plane, 190–191
  between points in space, 238–239
  as undefined term, 95
Distributive property, 156, 254
Divide and average, 90
Dividers, 125
Dividing segment in given ratio, 182–183
Dodecagon, 133
Dodecahedron, 226
Doubling a cube, 134
Duality, 278

Edges:
  of polyhedrons, 226–227
  of right prism, 224
Element, 4
Elevation, angle of, 178–179
Ellipse:
  base for cylinder, 229
  section of cone, 231
Elliptic geometry, 293
Empty set, 19
Endpoints of segment, 21
English system, 50–51
Equality of sets, 18
Equal vectors, 210
Equation(s):
  four variables, 240
  general form for line, 198
  for lines, 195–199
  for parallel planes, 237–238
  for plane, 237–238
  point-slope form for line, 198
Equation(s)—(continued)
  slope-intercept form for line, 198
  for sphere, 239
Equator, 242
Equilateral triangle, 70, 139–140
Equivalence, 12, 14–15
Equivalence relation, 46
Equivalent sets, 19
Error:
  greatest possible, 266–269
  relative, 269–270
Estimates, area of plane regions, 201
Euclid:
  assumptions of, 7, 60
  fifth postulate, 10
Euclidean geometry, axioms for, 25–30, 32, 48, 56, 58, 73, 103, 148, 152, 156–157, 188, 219, 220, 251–252
Euler, Leonard, 226
Euler's formula, 227
Even vertex, 66
Excenters, 141
Existence axiom, 26
Exponent, 149, 248
Exterior angle, 72, 140
Exterior of angle, 31–32
Extremes, 171

Faces:
  polyhedron, 226–227
  pyramid, 225
  right prism, 224
Fano's geometry, 291
Fermat, Pierre de, 189
Fibonacci numbers, 82
Fifth postulate of Euclid, 10
Finite geometries, 289–293
Finite sets, 18–19
Foot, 51
Forces, 210, 213–214
Formal deductive proof, 9
Foundations of Geometry, 8
Foundations of mathematics, 8
Four-color map problem, 287
Framework of geometry, 4
Function, 44, 194, 197
Fundamental Theorem of projective geometry, 282

Garfield, President, 91
Gauss, K. F., 132

General form, 198
Geometric figures, 2, 21
Geometry:
analytic, 194–201
finite, 289–293
framework of, 4
informal approach, 1–6
non-Euclidean, 293–296
non-metric, 17–25
projective, 275–285
Golden ratio, 82
Golden rectangle, 82
Graph:
points in plane, 187–200
points in space, 235–241
Great circle, 232, 262–263
Greatest possible error, 266–269
Greek problems, 134
Greenwich Observatory, 242
Grid to measure area, 162–163

Half-line, 23
Half-plane:
concept of, 22–23
definition of interior of angle, 31–32
graph of, 200
mathematical sentence for, 200–201
Half-space, 219
Harmonic set, 281
Heptagon, 84
Hexagon, 84, 132
Hexagonal prism, 224
Hexagonal region, 157
Hilbert, David, 8
Homothetic figures, 167–168
Homothety, 167–168
Horizontal axis, 188
Hyperbola, 232
Hyperbolic geometry, 293–296
Hypotenuse, 90–91
Hypothesis, 9

Icosahedron, 226
Ideal point, 277, 294
If-then form, 9, 15
Illusions in geometric drawings, 2
Image, 72
Implication, 12, 14–16
Improper subset, 20

Incenter, 137
Inch, 51
Incircle, 142
Included angle, 73
Independent axioms, 8
Indirect measurement, 176–180, 206–207
Infinite sets, 18–19
Informal approach to geometry, 1–6
Initial point, 210
Inscribed angle, 98
Inscribed triangle, 97
Instruments used in construction, 124–131
Interior:
angle, 31–32
circle, 96
simple closed curve, 65
simple closed surface, 229
International Bureau of Weights and Measures, 51
Intersection:
angles and lines, 35–36
dihedral angle and other figures, 221
lines in a plane, 34–35, 199
plane and line, 221
plane and simple closed surface, 231
rays and segments, 35
segments, 35
sets, 20
sets of points, 34–40
Intuitive approach to geometry, 1–6
Inverse, 15–16
Irrational number, 91
Isometry, 45, 118
Isosceles triangle, 70

Jordan Curve Theorem, 65, 285–286

Kilometer, 51
Klein, Felix, 288
Klein bottle, 288
Krypton lamp, 51

Lambert quadrilateral, 295
Lateral faces, 224
Lateral surface area:
cone, 258
cylinder, 253
prism, 246–247

Latitude, 241–244
Lemoine, E., 141
Lemoine point, 141
Length, 47–53
Leonardo da Vinci, 159
Limit, 161, 208
Line(s):
  equations of, 195–199
  intersections in a plane, 34–35, 199
  of latitude, 242, 263
  of longitude, 242
  notation for, 5
  parallel, 34–35
  in plane, 221
  skew, 219
  slope of, 193
  of symmetry, 116
  as undefined term, 5
Line segment, 21–22
Listing, J. B., 285
Lobachevsky, Nicolai, 293
Locating points on surface of earth, 241–244
Logic, 11–16
Longitude, 241–244

Magnitude of vector, 210
Mapping, 44, 173
Map problem, 287
Masheroni, 135
Mathematical laboratory, 136–145
Mathematical model, 2–3, 42–43, 218
Mathematical sentences, 174, 237–241
Mathematical space, 2, 234
Means, 171
Measurement(s):
  angle, 54–62
  angle between chord and tangent, 106
  approximate nature of, 265–274
  central angles, 102–103
  concept of, 41–46
  cylinders, 252–254
  cylindrical solids, 254–255
  English system of, 50–51
  indirect, 176–180
  length, 47–53
  in mathematical model, 42–43
  metric system of, 50–53
  physical objects, 42
  prismatic solids, 247–251
Measurement(s)—(continued)
  prisms, 245–247
  pyramids, 256–257
  spheres, 260–263
  standard units for area, 147–148
  unit of, 42–43, 47–48, 147–148
Measures, operations with, 271–274
Median of triangle, 136–137
Member of set, 4
Menelaus, 145
Meridians, 241–244
Meter, 51
Metric system, 50–53, 251–252
Metrology, 51
Micron, 52
Midpoint of segment, 47, 126, 189–190
Mile, 51
Millimeter, 51
Millimicron, 52
Minutes, 56
Model(s):
  mathematical, 2–3
  for polyhedrons, 227
Modern mathematics, 293
Modular arithmetic, 112–116
Modulus, 113–116
Moebius, A. F., 285
Moebius strip, 287–288
Monte Carlo Method, 201
Motions, 120
Multiply connected, 286

N-dimensional space, 240
Negation, 12–14
Non-convex polyhedrons, 226
Non-Euclidean geometry, 293–296
Non-intersecting lines, 294
Non-metric geometry, 17–25
Non-square units of area, 149–150
Non-standardized units, 50–52
North Frigid zone, 244
North Temperate zone, 244
N-tuples, 240
Number:
  cardinal, 19
  irrational, 91
Number line, 48–49, 187–188

Oblique circular cylinder, 230
Oblique prism, 224–225

Oblique prismatic solid, 249–250
Obtuse angles, 58
Obtuse triangle, 71
Octagon, 84
Octahedron, 226
Odd vertex, 66
Omega triangle, 294
One-dimensional, 234
One-to-one correspondence, 19, 48–49
Open half-plane, 23
Open segment, 22
Operations:
  with measures, 271–274
  on sets, 21
Opposite motion, 120
Order, axioms of, 8
Ordered pair of numbers, 187
Ordered triple, 235–236
Origin, 187, 235
Orthocenter, 137

Parabola, 231
Parallel lines:
  angles associated with, 60–61
  axioms of, 8
  concept of, 34–35
  in hyperbolic geometry, 294
  and similar triangles, 169
  slope of, 196
Parallelogram:
  altitude of, 151
  area of region, 151–152
  base of, 151
  boundary of region, 151–152
  concept of, 78
  properties of, 80–82
Parallel planes, 221, 237–238
Partition of set, 21
Pascal, B., 276
Pascal line, 283–284
Pascal's Theorem, 282
Path:
  between points on sphere, 262–263
  tracing puzzle, 65–66
Pentagon, 84
Pentagonal prism, 224
Percent of error, 270
Perimeter, 88–89, 246
Perpendicular:
  concept in plane, 58
Perpendicular—(continued)
  concept in space, 220–221
  at point on line, 126–127
  from point to line, 131
Perpendicular bisectors, 142
Perspective:
  from a line, 281
  from a point, 281
Perspective forms, 277
Pi, 110–111
Plane(s):
  equation for, 237–238
  notation for, 5
  parallel, 221
  reference, 235
  as undefined term, 5
Plane duality, 278
Plane motions, 118 120
Plane region, 67, 146
Planimeter, 161–162
Plato, 124–125
Point(s):
  coordinates in plane, 187–194
  coordinates in space, 234–241
  on line, 199
  in mathematical space, 2
  on plane, 238
  of symmetry, 117–118
  as undefined term, 5
Point of tangency, 97
Point-slope form of equation, 198
Polygon(s):
  circumscribed about circle, 98
  congruent, 86
  inscribed, 132–133
  more than four sides, 83–87
  regular, 85–86, 143–144, 157
  simple, 66
Polygonal region:
  area of, 155–159
  convex, 78, 156
Polyhedra (*see* Polyhedrons)
Polyhedral solid, 232, 250
Polyhedrons, 223–229
Poncelet, J. V., 275
Poncelet-Steiner Construction
    Theorem, 135
Postulates:
  of Euclid, 7
  for hyperbolic geometry, 293–294
Precision, 266–269

Prime Meridian, 242
Prism(s):
  measurements of, 245–247
  oblique, 224–225
  rectangular, 224
  right, 223–224
  triangular right, 224
Prismatic solid, 232, 247–251
Probability, 111
Projection, 276–277
Projective forms, 277
Projective geometry, 275–285
Projectivity, 281
Proof:
  of constructions, 126–130
  formal deductive, 9
  indirect method of, 9–10
  informal, 9
  rigorous, 9
  of theorem, 8–9
  two-column form, 29
Proper subset, 20
Proportion, 171
Protractor, 57–58, 179
Pyramidal solid, 232, 257
Pyramids, 225, 256–257
Pythagoras, 91
Pythagorean Theorem:
  distance formula, 190–191
  distances in space, 238–239
  equation of circle, 199–200
  proof of, 153–154, 159
  statement of, 89–94
  use of, 262
Pythagorean triples, 91

Quadrants, 189
Quadrature, 161
Quadrilateral:
  concepts of, 77–83
  congruence of, 81
  convex, 77
  cyclic, 143–144
  inscribed in circle, 9
  Lambert, 295
  Saccheri, 295

Radian, 112
Radius, 96
Random number pairs, 201
Ratio(s):
  basic ideas of, 171
  golden, 82
  measure of area of similar triangles, 181
  of similarity, 167, 173
  trigonometric, 203–207
Ray, 23–24, 31, 36–37
Real numbers, 48
Rectangle, 78, 82
Rectangular prism, 224
Rectangular region, 147–150
Reference planes, 235
Reflection, 120, 184–185
Reflex angles, 77
Reflexive property, 46
Region, 67, 146, 151–152
Regular polygons, 85–86, 131–132, 143–144, 157
Relation, 200
Relative error, 269–270
Reproducing angle, 128
Reuleaux triangle, 99–100
Rhombus, 78
Riemann, Bernhard, 285
Right angle, 58
Right circular cone, 230
Right circular cylinder, 229
Right prism, 223–224
Right triangle, 71, 169–170, 204
Rod, 51
Rotation, 119–120
Rotational symmetry, 120
Rounding measurements, 268
Rubber-sheet geometry, 285

Saccheri quadrilateral, 295
Scale, 148
Scale drawings, 173–176
Scalene triangle, 70
Secant, 97
Seconds, 56
Section, 231
Sectors, 159–160
Segment(s):
  closed, 22
  concepts of, 21–22
  congruent, 47
  directed, 210
  intersection of, 35
  midpoint of, 126, 189–190

Segment(s)—(continued)
- notation for, 22
- open, 22
- transferring, 125
- union of, 36

Semicircle, 101
Set(s):
- of constant width, 99
- convex, 25–27
- disjoint, 20
- element of, 4
- empty, 19
- equality of, 18
- equivalent, 19
- finite, 18–19
- infinite, 18–19
- intersection of, 20, 34–40
- member of, 4
- in non-metric geometry, 17
- operations on, 21
- partition of, 21
- points in space, 218–223, 237–241
- undefined term, 4
- union of, 21, 34–40
- universal, 19

Set-builder notation, 195, 237
Side(s):
- angle, 31
- right prism, 223
- triangle, 69

Side opposite, 74
Significant figures, 268
Similar figures:
- constructions of, 181–184
- examples of, 166–171
- in indirect measurement, 176–180
- surfaces and solids, 250–251

Similarity, 167
Similar triangles:
- concepts of, 167–170
- and hyperbolic geometry, 295
- and parallel lines, 169
- in trigonometry, 204–207

Simple closed curves, 64–68, 146
Simple closed surfaces, 229–233
Simple polygon, 66
Simply connected, 286
Simson, Robert, 143
Simson line, 143
Sine, 203–205
Skew lines, 219
Slope, 187, 191–193, 208
Slope-intercept form, 198
Small circle, 232, 241
Solid figures, 232–233
South Frigid zone, 244
South Temperate zone, 244
Space:
- description of, 2, 218
- mathematical, 2
- sets of points in, 218–223
- undefined term, 5

Space duality, 278
Sphere, 220–231, 239, 260–263
Spherical solid, 232, 261–263
Spheroid, 232
Square, 78, 132
Square centimeter, 147–148
Square foot, 147–148
Square kilometer, 147–148
Square meter, 147–148
Square mile, 147–148
Square millimeter, 147–148
Square region, 149
Square root, 90, 184
Square yard, 147–148
Squaring a circle, 134
Standard units:
- area, 147–148
- cubic, 251–252
- linear, 50–52

Statements, 11–16
Straight angle, 59
Straightedge, 125, 135
Straight line, 5
Subsets, 19–20
Subtend, 102
Subtraction of vectors, 211–213
Supplementary angles, 59
Surface(s):
- of cylinder, 253
- locating points on earth, 241–244
- one-sided, 287–288
- polyhedrons, 223–229
- right circular cone, 257–258
- simple closed, 229–233

Surface area:
- cone, 258
- cylinder, 254
- prism, 245–247
- pyramid, 256
- sphere, 260

Symmedian, 141
Symmetric property, 46
Symmetry, 116–123

Tangent, 97, 204–205
Terminal point, 210
Tetrahedron, 225–226
Theorem, 4, 8–9
Three-dimensional, 234
Tolerance, 270
Topological transformation, 287
Topology, 285–289
Torrid zone, 244
Tracing puzzles, 65–66
Transferring a segment, 125
Transformation:
- concept of, 72
- definition of, 44
- projection, 276–277
- of similarity, 167
- and symmetry, 117
- topological, 287

Transitive property, 46
Translation, 118–119
Transversal, 60
Trapezium, 78
Trapezoid, 78, 155–156
Triangle(s):
- acute, 71
- altitude of, 152
- angles of, 70
- base of, 152
- concept of, 69–76
- concurrent segments, 136–142
- congruent, 72–76
- construction of, 128–130
- definition of, 69
- equilateral, 70
- inscribed in circle, 97
- isosceles, 70
- of maximum area, 295
- obtuse, 71
- omega, 294
- right, 71
- scalene, 70
- similar, 167–170
- union and intersection of, 69

Triangular region, 152–154
Triangular right prism, 224
Trigonometric ratios, 203–207
Trigonometry, 203–207
Triples, Pythagorean, 91
Trisecting an angle, 134
Tropic of Cancer, 243
Tropic of Capricorn, 243
Truth tables, 12–16
Two-dimensional, 234

Undefined terms, 4–5
Union:
- half-lines and half-planes, 37
- rays, 36–37
- segments, 36
- sets, 21
- sets of points, 34–40

Unit angle, 54–55, 112
Unit circle, 203
Unit of measurement:
- area, 147–148
- concept of, 42–43
- for length, 47–48
- non-square, 149–150
- non-standardized, 50–52
- standardized, 50–52

Unit of precision, 267
Unit region, 149
Universal set, 19

Valid, 8
Variable:
- continuous, 149
- discrete, 149

Vectors, 210–217
Velocity, 210
Venn diagram, 38
Vertex:
- angle, 31
- cone, 230
- polyhedrons, 226–227
- right prism, 224
- tracing puzzle, 65–66
- triangle, 69

Vertical angles, 32, 59–60
Vertical axis, 188
Volume:
- concept of, 247
- conical solid, 258–259
- cylindrical solid, 254–255
- prismatic solid, 247–251

Volume—(continued)
  pyramidal solid, 257
  spherical solid, 261–262

Width of set, 98–100

x-coordinate, 194

Yard, 51

y-coordinate, 195

Zones, 244